Korean Science and Technology in an International Perspective

Jörg Mahlich · Werner Pascha
Editors

Korean Science and Technology in an International Perspective

Physica-Verlag
A Springer Company

Editors
Dr. Jörg Mahlich
University of Vienna
Department of Economics
Hohenstaufengasse 9
1010 Vienna
Austria
joerg.mahlich@univie.ac.at

Prof. Dr. Werner Pascha
University of Duisburg-Essen
Mercator School of Management and
Institute of East Asian Studies
Lotharstr. 65
47048 Duisburg
Germany
werner.pascha@uni-duisburg-essen.de

ISBN 978-3-7908-2752-1 ISBN 978-3-7908-2753-8 (eBook)
DOI 10.1007/978-3-7908-2753-8
Springer Heidelberg Dordrecht London New York

Library of Congress Control Number: 2011942158

Printed on acid-free paper

Physica-Verlag is a brand of Springer-Verlag Berlin Heidelberg
Springer-Verlag is a part of Springer Science+Business Media (www.springer.com)

This work was supported by the Academy of Korean Studies Grant, which is funded by the Korean Government (MOEHRD, Basic Research Promotion Fund).
AKS-2007-CB-2002

Contents

Contributors

Jörg Büechl Department of International Business, Tübingen University, Tübingen, Germany

Marcus Conlé Mercator School of Management, University of Duisburg-Essen, Essen, Germany

Bernhard Dachs Foresight & Policy Development Department of AIT Austrian Institute of Technology, Vienna, Austria

Markus K. Hödl Institute for International Business, Vienna University of Economics and Business, Vienna, Austria

Reinhold Hofer Policies – Centre of Economic and Innovation Research, Joanneum Research, Vienna, Austria

Jia Jin Investment Banking Division, Hanwha Securities Co, Seoul, South Korea

Michael Keenan Manchester Institute of Innovation Research, University of Manchester, Manchester, UK

Ji Hong Kim Korea Development Institute, Seoul, South Korea

Silvo Korez Austrian Federal Ministry of Science and Research, Vienna, Austria

Chol Lee Graduate School of Business, Sogang University, Seoul, South Korea

Keun Lee Center for Economic Catch-up, Seoul National University, Seoul, South Korea

Jörg Mahlich University of Vienna, Department of Economics, Vienna, Austria

Chan-Soo Park Sauder School of Business, University of British Columbia, Vancouver, Canada

Werner Pascha Mercator School of Management and Institute of East Asian Studies, University of Duisburg-Essen, Duisburg, Germany

Iain Paterson Department of Economics and Finance, Institute for Advanced Studies (IHS), Vienna, Austria

Jonas F. Puck Institute for International Business, Vienna University of Economics and Business, Vienna, Austria

Markus Pudelko Department of International Business, Tübingen University, Tübingen, Germany

Greg Scarlatoiu Business Issues of the Korea Economic Institute (KEI), Washington, DC, USA

Margot Schüller GIGA Institute of Asian Studies, Hamburg, Germany; GIGA German Institute of Global and Area Studies, Hamburg, Germany

David Shim GIGA Institute of Asian Studies, Hamburg, Germany; GIGA German Institute of Global and Area Studies, Hamburg, Germany

Bernhard Seliger Hanns-Seidel-Foundation, Korea office; University of Witten-Herdecke, Witten, Germany

Matthew A. Shapiro Illinois Institute of Technology, Chicago, USA

Jitendra Uttam Centre for East Asian Studies, School of International Studies, Jawaharlal Nehru University, New Delhi, India

Ilan Vertinsky Sauder School of Business, University of British Columbia, Vancouver, Canada

Eunsook Yang Complutense University of Madrid, Madrid, Spain

Chapter 1
Introduction: The Challenge of Innovation and Technology for Korea as a Newly Advanced Economy Revisited

Jörg Mahlich and Werner Pascha

Six years ago, we started a project to look into the evolution of the Republic of Korea's innovation and technology system that eventually led to the publication of a well-received book (Mahlich and Pascha 2007). In September 2009, we held a second conference at the Austrian Federal Economic Chamber in Vienna. By this time, it had become ever more obvious that Korea had indeed established itself internationally as a formidable newly advanced economy (NAE), having passed a "baptism of fire" by overcoming the global financial crisis with notable speed and effectiveness. The international community acknowledged this maturing of Korea by entrusting it with the chairing of the important G-20 meeting in late 2010.

Today, writing in early 2011, the issues are not so much whether Korea is indeed a newly advanced economy and the usual innovation and technology challenges for NAEs apply: those answers are obviously in the affirmative. The question to ask now is whether there is anything special about these challenges; is Korea simply facing normal issues associated with modernizing its innovation and technology system, or is there anything peculiar about this country's status? As peculiarity is an issue of degree, not a yes-or-no question, we should rephrase the question: is Korea only slightly different from other countries in facing these challenges or is there indeed something "peculiarly peculiar" in what Korea has to tackle?

To answer such a question, a comparative mind-set and approach is necessary. We have therefore asked the contributors to this second stage of our endeavour to apply an internationally comparative perspective to the topic of newly advanced economy status.

On the following pages, we will substantiate the suggestions made above somewhat more explicitly and offer a brief survey of the contributions to this book.

First, what does the recent global economic crisis of 2008–2009 imply for innovation and technology in Korea? Have former trends and considerations become obsolete, or has the crisis on the contrary strengthened Korean aspirations and contingencies for technological leadership? The Korean economy has recovered remarkably well from the shockwave of the Lehman collapse in September 2008. According to Bank of Korea figures for the third quarter of 2009, GDP had already increased again by 2.6% quarter-on-quarter, while the median estimate of the Bloomberg survey had

J. Mahlich and W. Pascha (eds.), *Korean Science and Technology in an International Perspective*, DOI 10.1007/978-3-7908-2753-8_1,

been 1.9%. While the current account had moved into negative territory in late 2008, it already registered a surplus of 4½% in the first quarter of 2009. This has been hailed as a dramatic turnaround. The Korea Times, for instance, noted that 2009 saw the fastest second quarter income growth in 21 years. Korean export strength and the international rebound have been major components of the recovery. Hyundai Motor, Samsung Electronics and other large exporters were already reporting surging profits during 2009. Government consumption had also increased significantly, but given the relatively small share of the government sector, exports have been the stronger force.

Four observations stand out in terms of lessons and consequences for Korean innovation and technology. The first and major lesson: for Korean policy makers, the immediate crisis is considered over. Symbolically, the 36th and last Crisis Management Meeting was held in December 2010, and the next one scheduled for January 2011 was named first Economic Policy Coordination Meeting. Korea has so quickly recovered from the global crisis that the pressure to change tack in the strategic outlook has remained small. Samsung Electronics and the other "high performers" that overcame the crisis are the well-known vanguards of Korean innovation; again, there is no strong incentive to change course. Second, like in many other advanced and emerging economies (Langhammer and Heilmann 2010), the Korean government's fiscal anti-crisis programmes were more geared towards short-term support measures, including a strong social policy component, than towards an opportunity to heighten long-term growth and development potential (for more detail, Pascha 2010). In that sense, a good chance to allocate considerable funds to reorient the strategic path of the economy was unfortunately missed. Rather, the support and, in almost all cases, any sense of recovery for industries like shipbuilding and automobiles as well as weak small and medium enterprises have tended to cement existing patterns. Third, the one major exception could be the government's "green growth" programme.[1] According to HSBC estimates of early 2009, Korea allocated the third-largest anti-crisis fiscal stimulus for green issues, surpassed only by China and the US and leading the entire EU, which lagged behind in fourth place; in terms of the percentage of green components among all fiscal measures, Korea actually took the No. 1 spot. The green stimulus programme has been evaluated in widely diverse ways in Korea.[2] The programme contains considerable portions devoted to construction work, for instance the cleaning and modernising of rivers, and it thus safeguards jobs for low-skilled workers, possibly for political reasons. The part of the programme that helps to develop and implement new technologies for green growth is thus considerably smaller than may appear at first sight. Nevertheless, despite some ironic remarks about a "green bubble", the programme is playing an important role in shaping the way new technologies emerge and lead Korea towards a brighter and cleaner future.

[1]Of course, this programme, both in terms of its background and its duration, goes well beyond its role as an anti-crisis fiscal response measure.

[2]For a more thorough analysis of the green growth programme, see the contribution of Greg Scarlatoiu in this volume.

The fourth and quite remarkable fact pertaining to Korea's handling of the global crisis is that Korea not only maintained its R&D activities during the crisis but has continued to expand them (data in this paragraph from KOITA 2011). For instance, between 2007 and 2009, R&D expenditure as a share of GDP was raised from 3.21% to 3.57%. Even Japan, long noted for its steadfast R&D investments throughout the difficult 1990s and early 2000s, saw R&D expenditures peak in 2007 at 3.44% of GDP, with some decrease in 2008 and beyond. An important source for Korea's consistent R&D support has been government policy. Despite the somewhat critical view of the fiscal programme outlined above, it should be recognized that the share of governmentally funded R&D expenditures rose considerably during the crisis, reaching 28.7% in 2009. Supported by this development, the share of research classified as "basic" research rose by 2% points from 2008 to 2009 as well, a sensible trend in a country aspiring for technology frontier status.[3]

This development is not accidental. In the midst of the crisis, Korean President Lee Myung-bak declared in April 2009 that the government would increase R&D spending by 10% each year until 2012 (Korea Times 2009). During the 1997/1998 Asian crisis, private Korean companies reduced R&D expenditures. This time, although the total share of public spending on R&D rose, the private sector also sustained higher R&D outlays. Private sector R&D spending increased from KRW239 trillion to KRW260 trillion and eventually to KRW282 trillion from 2007 to 2009 (KOITA 2011, p. 3). To further show his administration's determination to make technology a top national priority, President Lee announced creation of a National Science and Technology Council under his direct control in his 2011 New Year's Speech. It is to commence work in April 2011 and will coordinate some 75% of the country's R&D budget (Korea Times 2011a).

During its transformation into a country expanding technological frontiers, the Korean economy will face the kind of challenges outlined by the six points below that are typical for a newly advanced economy (cf. Mahlich and Pascha 2007, pp. 2–3):

1. Productivity and particularly total factor productivity increases become more important, as economic growth switches from an extensive to an intensive pattern.
2. The complementarities inherent in a catching-up style national innovation system are diminished; frictions appear among systemic components that change towards a technological frontier system at different speeds.
3. For economic policies as well, strategies that were relevant for a catching-up model of economic development, like sector-specific industrial policies, lose appeal and applicability.
4. Individualization, heterogenization of the social fabric, and democratization are typically associated with a maturing economy and lead to friction as well as changes in the political decision-making system and the execution of policies.
5. Social stabilisation and safeguarding the economic resilience gain importance.

[3]For the issue of whether all officially labelled "basic research" in Korea can truly be considered basic in the traditional sense, see the contribution of Michael Keenan in this volume.

6. Maturing economies typically undergo a process of ageing with the ensuing economic and social challenges.

Some of these issues can be illustrated through the profile of Korea included in the recent OECD science, technology and industry outlook for 2010 that focuses on science and innovation. As visualized in Fig. 1.1, Korea's profile is still quite different from the OECD average. Its R&D expenditures are very high compared to those of many other countries, but this is not yet reflected in a similarly pronounced R&D output performance, for instance as measured through triadic patents or the number of scientific articles (point 1). Several seemingly contradictory differences from OECD averages appear between various variables contained in the graph (point 2). For instance, the share of corporate R&D is still very much beyond the OECD average, but this does not seem to create a lot of potential for international collaboration: patents with international co-inventors and domestic R&D financed from abroad are still quite rare. With respect to economic policies (point 3), there are remarkable divergences in employment and education if the OECD average is, for simplicity, taken as the standard: despite Korea's strikingly high number of science and engineering graduates, the percentage of science and technology-related occupations is still rather small. The needed realignment of education and employment policies seems remarkable, particularly if the ageing of the population is taken into account (point 6).

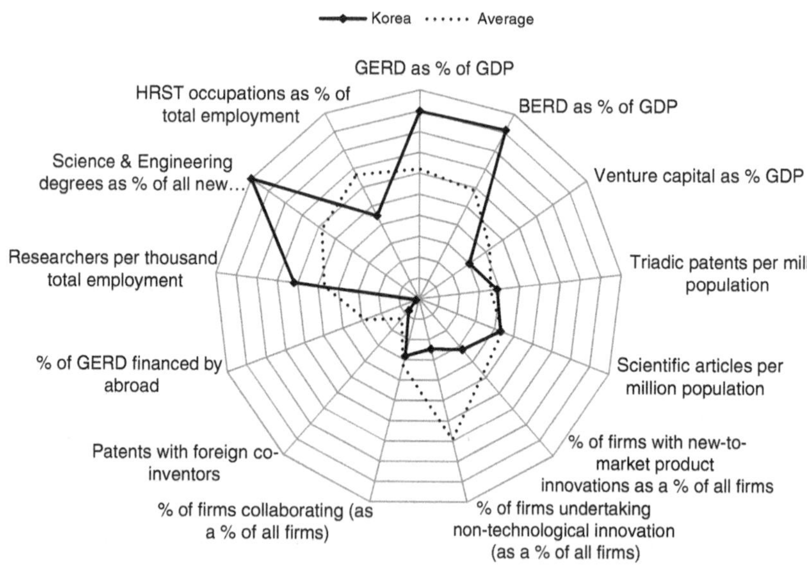

Fig. 1.1 Korea's technology and innovation profile (Source: OECD 2010, p. 199; statistical data from http://dx.doi.org/10.1787/888932334336, accessed in February 2011). *BERD* business expenditure on R&D, *GERD* gross expenditure on R&D, *HRST* human resources in science and technology

Other social challenges, like points 4 and 5 listed above, are less clearly visible in the graph. However, point 4, in particular, can be illustrated through an issue that surfaced with global and specifically Korean implications while making the last additions to this introduction, namely the Japanese Fukushima nuclear power plant disaster sparked by the 11 March 2011 earthquake and tsunami.

Given the cursory look necessitated by the publishing deadline, two observations about what Japan's nuclear accident means for Koreans seem particularly striking (Korea Times 2011b, c; Korea Herald 2011). First, the position of the government of Korea before the incident is very much characterized by an industrial policy approach reminiscent of the out-dated catching-up development model (point 3). Following the success in 2009 of a deal to build four nuclear reactors in the United Arab Emirates (UAE), valued at some USD2 billion, the Korean government stressed the role of nuclear power technology as a potential major exporting industry. It announced a plan to sell ten reactors abroad by 2012, to be increased to 80 by 2030, making Korea one of the top three global suppliers of nuclear technology. With pride, it is claimed that the current top-range APR-1400 model, sold to the UAE, features a fuel cost that is 23% lower than the French Areva's EPR-1600 type (Choi 2010). For domestic power generation as well, planning is very ambitious. While nuclear energy already accounts for about 40% of electricity supply, this is to be increased to 60% by 2030, which involves the construction of ten additional nuclear power plants. Second, following Japan's problems at Fukushima, the government is facing predictable concerns about nuclear safety issues. Important figures include lawmakers from the leading opposition party, the Democratic Party, but also concerned legislators from within the governing Grand National Party. Moreover, civic groups have become ever more active, and the Busan Bar Association has announced that it will file for an injunction to stop operation of an older reactor in Busan, the country's second-biggest city. While the government stresses that it has no plans to change its long-term strategy, within 1 month of the accident, it has promptly made considerable efforts to appease the opposition and confront the legitimate safety concerns. For instance, on 28 March, the nuclear power committee, chaired by the prime minister, announced that within 1 month all 21 nuclear reactors would be inspected. Moreover, in July the Nuclear Safety Commission, currently within the Ministry of Education, Science and Technology, is to become an independent agency headed by a ministerial-level appointee. It is unclear whether these safety-related measures could have longer-term implications that undermine the nuclear strategy. Nevertheless, the relevance of democratization in the ability of governments to install R&D policies in the traditional top–down manner has already become quite obvious (point 4).

Points 1–6 listed above are relevant for all advanced or semi-advanced economies, and it is legitimate to ask whether Korea faces any peculiar challenges. In the ensuing debate on nuclear power policy, Korean policymakers already compare the situation in their country with others and draw conclusions for handling the domestic issues. For instance, not only is Japan's radiation leak obviously of upmost importance for Koreans' current safety concerns, but the apprehensive policy response of a far-away country like Germany has also entered the local debate. The opposition Democratic Party chairman declared, "Like Germany, Korea must halt the operations of old reactors and instead focus on developing renewable and clean energy sources" (Korea Times 2011c). International gestation effects and the impact of multilateral monitoring thus lead to a certain levelling of issues and approaches.

So, what is peculiar about innovation and technology in Korea? Peculiarities with respect to the points raised above are certainly only peculiar by degree. For instance, it is well known that Korea is one of the countries, if not *the* country, with the lowest birth rate, leading in the forthcoming decade to a sharp decline in the size and a dramatic surge in the average age of the workforce (e.g., Poston and Davis 2009). This will have enormous consequences for Korea's future ability to innovate. Nevertheless, the challenge of societal ageing is by no means limited to Korea. Another topic of concern is the participation of women in the labour force, including their currently limited contribution as "human capital" in science and technology, but while gender equality may be particularly problematic in a male-oriented society like Korea's, almost all OECD countries face such concerns to some degree, some of them possibly more so than Korea.

Are there qualitative differences between Korea and other countries? While there is no easy way to reliably distinguish issues of degree from qualitative disparities, we propose the following list of Korean "peculiar peculiarities" that are relevant to innovation and technology:

- Economic Concentration
  Korea's economy is highly concentrated in a triple sense: the dominance of a few relatively large enterprises and the corresponding concentration in the external trade portfolio and R&D expenditures.
- "Korea Discount"
  The Republic of Korea faces a notable discount in its international profile. A major, but not exclusive, reason is its proximity to maverick North Korea.
- Closedness
  Korea is relatively closed towards international contacts and cooperation. While the same factor holds for some other countries, few of those approach Korea's size, development level and position.

Each of these deserves a closer look, starting with Economic Concentration. Based on the export-led development model, the Korean government has promoted a limited number of companies, organised in family-controlled business groups since the 1960s, widely identified as chaebol in English-language accounts. Despite considerable changes, for instance in terms of disentangling such groups and making them more accountable to outside shareholders after the 1997–1998 Asian financial crisis, a high degree of concentration still prevails. In contrast to the few globally successful electronics, automobile, shipbuilding and other assembly giants, many Korean small and medium sized enterprises (SMEs) lack the prowess and competence to compete internationally. Highly capable large enterprises even have to import critical semi-finished goods from elsewhere, and

Korea's production and trade portfolio is thus highly concentrated as well. Critical authors, like Michael Hobday, argue that striving for leadership in numerous fields and company sizes may be a wrongheaded strategy. In a sample of 25 leading Korean firms of various sizes, he finds that only a few are on or even reasonably close to the global R&D frontier. An over-enthusiastic endorsement of unconditioned global leadership may thus be inferior to pursuing a wider range of strategies, at least according to Hobday and his associates. They recommend to still including "behind the frontier" catching-up innovation strategies for some fields during the ongoing transition phase (Hobday et al. 2008).

Despite Korea's well-known export strength, in terms of key items that hold the world's top rank, the total number of globally dominant exports is only 52 (as for 2008), in the vicinity of much weaker exporters like Thailand or Turkey, and significantly lower than Taiwan (141 items) or even the city state Hong Kong (91) (Table 1.1). Diversification would probably have to move well beyond assembly-type industries, even beyond manufacturing and strive to establish a world-class service economy.

Consequences for R&D correspond to this structure. A few large enterprises dominate the technology landscape. Even among the top five companies, the differences are enormous. The R&D input of Samsung Electronics is bigger than numbers 2–5 combined (Table 1.2). The strong reliance on a few companies, export goods and R&D actors creates considerable dependencies and risks. In this respect, Korea faces some rather unique challenges. While it is nowhere easy to create

Table 1.1 Rank in terms of the number of top global export items, 2008

1. China	1,210
2. Germany	860
3. USA	587
4. Italy	288
5. Japan	213
6. Belgium	151
7. Taiwan	141
...	...
12. Hong Kong	91
...	...
17. Malaysia	53
18. Thailand	53
19. Republic of Korea	52
20. Turkey	40

Source: KOITA (2010, p. 143)

Table 1.2 Top five Korean companies according to R&D expenditures, 2009 (values in bil. KRW)

1. Samsung Electronics	7,042.5
2. LG Electronics	1,613.8
3. Hyundai Motor	1,277.0
4. Hynix Semiconductor	561.7
5. GM Daewoo Auto & Technology	533.1

Source: KOITA (2010, p. 47)

a multitude of small, innovation-friendly and competent SMEs, the challenges faced by Korea to diversify and spread risk seem almost formidable, despite the undeniable strength and success of its major enterprises.

Despite Korea's spectacular rise in recent decades, it still seems underrated on a global level, resulting in what McKinsey chairman Dominic Barton has referred to as the "Korea discount." While this relates to the undervaluation of Korea's brand image (Kim 2010), it is not pervasive. Samsung Electronics, for instance, has accomplished a stellar rise in two decades, moving from a cheap lower-end producer to one of the world's leading brands, ranked 19 in 2009. But apart from other household names like LG or Hyundai, the image of Korea in the world is still somewhat fuzzy and overshadowed by its two major-league neighbours, China and Japan. Nevertheless, this kind of Korea discount could well be a temporal problem, and given strategic efforts, there is no evidence to suppose that it will be an unassailable detriment.

A second, quite different Korea discount source is more important to consider: North Korea adds a certain level of risk for South Korea-related activities. For instance, the valuation of Korean joint-stock companies is usually somewhat lower than for comparable companies from elsewhere, which ceteris paribus implies somewhat higher refinancing costs for Korean enterprises. On one hand, a more peaceful situation on the Korean peninsula could help accelerate gains from innovation and technology there, but on the other, even with the prospect of an eventual confederation or even reunification, it is somewhat uncertain to what extent a "peace dividend" for innovation and technology could truly be realised.[4] Contemplating possible Korean unification scenarios based on Germany's experience in the early years after the 1990 reunification, considerable resources were needed to (re-)build the former East Germany, crowding-out at least some investment in and attention to technology- and innovation-related activities. Some observers believe this is one reason for the somewhat disappointing economic performance of Germany in the late 1990s and early 2000s. As for Korea, while it is frequently heard that in the longer run a unified Korea would profit from a wider resource base, particularly from natural resources and the labour pool in the northern parts of the peninsula, these resources would also influence the relative competitiveness of the Korean economy and dilute incentives to move quickly into knowledge-intensive and high value-added production, thus slowing the path towards intensive growth (point 1 from the above list of issues facing a newly advanced economy).

While South Korea is obviously not a closed society – having turned itself into one of the world's leading trading nations in recent decades – there is a conspicuous lack of deeply developed cooperation channels between Korea and the world outside, beyond the reach of more or less arms-length market transactions. This void, or closedness, is

---

[4]This is the topic of another book in this series on the modern Korean economy as part of the Strategic Initiative for Korean Studies of the Academy of Korean Studies (Seliger and Pascha eds 2011).

obviously related to Korea's caution with respect to superior former foes like Japan or current friends like the US. During the catching-up development phase, Korea found a viable model of selectively dealing with the world abroad, without losing control in overly dense cooperation arrangements that would be uncontrollable and thus dubious from a Korean perspective. The very viability and appealing success of this post-war model may paradoxically now create problems. While avoiding a mistaken sense of cultural determinism, this model has created a path of detained and selective opening that is difficult to leave. At the same time, Korea faces a considerable risk of not aligning itself with global trends and of falling behind.

Exemplifying the cost of this closedness in at least one respect, the Royal Society has recently undertaken an important study on the growing role of science collaboration on a global scale. It finds:

> Collaboration brings significant benefits, both measurable (such as increased citation impact and access to new markets), and less easily quantifiable outputs, such as broadening research horizons. The facilitation of collaboration, therefore, has a positive impact not only on the science conducted, but on the broader objectives for any science system. (Royal Society 2011, p. 6)

This trend creates new opportunities for regions that are not part of the traditional centres of science in the US, Western Europe and Japan, by creating an increasingly multi-polar scientific world. Countries like India and Brazil, but also those in the Middle East and elsewhere, and certainly including fast-rising China, make good use of these new opportunities, while Korea's role is still somewhat unclear. With respect to a key statistic used by the Royal Society study, international collaboration in research papers as a proportion of national output, Korea has made remarkably little progress during the period under study, 1996–2008.[5] Internationalising even more forcefully is thus one of the major challenges that Korea's innovation and technology system has to face.

The papers in this volume are divided into three parts. The first comprises six general surveys on Korea as a newly advanced economy. The contributions are accompanied by four brief comments that aim to open a broader discussion. The second part highlights two empirical papers on international knowledge flows to Korea both on an aggregate and on a firm level. In the third part, two contributions cover environmental issues of Korea's growth model. One comment in each part adds some insights from different angles.

Michael Keenan's opening paper elaborates on all six challenges Korea has to confront as a newly advanced economy and draws on the 2009 OECD Report on Innovation Policy in Korea, for which he was a principal author. He emphasizes that Korea's growth model used to be highly dependent on the exploitation of capital and labour. However, demographic problems and increased competition from newly industrializing countries such as China have rendered this model

[5]You can view a motion graph for this period and highlight the data for Korea at http://royalsociety.org/knowledge-networks-nations-graph/ (accessed in March 2011).

obsolete, and further growth can only be sustained by boosting innovation. While Korea has made remarkable progress in many fields, there are still many challenges ahead, particularly in education and public research. The author also points out that policy is in need of reform itself, as it must adapt to a more regulatory than developmental mode of intervention. Like several other contributions, he emphasizes the danger of closedness, one of the three peculiar factors noted above by us.

Jitendra Uttam's contribution relates to the third point listed above among the challenges facing a newly advanced economy, namely turning away from traditional industrial policies. He conceptualizes the distinctiveness of the Korean catching-up strategy and argues that the "developmental state" has transformed itself into a new "techno-scientific state". Instead of intervening in the market to promote exports, the Korean government is now trying to facilitate technological innovation. He reasons that the financial crisis of 1997 was a kind of inflection point, at which it was felt that Korea should broaden the social base of its science system to move beyond the old practice of technological catch-up. He evaluates recent policy initiatives such as BK-21/NURI, applying a Schumpeterian framework and deriving policy lessons that aim to widen the social base of innovation.

Ji Hong Kim's paper takes a similar view. The author mentions that the financial crisis of 1997 has shown the limitations of an input- and investment-driven growth paradigm, which is actually the No. 1 challenge on our list of issues facing a newly advanced economy. He summarizes Korea's catching-up strategy from the 1960s onwards, when Korea was at first still trapped in the vicious cycle of low savings, low investment and low growth. He identifies five development stages since then: industrialization, upgrading, rationalizing industry, liberalization, and knowledge economy. In his view, Korea's S&T policy has been well aligned to the respective development stages ranging from imitation to the establishment of a knowledge economy infrastructure. Ian Paterson comments on Kim's paper in the light of a recent World Bank ranking, which provides some useful international comparisons.

The contribution of Margot Schüller, Marcus Conlé and David Shim relates to our third and fourth points above. They present the outcomes of field research designed to shed light on the institutional and governance challenges that Korea currently faces as a newly advanced economy. Based on expert interviews with Korean policy makers and analysts, they evaluate the two latest S&T policy reforms that took place in 2004 and 2008, assessing the strengths and weaknesses of both while observing that the two followed different ideologies and methodological approaches. In her comment, Eunsook Yang adds some remarks on the cultural embeddedness of Korea's innovation approach and the influence of Confucianism.

Markus Pudelko and Jörg Büechl deal with the cultural and social foundation of innovation, which refers to the fourth challenge in our list. They analyze the Korean innovation system and highlight the differences between the Korean model and that of the US, Japan and Germany. They suggest that culture does heavily influence a country's capacity to innovate. Drawing on Hofstede's dimensions of culture, they propose that Korea shows particular strength in incremental innovations but will also increasingly need to build up its ability to generate transformational innovations – the latter being a strength of the US model. Due to social structure differences, however, it could be difficult for Korea to follow the US model. Instead, they suggest Germany as a more suitable role model for Korea, in particular with regard to Germany's small and medium enterprise sector. Markus K. Hödl and Jonas F. Puck discuss the paper within a more general and broader theoretical framework.

Bernhard Seliger addresses Korea's fifth challenge as a newly advanced economy by analyzing the need for social stabilisation and income redistribution by means of regional policy. He takes a close look at regional aspects of innovation and development and describes an initiative in Gangwon province that aims to foster innovation through the formation of innovative clusters and to alleviate regional inequalities in development. Although cluster policy is not a new policy instrument, it is by no means easy for Korea's regions to apply, given the nation's enormous centre-periphery problems. According to the author, the experience of Gangwon province shows that the attraction of companies to peripheral regions is possible only when subsidies are granted on a long term basis. Silvo Korez's comments emphasize the role of Korea as a de facto island economy. In his view, geographical constraints have a limiting impact on the flow of goods and ideas.

In the first paper from the second group, Keun Lee and Jia Jin focus on the mobile communication industry. Using patent data from the US Patent Office, they are able to trail cross border knowledge flows between Japan, Korea, Taiwan, and China. Specifically, they verify the proposition that the flow of knowledge follows the order of entry or catch-up in an industry, such that Japanese patents tend to cite American patents, Korean patents tend to cite Japanese patents, Taiwanese patents Korean patents, and Chinese patents Taiwanese. They conclude that Korea has yet to catch up with other advanced countries like Japan. In his comment, Bernhard Dachs poses the question if China's leapfrogging is the end of the story or if another emerging country after China is about to enter the stage.

Chol Lee, Chan-Soo Park and Ilan Vertinsky present results of a questionnaire based study that analyzes knowledge transfer between foreign firms and their Korean-based joint ventures with local companies. Their findings suggest that communication is the key to cross border knowledge transfer. Successful knowledge transfer in turn enhances local unit performance. Interestingly, the authors treat the opening up as a key issue for Korea.

The last group of papers is about environmental issues and demonstrates that Korea still has a long way to go when it comes to the first challenge of a newly advanced economy, namely switching from a resource-extensive to a sustainable-growth pattern. This issue is explicitly taken up in Greg Scarlatoiu's paper that deals with Korea's environmental policy and how it relates to the country's overall innovation path. Korea's growth has been heavily dependent on fossil-fuel and Korea has become one of the world's major emitters of greenhouse gases. Korea's recent commitment to "Low Carbon Green Growth" aims to decouple economic growth from environmental degradation. Korea has set up many programmes and measures to promote investment in environmentally friendly technologies. As large energy intensive industries, such as steel, still remain at the core of Korea's economic growth, the author concludes that the "greening" of South Korea's industry is probably the most difficult challenge ahead.

Mathew Shapiro empirically assesses the environmental Kuznets curve that proposes an inverted-U- shape relationship between pollution and economic development. In his contribution, he compares Korea with Japan, China, and Taiwan. Among these peers, Korea is the only country where higher income levels are associated with higher energy intensity, while in the three other countries the relationship is the other way round. This result underlines Korea's challenges in implementing its "green growth" strategy. In his comment, Reinhold Hofer calls for a further opening of the economy in order to benefit from a close relationship with foreign technology suppliers; this would both increase income and reduce carbon emissions at the same time.

As shortly mentioned in the opening, most of the contributions were first presented during the international conference "Innovation and Technology in Korea in International Perspective", held September 24–25, 2009, at the Austrian Federal Economic Chamber in Vienna. During the event, we invited commentators to discuss the presentations, and this as well as extra comments and reviews formed the basis for a fundamental revision of the papers included in this book. Several commentators offered input not only in regard to the respective papers, but including some added value on relevant questions. Moreover, we invited a number of additional contributions on topics and perspectives that we found particularly helpful to understand the presence and future of innovation and technology in Korea in an international perspective.

We would like to extend our thanks to all presenters, authors and commentators for their co-operation in focusing their attention on the issues at stake and for creating, as we hope, a valuable outlook on Korean innovation and technology in the early twenty-first century. Moreover, we would also like to thank the Austrian Chamber of Commerce for having hosted yet another conference on Korean innovation and technology, after the first such conference held in 2005. The 2009 event was combined with a "Korea Design Forum", with a fascinating view of what Korea (and the West) have to offer in terms of advanced design solutions, and a "Networking Korea Forum" of the Austrian–Korean Society that was held in the Austrian Parliament. To all the co-organizers of these meetings-in-conjunction, we are grateful for the additional opportunities they provided.

Our utmost gratitude is due to the Academy of Korean Studies, which under the Strategic Initiative for Korean Studies supported the conference and this publication. Without this support, the whole endeavor would not have been possible.

Bernhard Seliger, head of the Hanns Seidel Foundation, Seoul Office, was the important partner of Werner Pascha in the proposition of a small "book series on the modern Korean economy" as part of the Academy of Korean Studies Initiative mentioned above. Sarah Kohls, until recently also in the HSF office in Seoul, was an efficient, circumspect and simply invaluable support in organizing and managing the conference as well as the publication. David Kendall of Kendall Editing helped to bring the texts into a much more readable form. To all of them, we owe our sincere gratitude.

References

Choi Y-H (2010) Korea boasts global competitiveness in nuclear tech. *Korea Times*. http://www.koreatimes.co.kr/www/news/biz/2010/05/291_64190.html. Accessed Mar 2011

HSBC Global Research (2009) A climate for recovery. The colour of stimulus goes green, 25 February 2009. http://www.globaldashboard.org/wp-content/uploads/2009/HSBC_Green_New_Deal.pdf. Accessed in Mar 2011

Hobday M, Rush H, Bessant J (2008) Approaching the innovation frontier in Korea: the transition phase to leadership. Presentation for the Sabanc University Competitiveness Forum, Istanbul. http://ref.advancity.net/tr/dokumanlar/Hobday_presentation.pdf. Accessed Mar 2011

KOITA (Korea Industrial Technology Association) (2011) Major indicators of industrial technology, Seoul

Kim J-K (2010) Build on G20 role to address Korea discount. *Korea Times*. http://www.koreatimes.co.kr/www/news/biz/2010/11/301_71327.html. Accessed Mar 2010

Korea Times (2009) Korea to increase R&D spending 10%. http://www.koreatimes.co.kr/www/news/nation/2009/10/113_43582.html Accessed Mar 2011

Korea Herald (2011) Korea may close old reactors. http://www.koreaherald.com/national/Detail.jsp?newsMLId=20110328001033. Accessed Mar 2011

Korea Times (2011a) National science, technology council to be launched. http://www.koreatimes.co.kr/www/news/nation/2011/03/113_83887.html. Accessed Mar 2011

Korea Times (2011b) Focus more on safety. http://www.koreatimes.co.kr/www/news/opinon/2011/03/167_83938.html. Accessed Mar 2011

Korea Times (2011c) Growing calls to overhaul energy policy. http://www.koreatimes.co.kr/www/news/nation/2011/03/113_83232.html. Accessed Mar 2011

Langhammer RJ, Heilmann S (2010) Managing the crisis: a comparative assessment. In: Bertelsmann Stiftung (ed) Managing the crisis. a comparative analysis of economic governance in 14 countries. Verlag Bertelsmann, Gütersloh, pp 9–30

Mahlich J, Pascha W (eds) (2007) Innovation and technology in Korea. Challenges of a newly advanced economy. Physica, Heidelberg/New York

OECD (2010) OECD science, technology and industry outlook 2010, Paris

Pascha W (2010) South Korea's economic policy response to the global financial crisis. In: Köllner P et al (eds) Korea 2010: politics, economy and society. Brill, Leiden/Boston, pp 135–164

Poston DL, Davis MA (2009) Population aging in the Koreas. In: Uhlenberg P (ed) International handbook of population aging. Springer, Heidelberg, pp 173–190

Royal Society (2011) Knowledge, networks and nations. Global scientific cooperation in the 21st century, London. http://royalsociety.org/uploadedFiles/Royal_Society_Content/Influencing_Policy/Reports/2011-03-28-Knowledge-networks-nations.pdf Accessed Mar 2011

Seliger B, Pascha W (eds) (2011) Towards a Northeast Asian security community: implications for Korea's growth and economic development. Springer Science/Business Media, New York

Chapter 2
Moving to the Innovation Frontier: Lessons from the OECD Review of Korean Innovation Policy

Michael Keenan

Introduction

During 2007–2008, the OECD carried out a review of Korea's innovation policy (OECD 2009), offering a comprehensive assessment of the innovation system with a special focus on the role of government. The Korean review is one of a series of national innovation policy reviews being carried out by the OECD[1] and was intended to assess government policy in the area of science, technology and innovation (STI). The review paid particular attention to the science base, but also covered the wider innovation system, including the activities of firms and the 'framework conditions' conducive to supporting their innovation performance.

At a strategic level, the main task of Korea's innovation policy is to contribute to the country's economic convergence with the more advanced OECD economies. Historically, Korean growth can largely be attributed to the exploitation of capital and labour, driven by the highest rate of business investment in the OECD area, a growing population, long working hours and rising participation in the labour force. However, since growth of inputs of labour and capital have begun to slow, continued convergence to the OECD average through input-based growth will become more difficult. Indeed, with the lowest fertility rates in the OECD area, combined with increased competition from newly industrialising countries, particularly China, sustaining productivity growth will increasingly depend upon boosting innovation.

It should be stressed at this point that Korean government policy has long recognised the importance of innovation for development. Even the earlier technological imitation activities of firms and public technological development centres, often promoted by public policy, involved significant amounts of technological and

[1]In addition to Korea, the innovation policies of the following countries had been reviewed by 2010: Luxembourg, Switzerland, South Africa, New Zealand, China, Chile, Norway, Hungary, Mexico, Greece and Russia. Further information can be found at http://www.oecd.org/sti/innovation/reviews

J. Mahlich and W. Pascha (eds.), *Korean Science and Technology in an International Perspective*, DOI 10.1007/978-3-7908-2753-8_2,

organisational innovation. However, as Korea has approached technological frontiers in several areas over the last decade or so, the scope for relying solely on adaptation and exploitation of existing technologies has diminished considerably. A different type of innovation is now required, one characterised by greater levels of high-risk exploration in addition to a more familiar exploitation.

This calls for change in the organisation and dynamics of the national innovation system, change that has been framed by government policymakers as a shift away from a "catch-up" towards a "creative" mode of development.[2] The catch-up mode centred upon large-scale strategic technology development with government-affiliated research institutes and large family-owned business conglomerates (chaebol) taking a leading role. It harboured little space for the creation of innovative start-ups, the development of technological capabilities among local independent SME suppliers, and the building of basic (fundamental) research capacity. The creative mode, by contrast, emphasises such actions and calls for greatly increased spending on R&D – by both the public and private sectors – with the aim of improving knowledge generation and diffusion across the national innovation system.

Shifting from a catch-up to a more creative mode of development requires changes in the institutions and ideas that structure the actions and interactions of national innovation system actors. In many instances, arrangements that worked well during earlier waves of catch-up need to be dismantled and reconstituted. This is far from being a straightforward task and, in today's Korea, cannot be centrally coordinated by the government to the same extent as in the past (Mahlich and Pascha 2007). Instead, new institutional arrangements will emerge through conflict and cooperation among existing and new protagonists that seek to pursue their interests and to promote particular ideologies of development.

This chapter sets out to provide a snapshot of Korea's progress in its transition to a more creative mode of development. It focuses on four key areas where government policy continues to play an important role in shaping the Korean innovation system, namely education and skills, SME development, public research, and international knowledge linkages. With reference to these four areas, a national innovation system conducive to creative exploration activities requires, among other things, a ready supply of advanced and adaptable skills, the participation of both large and small firms in frontier knowledge generation and utilisation, a strong basic (fundamental) research base, and strong linkages to international sources of frontier knowledge.

While Korea has made remarkable progress on all of these fronts, there is still much left to do. Structural and institutional 'lock-ins' along catch-up development paths and trajectories, particularly in education and public research, mean that a lot of old ideas and institutions remain, as do old habits and routines, co-existing with a lot of new initiative. Policy is trying to accelerate the shift, but is also in need of reform itself – it, too, can no longer work in a catch-up mode and must adopt a more regulatory than developmental mode of intervention, not only because of the

[2]For example, the 2004 Implementation Plan for the National Innovation System draws this distinction, as do more recent policy strategies.

independence and strength of other actors in the system, but also because of the very nature of the changes that must be engineered.

Policy statements have acknowledged this: for instance, as far back as 1999, the Vision 2025 initiative proposed a number of significant shifts in S&T policy, arguing for the development of a more globally open, privately-led and diffusion-oriented innovation system driven by longer-term S&T considerations. This was followed up in 2004 by the Implementation Plan for the National Innovation System, where the contrast between a catch-up and creative innovation system was first articulated in detail. The Plan paid particular attention to developing the innovation capabilities of SMEs, universities and government research institutes and to improving the linkages between them. It also emphasised various inputs, including human resources and finance, and the need for the conditions of their production and availability to be reformed. More recently, the Lee Myung-bak administration has launched its so-called "577 Initiative", which includes several ambitious targets: to reach an R&D intensity of 5% by 2012 (it stood at an impressive 3.36% in 2008); to focus upon seven key areas of R&D and seven systems; and to become one of the seven major S&T powers in the world (MEST 2008).

The sections that follow focus on the aforementioned key areas for policy action, namely the development and utilisation of appropriate skills, strengthening the innovation capabilities of SMEs, increasing and realigning R&D spending, and promoting a new internationalism. The chapter is then rounded off with some concluding remarks that include an overall assessment of the strengths, weaknesses, opportunities and threats facing the Korean innovation system.

Development and Utilisation of Appropriate Skills

Throughout Korea's industrialisation process, an adequate supply and mobilisation of qualified S&T human resources has been a cornerstone of industrial and S&T policy. In an earlier period, when Korean exports were still relatively simple, the workforce required only minimal skills. But as Korea entered new high-tech sectors of production, the education system had to expand in scope and scale to meet new demands for higher level skills. With the population almost doubling in size in the period 1960–2000, up-skilling efforts centred on young people. These efforts saw a massive expansion in tertiary education provision, with new universities and colleges springing up all over the country. Today, more than 80% of young Koreans are participating in tertiary education programmes, the highest level in the world. The proportion graduating with science and engineering (S&E) degrees is also the highest in the OECD (Fig. 2.1).

These achievements are undoubtedly impressive and are evidence of the definitive steps taken to develop the necessary skills base for a more creative mode of innovation-led development. Yet, there are also concerns. Firstly, as in many other OECD countries, the 'massification' of tertiary education has been accompanied by anxieties of falling standards. Furthermore, the conventional lecture-dominated

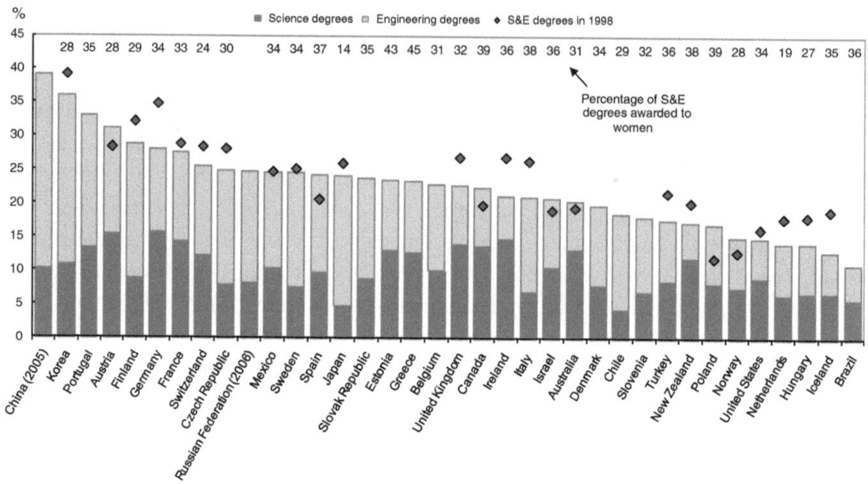

Fig. 2.1 Science and engineering degrees as a percentage of total new degrees (2007) (Source: OECD 2010)

teaching style that is common to Korean universities has been criticised for failing to nurture creative thinking. Korean firms regularly complain about the standard of graduates entering the employment market, arguing that curricula are outdated and in need of renewal. Similar complaints can be heard in most OECD countries, but they likely have some merit in the Korean context. Recent government policy has sought to tackle such problems by subjecting universities and colleges to stronger and more systematic evaluation arrangements and by promoting their specialisation in areas where they can better serve the needs of local employers (the latter applies particularly to universities and colleges outside of the Seoul metropolitan region). Stronger linkages to other innovation system actors, particularly firms, are also being actively encouraged through a variety of incentives. Such measures offer scope for quality improvements in tertiary education, though they are unable to address perhaps the most significant concern facing Korean education, which lies with secondary schools. The extensive use of rote learning and cramming, done in support of preparations for the university entrance exam, is less than conducive to the development of creative skills and continues to be a major weakness of the Korean education system.

Secondly, and perhaps more worrying, there is strong evidence of a mismatch between the skills required by the labour market and the qualifications of tertiary education graduates. Overall, there are too many people with bachelor's degrees and too few with master's or doctoral degrees in high-technology fields. Data from the Korean Educational Development Institute (KEDI) show that only two-thirds of new graduates of universities and colleges find employment within 6 months of graduating (KEDI 2006). In a further study, Grubb et al. (2006) conclude that the expansion of tertiary education seems to have outpaced the demand for jobs requiring high levels of schooling, and that the Korean system is preparing too few technicians in colleges, while there are too many individuals with bachelor's degrees who cannot find appropriate employment. It should be said that this problem is hardly unique to Korea – in most OECD countries, there have been declines in uptake of vocational courses as more and more young people are attracted to the higher prestige of university degrees but ultimately face problems finding appropriate employment. The problem is, however, perhaps more acute in Korea given the very high university enrolment rates. Recent government policy has sought to tackle the issue by attempting to raise the prestige of vocational colleges and qualifications and to offer them as routes for obtaining degrees.

Efforts to raise the skills base must also take account of plummeting fertility rates. Korea currently has one of the lowest total fertility rates in the OECD; the flow of young people into the workforce will fall sharply over the next decade or so. The focus of skills upgrading efforts must therefore shift increasingly to the existing workforce, i.e. to lifelong learning provision. The Korean government has acknowledged this and has in recent years established important elements of a system of lifelong learning. However, these elements are insufficiently connected and inadequately linked with the employment system, and too few resources are allocated to adult learning (OECD 2005). On the supply side, the continuing emphasis upon full-time degree programmes in the Korean education system is generally ill-suited to the needs of adults already in employment (Grubb et al. 2006). But the most significant barriers are to be found on the demand side, where the dualistic structure of the Korean labour market[3] offers too few incentives for employers and employees to engage in lifelong learning activities. Specifically, individuals in regular employment are paid in accordance with seniority rather than qualifications, while the lack of mobility prospects for irregular workers and workers in SMEs all mean that there is little financial incentive for many workers to invest in their own skills development. Labour market reforms are therefore a likely necessary condition for lifelong learning initiatives to be widely embraced.

If reforms of labour markets and education systems are not big enough challenges for policy makers to tackle, perhaps the biggest concern is the rather low rate of female employment, which can at least partly be ascribed to Korean societal values and tradition. Unusually, women with university level educational attainment have lower rates of participation in the labour force (57%) than those

[3]The Korean labour market is divided into regular and irregular/non-regular workers. Wages and fringe benefits of the latter are substantially lower than those of the former. The labour market for regular workers is characterised by lifetime employment, low risk of layoffs, good social security, trade-union representation (at least in larger manufacturing companies), and minimal intercompany mobility. By contrast, the market for irregular workers can be described as a competitive market with almost no restrictions on hiring and firing. The labour market is also segmented by company size, gender and levels of qualification. In smaller companies wages are substantially lower than in bigger ones. Furthermore, there is little opportunity for workers in small companies to move to larger ones. A far higher proportion of women than men are in irregular employment. Higher percentages of irregular workers are found among the less skilled than among the more highly skilled (OECD 2005).

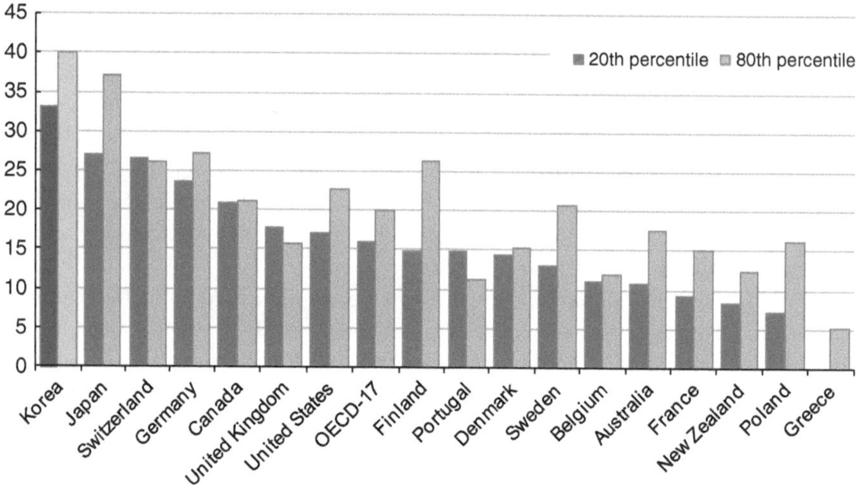

Fig. 2.2 Gender gap in full-time earnings at the top and bottom of the earnings distribution, 2003 or latest available year (Source: OECD 2006a)

who have only compulsory education (59%) (OECD 2007a).[4] An important reason for the low activity rate among well-qualified women lies in the duality of the labour market, which manifests itself most clearly in the significant gender wage gap. As Fig. 2.2 shows, while there are still wage differentials between women and men across all OECD countries, the situation in Korea is particularly acute. Other unfavourable labour market outcomes for women include a large proportion (one-third) of women holding temporary work contracts (with few prospects for career advancement) and a working culture geared towards long hours.[5] These are structural problems in the economy that will be difficult to solve, yet there is a clear need for greater female participation in the labour force, given the projected decline in the Korean labour force (OECD 2005).

Strengthening the Innovation Capabilities of SMEs

Korea's catch-up model has been dominated by the chaebol and has, until recently, largely neglected SMEs. The chaebol-dominated industrial structure that created economies of scale and scope and opportunities for substantial cross-subsidisation

[4]An important reason for this anomaly lies in traditional family values that expect women to remain at home. Well-qualified men can normally earn a wage that allows them to finance the traditional family model. Since highly qualified Korean women tend to marry well-qualified men, they are more likely to participate in the traditional family model (OECD 2005).

[5]Almost 90% of the male workforce spends over 40 h a week at work as compared with the OECD average of 76%. Although fewer Korean women (77%) work over 40 h a week, this is significantly higher than the OECD average of 49% (OECD 2007a).

proved ill-equipped to foster a dynamic SME sector that could provide key components and critical complementary support services to the chaebol (Ernst 2000). Taking the electronics industry as an example, this situation differs markedly from that found in Chinese Taipei and Japan, where innovative SMEs have played an active role in industrial upgrading.

The growth of a more dynamic SME sector has been strongly associated with the shift to a more creative mode of development. It is seen to offer the possibility of opening new growth paths based on radical innovations that larger firms might otherwise disregard. This is because radical innovations often make obsolete the extensive technological fixed assets of large firms and their accumulated capabilities. By contrast, SMEs are less burdened by such sunk costs and benefit from a greater flexibility to develop and exploit radical innovations. This 'New Economy' vision of the central role of SMEs in innovation has been widely adopted by OECD governments, including by Korean policy makers. Yet it tends to underplay the continuing dominant roles of large firms in innovation.[6] It also provides only a partial account of the contributions SMEs can make to industrial dynamism and pays, in particular, too little attention to the prominent role played by linkages between large and small firms in promoting innovation, e.g. through supply chains. As implied above, these relations have, historically, failed to foster the development of strong innovation capabilities among Korean SMEs. But there are signs that this is changing, with the chaebol looking to play more supportive roles in the development of a vibrant SME sector. This is in their self-interest, of course, since their technological leadership positions increasingly require supplier companies to have substantial innovation capabilities. If these capabilities cannot be found among Korean suppliers, then the chaebol can find them elsewhere, a scenario that could see extensive industrial 'hollowing-out'. A particular case of the influence of a major chaebol in driving supplier and associated companies is provided by the rapid development of a mobile telecommunication cluster in Gumi, driven by the outsourcing strategy of Samsung Electronics (Lee 2006). Increases in telecommunications sales in the global market have driven the rapid expansion of the local agglomeration of mobile companies with most of the contract suppliers accumulating core technologies through their Samsung contracts. At the same time, recognising their vulnerability to sudden changes at Samsung, they have also sought to develop independent strategies and customers.

One of the most important public agencies promoting innovation capacities in SMEs is the Small and Medium Business Administration (SMBA). Founded in 1996, it operates several schemes to support SMEs, covering areas such as entrepreneurship, human resources, financing, marketing and innovation. A selection of its innovation promotion support schemes, aimed largely at existing (as opposed to start-up) SMEs, is provided in Box 2.1. The government has

[6] The performance of some of Korea's chaebol is testimony to this, with the likes of Samsung Electronics going from strength to strength over the last decade while working at technological frontiers.

also initiated a broad array of incentive schemes targeted at promoting R&D in SMEs, including direct R&D funding, tax waivers, tariff exemption for R&D equipment, and exemption from military service for research personnel. It would seem that these efforts have met with some success. For example, R&D expenditure by SMEs grew almost fivefold from 1997, when it amounted to 12% of total firms' R&D expenditure, to 2006, when it reached 24%. Furthermore, in 1997, only 24% of researchers worked in SMEs, but by 2006 43% did so, for a fourfold increase in actual numbers of researchers to almost 75,000. This impressive growth of R&D activities in Korean SMEs can be accounted for by both the 'traditional' SME sector and the so-called 'venture firm' sector.[7] The latter has performed particularly strongly, and now accounts for a higher proportion of business expenditure on R&D and number of researchers than the traditional SME sector.

Box 2.1: Main SMBA Programmes in Support of Technological Innovation in SMEs

To strengthen the innovation capabilities of SMEs, SMBA is pursuing various policies to foster innovative SMEs; reinforce the networking of industry, academia and research institutes; and promote the commercialisation of developed technology. Some of its main programmes involve the following:

Fostering Innovative SMEs (Inno-Biz): Prospective SMEs with technology development and innovation capabilities are designated as "Inno-Biz" and fostered as a core engine of growth. Under this scheme, the SMBA identifies innovative SMEs with superior technologies which are able to raise their technological level through their own technological innovation system. The purpose is to ensure that they will develop into global blue-chip firms by providing comprehensive support through schemes such as technology assurance and preferential treatment for credit loans.

Korea Small Business Innovation Research (KOSBIR) system: Under this system, government ministries and government-financed institutions are required to allocate at least 5% of their R&D budget to support SMEs' technology development and to cover R&D expenses of SMEs capable of separately developing technology. SMBA spent approximately USD920 million in 2005 on this system.

SMEs' Technology Innovation Programme: SMEs capable of developing technologies without support can recover up to 75% of the expense of developing new products or enhancing product quality. The financing ceiling

[7] A venture firm can be generally defined as an early-stage innovative SME. However, there is a more specific legal definition for a Korean venture firm that classifies it as either an SME in which a venture capital firm has invested, or which has a relatively large R&D expenditure as a percentage of sales, or whose business stems from technology, as indicated by a patent.

is USD300,000 for 2 years for strategic tasks or USD100,000 for 1 year for general tasks. Under this programme, the SMBA supported 1,912 SMEs in 2005 alone.

Industry-University-Research Consortium Programme: Various policy measures have been put in place to reinforce networking for technological innovation among enterprises or among industry, academia and research institutes. The measures include the Industry-University-Research Consortium Programme, which seeks to boost the technological capabilities of manufacturing SMEs through collaborative technological development with universities or GRIs. About 220 consortia were formed in 2005, to support 2,700 SMEs in developing new technologies.

Transferred Technology Development Project: This project aims to prevent superior technologies from being discarded and to enhance the technological innovation capabilities of SMEs. To this end, the SMBA covers the additional development costs required to commercialise transferred technologies owned by universities, research institutes and businesses. In 2005, the SMBA supported about 90 such tasks.

Source: SMBA website (http://www.smba.go.kr/smba/main/english/index.jsp), accessed December 2007.

The availability of finance has naturally played an important role in these developments. As is common in most countries, SMEs have difficulties in borrowing from traditional lenders, largely on account of the information asymmetry between lending financial institutions and borrowing SMEs and a lack of tangible collateral to secure creditors' confidence. To address this market failure, the government employs loan financing and loan guarantee programmes for technological innovation in the private sector. In addition, SMEs can borrow "policy funds" from the SMBA at low interest rates to promote investment in facilities, restructuring, commercialisation of new technologies, and to assist start-up activities. The total policy funds allocated in 2005 amounted to more than USD3 billion. For indirect financing services, the SMBA also provides a security assurance service for SMEs ineligible for bank loans owing to a lack of collateral and technology. This service allows these SMEs to borrow needed funds from the Korea Credit Guarantee Fund and its local offices and from the Kibo Technology Fund.[8] Significantly, these organisations conduct technology appraisals, thereby increasing the transparency of opportunities and risks associated with new technology investments.

[8]The mission of the Kibo Technology Fund is to contribute to the national economy by providing credit guarantees to facilitate financing for new technology-based enterprises while promoting the growth of technologically strong SMEs and venture businesses. It has three main functions, namely technology guarantee, technology appraisal and business consultation.

Besides providing loan financing and loan guarantee programmes, the Korean government has sought to stimulate private venture capital markets with a view to enhancing high-technology venture firms and start-ups in Korea. It has revised laws related to financial markets and provided funds to be injected into venture capital investment funds. For example, to build a foundation for the stable growth of venture capital, the SMBA created a Fund of Funds of around KRW1trillion (for the period 2005–2009) for financing investment funds for early-stage venture businesses. The objective is to provide stable financing that is able to meet the needs of the capital market in the long term. Furthermore, the SMBA has introduced streamlined procedures for mergers and acquisitions and business transfers thereby making it easier for venture capital to exit venture businesses.

Increasing and Realigning R&D Spending

An important move towards a more creative mode of development involves increasing levels of spending on R&D. Korea has performed exceptionally well in this regard and has one of the highest R&D intensities in the OECD (Fig. 2.3). Furthermore, around three-quarters of Korean R&D is carried out by the private sector, a proportion similar to other leading manufacturing economies, such as Japan and Germany.

Increases in R&D intensity cannot alone be taken as evidence of a shift towards a more creative mode of development – the sorts of R&D performed should also have qualitatively shifted. The OECD's Frascati Manual distinguishes between basic research, applied research, and experimental development, which, while somewhat imperfect, can nonetheless be used as a proxy to estimate any qualitative shifts.[9] To recall, a more creative mode of development translates into more exploration, which, among other things, calls for more basic research to be performed. The proportions of gross domestic expenditure on R&D (GERD) spent on basic research have indeed increased over recent years, reaching 16.1% of the total in 2008 (compared to 12.6% in 2000), which is more or less in line with the proportions seen in other advanced economies.[10] More unusually, the majority of this basic

[9]The Frascati Manual defines these forms of research as follows (OECD 2002):

- *Basic research* is experimental or theoretical work undertaken primarily to acquire new knowledge of the underlying foundation of phenomena and observable facts, without any particular application or use in view.
- *Applied research* is also original investigation undertaken in order to acquire new knowledge. It is, however, directed primarily towards a specific practical aim or objective.
- *Experimental development* is systematic work, drawing on existing knowledge gained from research and/or practical experience, which is directed to producing new materials, products or devices, to installing new processes, systems and services, or to improving substantially those already produced or installed.

[10]For example, the proportion of GERD devoted to basic research in 2006 in the United States was 18.7% and in Japan 14.3%.

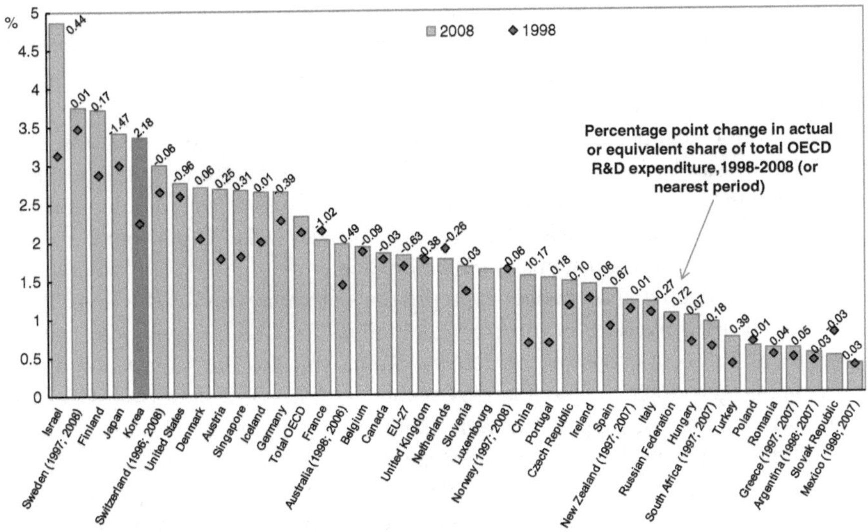

Fig. 2.3 R&D intensity (gross domestic expenditure on R&D as a percentage of GDP), 2008 or latest available year (Note: In Israel, defence R&D is not covered. Furthermore, humanities and law are only partially covered in the higher education sector. Due to the lack of a comprehensive business register for South Africa, R&D expenditure may be underestimated by 10–15%) (Source: OECD 2010)

research – around 60% in 2006 – is performed by firms. The share of business expenditure on R&D (BERD) given over to basic research has almost doubled, from 6.5% in 1998 to around 12% since 2004. Around 90% of this is performed by large firms, reflecting their concentration in high-technology manufacturing industries, particularly in the information and communication technology (ICT) and automotive sectors. The large firms in these industries are increasingly working at the technological frontier in many product areas and have had to invest in more fundamental research to support their leadership positions. The remaining 10% is performed by venture firms and traditional SMEs in roughly equal amounts. Although this proportion appears relatively low, it should be borne in mind that the level of basic research performed by SMEs has increased rapidly in recent years from a very low starting base.[11]

As for the other Frascati Manual categories of R&D, experimental development is of particular interest. Korea has a long and fruitful tradition of performing this type of R&D in its efforts to exploit and adapt existing technologies. It still accounts for the majority of R&D performed in Korea, as it does in other advanced manufacturing economies. What is of particular interest, however, is its increasing share of GERD – up from 60.7% in 1999 to 65.0% in 2006. During this period, the

[11]For example, between 2005 and 2006, expenditures on basic research in venture firms increased by an impressive 60% (MoST and KISTEP 2007).

share of BERD given over to experimental development has remained static, so the increase can only be accounted for by the changing composition of R&D performed in government research institutes (GRIs) and higher education institutes (HEIs). Korean R&D data confirms this, showing marked increases in the share of experimental development performed by these institutes while the share of basic research has fallen by similar amounts. Funding of GRIs' and HEIs' R&D by the business sector is not the explanation for this evolution – in fact, firms account for a diminishing proportion of funding of GRIs' and HEIs' R&D activities. This evolution must therefore reflect changing patterns in public funding.

In this regard, civil gross budget appropriations or outlays on R&D (GBAORD), a commonly used measure of government funding for R&D, shows little change in terms of its socio-economic objectives. Korea still has the largest proportion of economic development programmes within its GBAORD in the OECD area (Fig. 2.4), a pattern of support that reflects the legacy of the government's role in facilitating the country's technological 'catch-up'. Rather, the change lies in the nature of this support: public R&D programmes that historically targeted mostly large-scale industrial technology (with the intention of supporting near-immediate industrial development by the chaebol) have been largely superseded by public R&D programmes with a greater focus on generic technology development (see Box 2.2). Furthermore, greater attention is being given to support for R&D commercialisation and for technological developments in SMEs, as outlined already. These changes have implications for the GRIs and HEIs, who have come to be viewed as increasingly important players in creating and contributing to new high-tech industries. This perspective has strongly shaped policy incentives and accountability measures and appears to have shifted GRIs' and HEIs' R&D activities more 'downstream' as they seek to better commercialise their R&D results.

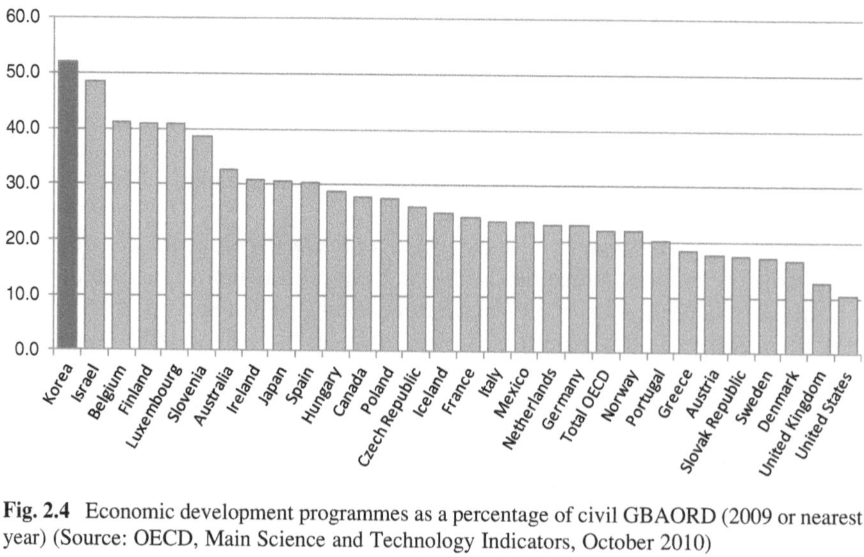

Fig. 2.4 Economic development programmes as a percentage of civil GBAORD (2009 or nearest year) (Source: OECD, Main Science and Technology Indicators, October 2010)

Box 2.2: Major Public R&D Programmes Focused on Economic Development

The high proportion of GBAORD given over to economic development programmes reflects the fact that much public R&D spending is channelled through public-private R&D collaboration programmes funded by the Ministry of Knowledge Economy (MKE), and to a lesser extent by the Ministry of Education, Science and Technology (MEST).

The MKE has the largest support measures for industrial technology development – including sectoral technology programmes to support "flagship industries". In a more targeted manner, the MKE is also responsible for the Next-generation Growth Engines R&D Programme, a scheme started in 2003 which targets ten strategic "growth engine" industrial sectors. The programme is further supported by the MEST's Human Resources Development Plan for Next-generation Growth Engines Sectors to ensure a supply of appropriately skilled human resources.

The MEST is also a major player in industrial technology development, though it has sought to take a longer horizon than the programmes supported by the MKE. For instance, its 2006 Total Roadmap included a foresight project that identified 90 strategic technologies with the potential for considerable impact upon Korea's economic growth. More recently, the 577 Initiative has identified 50 "critical" technologies and 40 "candidate" technologies in seven major technology areas. The MEST is also responsible for funding several national R&D programmes that seek to improve national competitiveness. For example, the twenty-first Century Frontier R&D Programme is a medium-to-long-term funding programme to develop a selection of future technologies which could allow Korea to exploit its technological capabilities to achieve global competitiveness. The programme aims to develop these technologies within 10 years with a view to their rapid contribution to economic growth. The likely economic impact and marketability of the future technologies constitute major selection criteria for this programme.

It could be argued that this downstream movement is, in fact, in the wrong direction and that Korean innovation policy should be encouraging the GRIs and HEIs to conduct more explorative basic research. Even if firms are conducting an increasing amount of basic research, the vast majority of this is still likely to be strategically targeted. The public sector should therefore be performing the more explorative basic research that private firms are unwilling to perform themselves. An assessment of the merits of this argument necessitates a closer examination of the history and activities of the GRIs and HEIs.

Starting first with the GRIs, these institutes played a crucial part in the technological upgrading of Korean industry during the catch-up era, working closely with the chaebol. But as the research capabilities of Korea's largest firms deepened and reached knowledge frontiers, the role of the GRIs became less clear-cut.

Nevertheless, by budget expenditure, the GRIs remain the largest performers of research in the public sector, accounting for 53% of publicly performed R&D in 2008, although their leading position is increasingly challenged by HEIs, who accounted for 47% of publicly performed R&D in 2008.[12] Roughly speaking, around half of the GRIs' budget comes from a government core grant, while the other half comes from contract research for various organisations, including a range of central government ministries (the main purchasers of research) and local governments, plus private companies. The GRIs still conduct much of the "big science" carried out by the public sector in Korea and HEIs cannot match their facilities. This is not an unusual situation and international experience shows that such government sector labs often carry out large-scale research that would be difficult to conduct in HEIs.

Data for the types of research performed by GRIs shows that only around 20% of R&D expenditure was devoted to basic research in 2006, down from 27% in 1998. Over the same period, the share of R&D devoted to experimental development increased from 38% to 44%, while the share of applied research remained at around 35% (MoST and KISTEP 2007). There are several possible explanations for this profile and its evolution. One of these is likely to be path dependence from an earlier era of technology development, which continues to favour applied research and experimental development. But the introduction of a project-based funding system in the mid-1990s, together with the more recent deployment of a performance evaluation system skewed towards counting patents registered by GRIs, have probably also contributed to an increased outcomes-based short-termism and to the relative decline of basic research.[13] It is important to consider how well the GRIs perform their significant role of experimental development. The number of patents registered has shown impressive increases in recent years, suggesting some success, but this should be seen in light of relatively weak patent exploitation (Lee 2007). This raises the suspicion that researchers in the GRIs are perhaps over-patenting in order to meet certain performance indicators. It also calls into question the ability of the GRIs to successfully commercialise the results of their R&D themselves, with several critics concluding that this is best done by firms.

Turning to the HEIs, the Korean government has in recent years favoured the strengthening of R&D capabilities in universities, which are considered the "natural" sites of skills development and knowledge transfer. Nevertheless, the proportion of R&D performed in the HEI sector remains small by international

[12]A decade earlier, the GRIs accounted for 62% of publicly performed R&D, the HEIs for 38%.

[13]The project-based funding system (PBS), introduced to improve R&D management and performance, has been particularly detrimental to the conduct of basic research, introducing a stronger short-termism into the GRIs, reducing the scope for knowledge deepening as GRIs compete for a wide spectrum of project funding, and reducing the capacity for GRIs to commit to long-term employment contracts (for example, in 2006, special service interns, most on temporary short-term contracts, outnumbered regular employees by almost two-to-one in one of the leading GRIs, the Korean Research Institute for Bioscience and Biotechnology). On a more positive note, the PBS has been recently reformed in an attempt to reduce these adverse effects.

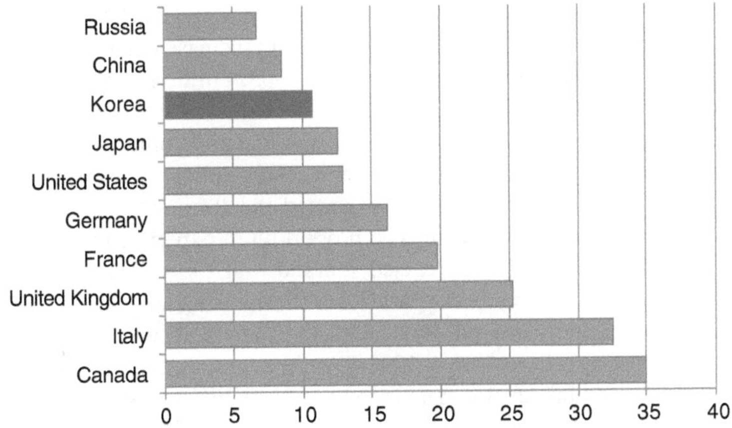

Fig. 2.5 Percentage of gross expenditure on R&D (GERD) performed by the HEI sector in selected countries (2008 or nearest year) (Source: OECD, Main Science and Technology Indicators, October 2010)

standards (Fig. 2.5) and is closest to that of China and the Russian Federation, both of which have traditionally separated research and teaching activities between national science academies and universities, respectively. This has also been the case in Korea, as the HEIs were concerned primarily with producing skilled manpower through their teaching activities during the catch-up years. In pursuing a more creative mode of development, current policy frameworks have taken a broader view of the role of HEIs (especially universities) and see them as places where new scientific and technological principles should be discovered. On paper at least, such a reorientation would seem to be well-overdue: despite performing a relatively small proportion of Korea's GERD, HEIs employ a little over one-quarter of all researchers in Korea and about four times as many researchers as the GRIs. Furthermore, they employ around 70% of all PhDs in Korea. Up until recently, much of this human resource was relatively inactive in research, having been marginalised by public funding regimes that focused on supporting large-scale, mission-oriented R&D programmes unsuited to the participation of single researchers or small research groups. In addition, the R&D facilities of most universities are less well equipped and their research teams are less well organised than those in the GRIs. Acknowledging the problems, the MEST earmarked some KRW500 billion for promoting grassroots efforts in basic research in 2009, an increase of 37% on 2008. At the same time, 7,000 individual or small-group researchers were granted research fees during 2009, and an expanded KRW255 billion was invested in the general researcher support project. The latter puts particular emphasis on facilitating basic R&D activities by young university faculty, general professors, female professors and faculty at local universities.

At the same time, the policy emphasis on promoting basic research in HEIs has been coupled to an exploitation agenda concerned with transforming discoveries into innovations. This has led to a policy focus on intellectual property rights, patenting and technology transfer from HEIs. Against this backdrop, the top ten patenting universities have shown an almost tenfold increase in the total number of patent applications filed over the period 1997–2005. Furthermore, there has been a marked diversification in the types of universities filing patent applications: in the 1990s, applications were mostly confined to the technical universities of KAIST and POSTECH, but this is no longer the case. This policy focus on commercialisation has, as in the case of the GRIs, shifted the R&D activities of HEIs, in their aggregate, further downstream. The proportion of experimental development performed in universities has increased markedly at the expense of a relative decline in basic research in recent years. Data for the types of research performed by HEIs shows that only around 33% of R&D expenditure was devoted to basic research in 2006, down from 40% in 1998. Over the same period, the share of R&D devoted to experimental development increased from 26% to 34%, while the share of applied research remained at around 32% (MoST and KISTEP 2007).

It is likely that the government's renewed commitment to and increased investments in basic research will see some reversal in this dynamic in the coming years. Furthermore, it is important to bear in mind that these figures represent proportions of a fast-growing level of expenditure on R&D. This means that the absolute levels of basic research being performed have increased markedly, though less rapidly than the levels of experimental development and applied research. This is perhaps best indicated by the increasing number of papers in SCI publications by Korean universities, which rose from 3,765 in 1998 to 7,281 in 2005. The SCI impact factor of university lecturers and new researchers in science and technology fields also increased from 1.9 in 1999 to 2.43 in 2005 (KRF BK21-Nuri Committee 2007).

Clearly, a balance is required between the performance of different types of research in GRIs and HEIs. In a more creative mode of innovation, these institutes will need to play a variety of roles requiring a mix of activities and linkages with other actors in the innovation system. At the same time, it needs to be acknowledged that trade-offs are likely between different types of research, on account of opportunity costs and conflicting incentives.

Increased spending on basic research that the private sector is unlikely to fund itself should be a priority. The 577 Initiative commits the government to expanding its investment in basic research to 50% of the public R&D budget by 2012, up from around 25% in 2007, a very ambitious target by any standard. To reach this goal several sub-targets have been set:

- Expanding research grants for individual investigators (including small groups) from KRW368 billion in 2008 to KRW1.5 trillion in 2012
- Increasing the ratio of university professors in S&T fields receiving basic research grants from around 25% in 2006 to 60% in 2012
- Increasing the ratio of young researchers in their 20s–30s receiving basic research grants from around 18% in 2006 to 25% in 2012
- Expanding support for basic research in GRIs
- Expanding research support for high-risk high-return projects

The reorientation to more basic research should be accompanied by appropriate expectations for the potential results. In this regard, the innovation-studies literature has demonstrated that the development of embodied skills and understanding from basic research tends to have a more significant and lasting impact than the generation of codified results. Acknowledging this has consequences for the way such research is evaluated and ultimately governed, since many of the returns to investment are unlikely to be immediate and can thus create attribution problems. Understanding these benefits of basic research also highlights the need to put in place policies and programmes that will facilitate the flow of skills and knowledge throughout the innovation system.

Promoting a New Internationalism

The Korean economy is often characterised as being relatively closed on account of its low levels of (1) import penetration, (2) inward foreign direct investment (FDI), and (3) foreign workers. These conditions are often viewed unfavourably by international governmental organisations that champion inward FDI and trade-openness as channels for improving productivity and competitiveness. Korean policy has seemed generally suspicious of these tenets of development, however, and has instead sought to nurture alternative channels of technological upgrading in efforts to drive economic catch-up. Significant among these have been the contracting arrangements of the chaebol which saw them enhance their technological capabilities via technical licensing arrangements with foreign firms, the importation of capital goods and components, and the exporting of original equipment manufacturing (OEM). In particular, OEM arrangements proved to be one of the most cost-effective channels for acquiring core capabilities in production and investment.[14] Another important channel promoted by the government for acquiring foreign knowledge has been overseas education and training. Much of Korea's top talent has benefited greatly from advanced education and training in the United States and elsewhere. Crucially, most of this talent has traditionally returned to Korea after training, in contrast to the situation in many other developing economies, and has contributed to the country's successful catch-up.

Taken together, these channels demonstrate that the Korean economy was in fact well-linked to global knowledge sources during the catch-up period. They allowed for the transfer of know-how and technology that have enabled Korea's transition to a newly advanced economy. The question is whether they are sufficient or even appropriate for a more creative mode of development. Put another way, while Korean

[14]The often tedious and gruelling qualification process that any potential supplier has to complete successfully in order to compete for OEM contracts opens up a variety of learning possibilities with regard to business organisation and the use of technology. In addition, customers often provide technical assistance in engineering and manufacturing processes in order to ensure quality and cost efficiency (Ernst 2000).

development has been traditionally based upon the exploitation of internationally-sourced knowledge, to what extent does the shift towards a more creative mode of development change the nature of knowledge to be acquired and the channels through which it is to be sourced? The manner through which the chaebol accumulate technological capabilities certainly changes as they reach knowledge frontiers. They have become significantly more reliant on R&D to support their exploration activities, an increasing amount of which is performed outside of Korea. For example, the Korea Industrial Technology Association (KOITA) reports that Korean firms had set up around 60 overseas R&D centres by 2005. Samsung Electronics, LG and Hyundai Motors have been particularly active in establishing such centres from the mid-1990s onwards, especially in the United States, Europe, Japan and, increasingly, China. The rationales for performing R&D in overseas centres are multiple and range from the need to adapt technologies for local markets to the exploitation of knowledge held by highly-skilled researchers.

As for overseas education and training, the other knowledge accumulation catch-up channel highlighted above, this continues to be important but has also changed over time. Numerically, more Koreans than ever study abroad, increasingly so for undergraduate degrees. They tend to be self-financed rather than sponsored and, crucially, many are remaining overseas after graduating. For example, the number of doctoral students returning to Korea after graduation has dramatically declined over the last decade, leading some to declare an apparent "brain drain" crisis. The reasons for (immediate) non-return tend to be complex and have yet to be fully researched in the Korean context. But preliminary findings seem to point to higher quality research and work environments in the United States and lack of suitable post-doc positions in Korea, especially in universities. This relates to the discussion in the previous section and the need to further develop basic research capacities in HEIs.

These brain drain fears, together with the possibility that the leading chaebol could perform their most cutting-edge research work in overseas R&D centres, have prompted the Korean government to accelerate the development of the country's science base in the same way the rate of technological learning was accelerated in earlier times, i.e. by accessing international sources of knowledge. However, some of the available channels for strengthening the science base in this way remain weakly-developed. For instance, very little R&D carried out in Korea is financed from abroad, research linkages with foreign firms and institutions are relatively weak, and few foreign researchers and students come to Korea to work or study. The language barrier and a traditional emphasis on more proprietary industrial research have probably contributed to the low level of international R&D activities; the situation appears to be similar to that of Japan. A range of policy initiatives have been subsequently launched to increase the internationalisation of the Korean research base. These include promoting international R&D collaboration, attracting foreign R&D centres to Korea, and attracting human resources in the form of students and skilled researchers to study and work in Korea.

The main policy actors in this regard are the MEST and MKE. Both have dedicated programmes and sub-programmes in support of S&T internationalisation. For example, international R&D collaboration is supported by the MEST's Global

Joint R&D programme, with more than 60% of funding directed towards cooperation involving the EU, Japan or the United States. The advanced level of R&D in these regions provides a strong incentive to collaborate on account of the learning benefits. One important component of this programme is the Global Research Laboratory scheme, which provides funding for Korean laboratories to collaborate with leading research centres overseas on fundamental research. The aim is to raise national research capacities to world-class levels. In 2010, more than 30 Korean laboratories benefited from the scheme. The MKE has similar programmes to promote R&D cooperation but targeted at industrial R&D. Korea is also participating in international S&T programmes such as the EU Framework Programme, CERN and ITER.

Besides providing support to R&D collaboration with overseas research centres, Korean policy also seeks to attract foreign R&D centres to Korea. Again, both the MEST and MKE are active. The MEST established the Korea Foundation for International Co-operation of Science & Technology (KICOS) in 2004 with the aim of making Korea the Northeast Asia R&D hub by attracting prestigious foreign research institutes. There have been some successes, notably the attraction of the Institut Pasteur to establish a research centre in Korea, lured by the offer of new buildings and generous subsidies. For its part, the MKE is working to attract private firms' R&D centres and has hosted 18 new R&D centres over the period 2003–2006, including from the likes of Microsoft, Siemens and DuPont. Crucial to the continuing attraction of such R&D investments will be the establishment of a world-class science system with high-performance public research units, a highly developed infrastructure and a supply of excellent human resources (OECD 2006b).

The brain drain of Koreans to the United States and Europe would be less of an issue if Korea were more successful in attracting highly-skilled workers from overseas. The number of foreign highly skilled workers in Korea has stagnated at around 25,000 since 2000, accounting for about 6% of the total foreign labour force. In contrast, skilled labour accounts for 19% of foreign labour in Japan. The low share of highly skilled workers is largely a result of problems in the business and living environment and administrative regulations that make it difficult to work in Korea (OECD 2007b). The Korean government has begun to provide preferential treatment and incentives to attract highly skilled foreign workers. These include preferential visa programmes for foreign scientists and engineers, and attractive packages aimed at encouraging leading scientists to spend time in Korean universities. A recent MEST-funded policy measure in this regard is the World Class University programme, which is designed to attract top researchers from around the world to collaborate with Korean scientists in key fields. Announced in 2008 with a 5-year budget of around USD800 million, the programme supports universities to build new research departments around leading foreign academics and to employ Nobel Prize winners and the like to take up visiting posts.

If these schemes were not ambitious enough, the Lee Myung-bak administration has announced plans to build large-scale scientific infrastructure, including a heavy-ion accelerator, in a new city of science and culture, known currently as the International Science and Business Belt (ISSB) initiative. Box 2.3 sets out the Korean government's vision for the Belt, which is expected to make connections with several overarching goals on the current Korean STI policy agenda. These include a shift towards more fundamental and interdisciplinary research, efforts to further internationalise research, and support for national competitiveness through the commercialisation of research findings and the development of highly skilled human resources.

Box 2.3: The Promise of the International Science and Business Belt (ISSB) Initiative

The International Science and Business Belt (ISBB) will be Korea's showpiece for a science nation of the future. With the ISBB as an exemplar, Korea as a whole will become an international focus for science and business. Korea will become a leading country in the world as it establishes a Global Knowledge Platform where the applied sciences flourish on the foundation of strong basic science. In this way, the ISBB will be the heart that pumps the economic lifeblood of the country to become one of the Big Seven Powers. More specifically:

The ISBB promises to add much value to academia and to other related industries. The leading researchers gathered in the ISBB from across the globe will drive a "Korean Wave" in science, much like the cultural wave that has swept Asia in recent years. The country that used to send a great number of students overseas will bustle with students from abroad.

As a city where people desire to live and visit, the ISBB will be a showpiece for twenty-first century cities. The city will enable science, art, culture and industries to merge creatively and produce synergy between the basic and applied sciences. A top-class basic science research institute comparable to the world-renowned Brookhaven National Laboratory in the United States and the Max Plank Institute in Germany will be established at the centre of the city.

The ISBB will be the hub of a "twenty-first century creative network", where science, art, culture and industries converge. It will be Korea's "twenty-first century Silicon Valley", where the latest scientific knowledge in service, medical, banking, manufacturing, communications, transportation, real estate, architecture, and many other industries can be readily translated into business.

At the ISBB, experts will meet and exchange ideas with other experts in the same field as well as with those in different fields. Research preparation, knowledge creation and propagation/transmission will be carried out systematically. The city will thus become the leader of international knowledge distribution as the central axis of science shifts to Asia, following similar shifts in industry.

As more than a city for scientists, the ISBB will be a global nexus where research and industry, East and West, and traditional Asian culture and modern culture converge. Not only will it bring about differentiated scientific competitiveness to participating organisations, it will also become a place where science and the humanities, and industry and art come together for lively exchange and co-operation.

Source: Excerpt from the report of the International Science and Business Belt Task Force (2008).

As well as trying to attract foreign researchers to Korea, the government is looking to make better use of Korean scientists and engineers working in different parts of the world. In this regard, the MEST is providing funding for the Korean Scientist and Engineers Network (KOSEN), a virtual network of almost 70,000 scientists and engineers, including over 5,000 expatriate researchers. The MEST has also organised societies of Korean scientists and engineers in 11 countries, which have undertaken, among other things, to collect information on Korean scientists and engineers. This is an important set of initiatives, as the Korean diaspora offers a unique resource that could be drawn upon to foster international collaboration and to provide an independent voice for domestic evaluation and assessment.

In addition to attracting established researchers, the MEST has also sought to attract more foreign students to Korean universities in order to increase the country's pool of foreign talent. The Study Korea Project, launched in 2004, provided scholarships to 1,500 students from developing countries in 2008, a figure that is set to double by 2012. In addition, the Study Korea Project Development Plan seeks to double the number of foreign students in Korea to 100,000 by 2012 (a previous target of 50,000 by 2010 was already exceeded by 2007) and to broaden their countries of origin. International students in S&T fields are now mainly from Southeast Asian countries. The MEST hopes these policies have a positive impact on Korea's economic and diplomatic development and expects foreign graduates to serve as a driving economic force by helping to counter low fertility rates and rapid population ageing.

Thus, the Korean government has put in place an impressive array of initiatives to improve the internationalisation of its science and innovation base. Flagship projects, such as the International Science and Business Belt (ISBB) and the appointment of Nobel laureates to Korean universities, will strongly signal Korea's intent to become a place for leading-edge science. However, the success of these and similar projects cannot be taken for granted. Indeed, a number of fundamental barriers to the internationalisation of Korean science remain. They include language and cultural barriers, which make Korea a less attractive destination for foreigners. The increasing use of English in Korean research centres and a growing international awareness of the richness of Korean arts and culture should improve the situation, but it will obviously take time to see a major impact. A further barrier concerns the education system, which is perceived negatively – an important consideration for foreign scientists with children – as evidenced by the growing tendency of Korean parents to send their children to schools outside of Korea. Finally, the vast majority of universities still have weak research capacities and thus have a low international profile. This makes them relatively unattractive destinations for foreign researchers.

While these challenges in no way undermine the potential value of the flagship projects announced, they highlight the need for *systemic* solutions to the internationalisation issue. Moreover, they also point to the need for broad grassroots efforts to improve the science base rather than efforts in a few selected sites. International scientific exchanges and mobility are more likely in basic science and are often concentrated in universities. They should be boosted by better research conditions in a broad set of universities and by increases in the levels of basic research performed.

Concluding Remarks

Table. 2.1 summarises some of the strengths and weaknesses of Korea's innovation system, as well as some of the opportunities and threats that are likely to be faced over the next few years. Sustaining the impressive productivity growth seen over the last four decades will increasingly depend upon boosting innovation, which in turn requires a reorientation towards more explorative search and selection, e.g. through basic research, as Korean firms reach knowledge frontiers. The transformation of the Korean innovation system – from one based almost solely on the goal of catch-up through exploitation of existing knowledge to one that extends to creative exploration – has been underway for some years now. It was accelerated by the 1998 Asian financial crisis, particularly by the government's strong intervention that followed, but also by several chaebol rapidly approaching knowledge frontiers. An innovation system geared to more creative exploration requires different skills and education, firm-level capabilities, public research, and international knowledge linkages than those required by one based almost solely on exploitation of existing technologies. These new requirements entail changes in the organisation and dynamics of the national innovation system, in particular, changes in the institutions and ideas that pattern the behaviours of and linkages between various innovation system actors (Mahlich and Pascha 2007). By their very nature, such changes are gradual, most likely uneven, and can be only partly engineered by policy intervention. The chapter has explored some of these changes in the context of four key areas where government policy continues to play an important role in shaping the Korean innovation system, i.e. education and skills, SME development, public research, and international knowledge linkages. The paragraphs that follow summarise some of the chapter's main arguments and draw a few tentative conclusions.

Table 2.1 SWOT analysis of the Korean innovation system

Strengths	Weaknesses
• Strong, mobilising national vision	• Under-developed fundamental research capabilities and weak research capacity in universities
• High growth rates in GDP	• Weak linkages between GRIs and higher education institutes
• Strong government support for innovation and R&D	• In education, prevalence of rote learning, overemphasis upon university entrance exam, and high cost of private education
• Good and improving framework conditions for innovation	• Under-utilisation of female labour
• High ratio of gross domestic expenditure on R&D (GERD) accounted for by significant business enterprise expenditure on R&D (BERD)	• Low productivity in the services sector
• Highly educated workforce	• Relatively weak SME sector
• Good supply of human resources for science and technology (HRST)	• Uneven development across regions and sectors
• Ready early adopters of new technologies	• Relatively small domestic market (compared to China, Japan, United States)
• Strong ICT infrastructure	
• Exceptionally good fast followers	
• Strong and internationally competitive firms	
• Learning society with a capacity to learn from failures and international good practices	• Policy co-ordination problems between ministries
Opportunities	**Threats**
• Geopolitical positioning in one of the most dynamic regions of the world	• Low fertility rates and an ageing society
• Globalisation, including of R&D	• Arrival of strong new competitors in fields in which Korea excels, e.g. ICTs, particularly from China
• Growing Korean S&T diaspora	• Geopolitical developments in the region
• Developments in S&T (technological change), particularly in information technology, nanotechnology, biotechnology and environmental technology – and their possible fusion	• Disruption in the supply of imported natural resources and energy upon which the Korean economy is highly dependent
• Growth of China and other newly industrialising economies, both in the region and worldwide, offering new markets for Korean exports	• Global economic outlook and its consequences for export-oriented economies

Source: OECD (2009)

In education and skills, changes have perhaps been slower to come about than in other areas. For example, the education system is still heavily geared towards young people and their attainment of bachelor degrees and relies too heavily on rote-learning. It could be argued that Korea has been too successful in producing large numbers of bachelor degree holders, while concerns remain over the appropriateness of the skills obtained from many degree courses. Shifting education systems in new directions is notoriously difficult, though in Korea, demographic change makes it inevitable. The government recognises the need for an urgent broadening of focus beyond young people to include greater lifelong learning provision, but HEIs and the business sector have yet to fully embrace the transition. At the same time, labour market rigidities hinder the uptake of lifelong learning opportunities, as well as the participation of more women in the workforce. Yet both will be crucial in the coming years as the population increasingly ages. These multiple considerations demonstrate well the complexity of issues around education and skills and highlight the need for sustained, coordinated action on several fronts.

Moving on to SMEs, the chapter presents some evidence of a population of smaller firms emerging with enhanced technological capabilities – a promising sign for an industrial ecosystem that has been for so long (and continues to be) dominated by a few large firms. The chaebol have supported this emergence in some instances, no doubt reflecting their self-interest in nurturing a cadre of local suppliers with strong technological capabilities. But the government has perhaps lent even greater support. Guided somewhat by New Economy thinking, it has provided incentives and funding for increased R&D activity in SMEs and has worked hard to introduce a wide variety of opportunities for innovation-friendly financing. Even though these initiatives seem to have had some success, nobody pretends that innovative SMEs will become the dominant industry players in Korea anytime soon, though to be fair, this is not the aim. Rather, the policy aim is to avoid an industrial 'hollowing out' scenario, driven by the chaebol sourcing increasingly from overseas suppliers, particularly from China. If Korean SME suppliers can accumulate greater technological capabilities, then the risk of this scenario being played out should, it is reasoned, be reduced.

R&D can be an important mechanism to aid exploration. Any shift to a more creative mode of development, particularly in manufacturing economies, would normally be accompanied by increases in R&D spending. The chapter has shown that growth in Korean GERD has been one of the highest in the OECD area and that an increasing share of the R&D performed is devoted to more explorative basic research. Much of the latter is, however, performed by the business sector while the proportions of basic research carried out in HEIs and GRIs have declined in relative terms over recent years. An explanation for this lies in the way government policy has assigned roles to the GRIs and HEIs as drivers of future engines of growth – again, reflecting a New Economy ideology. One outcome is a greater emphasis on commercialisation activities, even though they may not be fully congruent with a parallel focus on promoting basic research. This is because expectations around commercialisation – particularly if success is measured in terms of crude indicators, such as patent registrations – tend to place less value on many of the other, often more important, yet difficult to measure, benefits of public research, such as human resource development. Instead, they focus on the measurable and shorter-term pay-backs, an orientation that can be unhelpful to a more explorative type of research, where uncertainties are much higher and the risks are greater. There is evidence that Korean S&T policy, with a legacy of results-oriented programmes aimed at catch-up, still needs to draw realistically upon the potential of more explorative basic research.

Accessing international sources of knowledge has always been a central part of Korean development, but the nature of the knowledge and relationships sought has changed as the Korean economy has reached knowledge frontiers in several fields. Much greater emphasis is now placed on more explorative basic research linkages with leading global research centres with a view to accessing state-of-the-art knowledge. Korea has little track record in sourcing this type of knowledge and in fostering these sorts of relationships on account of its previous R&D profile, which was dominated by proprietary R&D activities. Crucially, the ability to access such knowledge depends upon having similar centres of excellence at home that have the necessary credentials to participate in global knowledge networks and the absorptive capacity to benefit from them. While the leading chaebol have developed such centres or have acquired them overseas, the public sector still lags behind. The government has introduced various schemes to accelerate the accumulation of basic research capabilities, e.g. through schemes to attract leading international talent to universities, the International Science and Business Belt, etc. Their long-term success, however, depends ultimately on Korea developing a deeper research milieu that extends well beyond a handful of islands of excellence to encapsulate a broad set of leading-edge research centres operating across a spectrum of GRIs, HEIs and the corporate sector.

The dynamics of change outlined above have consequences for the roles to be performed by the main actors in Korea's innovation system. For example, the government still performs more than an enabling role and remains rather active in designing and implementing a directive industrial policy. The sets of ideas guiding policy intervention have changed, however, and reflect new thinking informed by a national innovation systems perspective. The levels and nature of interventions suggest a certain policy impatience, which has been typical of Korean policy over the years. While this urgency might be warranted, it also leads to demands for demonstrable effects, which might be problematic, since many benefits may take years to appear and are likely to be only partially attributable to policy interventions. As for the chaebol, their needs regarding education and research systems have changed as their work increasingly occurs at knowledge frontiers. But they also have responsibilities to the Korean innovation system if they intend to continue to draw upon its knowledge assets. This applies particularly to SME suppliers, but also to education providers who could benefit from a stronger strategic partnership with the chaebol. Finally, the public sector research performing institutes, i.e. the GRIs and HEIs, have important roles to play in increasing the amounts of more explorative research carried out in Korea. Yet, recent years have seen perhaps too much emphasis placed on certain types of commercialisation-oriented activities, e.g. patenting. Such institutes should play a variety of roles, but a better balance is required that puts greater value on the explorative search activities of the GRIs and HEIs and on their contributions to the skills base of the Korean innovation system.

Acknowledgements This chapter is based upon the report *OECD Reviews of Innovation Policy: Korea*, to which there were many contributors. The author would like to acknowledge the inputs of OECD colleagues, namely Jean Guinet, Gernot Hutschenreiter, Dong-Hoon Oh and Tae-Seog Oh. The Review also benefited greatly from a background report prepared by a team of Korean experts and edited by Kong-Rae Lee of the Science and Technology Policy Institute (STEPI) in Seoul. The team's experts included Tae-Kyung Sung (Jeonju University), Woo-Sung Lee (STEPI), Sang-Won Ko (Korea Information Society Development Institute) and Jung-Tae Hwang (STEPI). Last, but by no means least, Ron Johnston (University of Sydney) played an important role in drafting certain sections of the Review. As for the form and content of the present chapter, the usual disclaimer applies.

References

Ernst D (2000) Catching-up and post-crisis industrial upgrading: searching for new sources of growth in Korea's electronics industry. East-west center working paper, No. 2, East-West Center, Honolulu

Grubb WN, Sweet R, Gallagher M, Tuomi O (2006) Thematic review of tertiary education: Korea country note. Directorate for Education, OECD, Paris

International Science and Business Belt Task Force (2008) International science and business belt. Presidential transition committee report, Seoul

KEDI (2006) OECD thematic review of tertiary education country: background report for Korea. Korean Educational Development Institute, Seoul

KRF BK21-Nuri Committee (2007) Brain Korea 21. Korea Research Foundation, Seoul

Lee J-H (2006) Overcoming the legacies of the developmental state model: the case of the gumi mobile cluster. Presentation at the KDI & world bank senior policy forum on innovation policies and institutions for knowledge economy, Seoul, Nov 29–Dec 1 2006

Lee C-W (2007) Challenges and issues to upgrade government-sponsored research institutes in science and technology in Korea. Paper presented at the annual conference of the Korean society for innovation management and economics, Jeju Island, 20–21 July 2007

Mahlich J, Pascha W (2007) Introduction: Korea as a newly advanced economy and the role of technology and innovation. In: Mahlich J, Pascha W (eds) Innovation and technology in Korea: challenges of a newly advanced economy. Springer, Heidelberg

MEST (2008) Becoming an S&T Power Nation through the 577 Initiative, Science and Technology Basic Plan of the Lee Myung Bak Administration, Ministry of Education, Science and Technology, Seoul

MoST, KISTEP (2007) Report on the survey of research and development in science and technology. Ministry of Science and Technology and Korea Institute of Science and Technology Evaluation and Planning, Seoul

OECD (2002) Frascati manual: proposed standard practice for surveys on research and experimental development. OECD, Paris

OECD (2005) Thematic review of adult learning: Korea country note. OECD, Paris

OECD (2006a) Society at a glance. OECD, Paris

OECD (2006b) Science, technology, and industry outlook 2006. OECD, Paris

OECD (2007a) Babies and bosses – policies towards reconciling work and family life. OECD, Paris

OECD (2007b) OECD economic surveys – Korea. OECD, Paris

OECD (2009) OECD reviews of innovation policy: Korea. OECD, Paris

OECD (2010) Science, technology and industry outlook 2010. OECD, Paris

Chapter 3
Korea's New Techno-Scientific Strategy: Realigning State, Market and Society to Move Beyond Technological Catch-Up

Jitendra Uttam

Being a late-late industrializing country, Korea carefully designed a unique model of technological catch-up led by a 'developmental state' (Johnson 1982) and spearheaded by *chaebol*. The model became operational in two distinct but mutually reinforcing phases: first, an imitation-driven GRI System based on reverse engineering of foreign technologies; and second, a *chaebol*-led, private sector funded system of corporate R&D. During both phases, an uninterrupted inflow of foreign technology to Korea was supported by Cold War era 'strategic constraints' to maintain the U.S.-led 'alliance system' aimed at containing the perceived Soviet threat. The dynamic interplay between foreign technology and indigenous R&D efforts paved the way for Korea to substantially bridge its technological divide with the developed world, particularly in the manufacturing sector.[1] Korean products from ships to chips, from computers to semiconductors acquired global recognition. Reflecting fierce competition with technologically advanced economies, the R&D budget of Korea has witnessed a phenomenal rise.[2]

Despite the success, the Korean model of technological catch-up (KMTC) built by an interventionist state to reverse engineer foreign technologies has been constrained by its persistent narrow social base, and has thus become unsustainable in the long-run. The dramatic end of the Cold War removed 'strategic constraints' responsible for facilitating smooth technology flow to Korea, which greatly exposed vulnerabilities inherent in the reverse engineering-based KMTC. The financial crisis of 1997 provided a grim reminder that Korea had to move beyond

[1]Korea launched its Highly Advanced National (HAN) Project in 1992, which was supposed to be a springboard for Korea to catch up with advanced nations. It funded 18 teams conducting research on areas diverse as agrochemicals and nuclear fusion. Although the results have helped Korean companies to commercialize such products as high-definition television sets and 256-MB DRAM chips, the project fell short in other areas, including fusion and high-speed rail transportation.

[2]Korea spent about KRW34.5 trillion (USD28 billion) in 2008 on research and development, up 10.2% from 2007. R&D spending now accounts for 3.37% of gross domestic product, placing South Korea sixth among 30 OECD member states. For details see, Song Sang Ho, "R&D spending jumped 10% last year", Korea Herald, 16 September 2009.

J. Mahlich and W. Pascha (eds.), *Korean Science and Technology in an International Perspective*, DOI 10.1007/978-3-7908-2753-8_3,

the technological catch-up and construct a new model aimed at achieving techno-scientific leadership. Faced with twin competitions – from the top, Japanese high-tech industrial products and from the bottom, Chinese mass produced cheaper low-tech items – Korea has been left only with the high-tech option. In the real world it meant that to avoid Chinese competition Korea had to move to compete directly with advanced economies such as Japan, Germany and the United States.

The down swing precipitated by the massive financial crisis pushed Korea into an uncertain era of experimentation. The malfunctioning of the system put more and more pressure on the state to find new means of stimulating innovation and entrepreneurship to better manage the economy. Korea's search for systemic transformation of innovation and entrepreneurship found support in Schumpeter's argument that cyclical behavior of the capitalist economy has a single root cause: innovation which is in turn due to a privileged agent: the entrepreneur. By taking Schumpeterian thought to a logical conclusion, the post-developmental state in Korea initiated bold programs to move the nation beyond the set limits of the KMTC based on the commercialization of imitated technologies.

The tectonic shift in Korea's political landscape made the re-making or transforming of the KMTC an easier task.[3] Kim Dae-jung's 'participatory democracy,' unlike the 'exclusionary democracy' of the earlier era, paved the way for the rise of a 'third sector' of civic organizations to complement the catch-up era of state-market interaction and play the role of 'whistle-blower' in overseeing the symbiotic often corrupt links between state and business. At this crucial juncture, Korea's earlier era experience of establishing a 'plan-rational, capitalist developmental state,' which effectively converged state, market and narrow societal interests to vigorously push for technological catch-up and the creation of a domestic entrepreneurial class, i.e. the *chaebol*, turned out to be very useful. It's an established fact that by controlling the financial nerve, the Korean developmental state was able to construct state, market and societal relations (SMSRs) conducive for technological catch-up, which provided backbone to the decades of double digit economic expansion and transformed a poor agrarian economy into a full-fledged industrial economy. The Korean state's proven capacity to construct SMSRs for technological catch-up was once again needed to mediate collaboration between the market and society to facilitate interest convergence between government agencies, universities and industrial houses to move the nation beyond the imitation trap created by the KMTC.

Understanding the seriousness of the task at hand, the Korean state committed a large amount of financial resources under programs such as Brain Korea-21 (BK-21), New University for Regional Innovation (NURI), etc., which utilized

[3]The Korean financial crisis coincided with the inauguration of the Millennium Democratic Party (MDP) candidate Kim Dae-jung's government, which came to power based on policies committed to rectifying wealth concentration and regional development imbalances. As 'DJnomics' took center stage in Korea, the locus of political power shifted away from its traditional Gyeongsang-do seat in the southeast part of the peninsula to Jeolla-do in the southwest.

the state's monetary support based on individual/institutional performance to empower firms to face the increasingly high-tech based, ever competitive market. Taking performance as the key variable, similar to the earlier era *chaebol*'s performance in the world export market, Korean leadership is trying to link monetary resources with the innovative capabilities of individuals/groups. Instead of subsidizing *chaebol*, the post-developmental state in Korea has taken a firm step to fund the nation's innovative skills. Though, the post-developmental state still uses the developmental state's performance criteria, the earlier era's 'export performance' has now become 'publication-patent performance.' Recent policy initiatives under the BK-21 and NURI projects are aimed to move Korea into the new techno-scientific era, bearing similarities with the earlier era developmental state's heavy and chemical industries (HCI) project, a big push by the state to augment industrial progress and technological competence. Also, Korea's current policy instruments resemble the earlier era developmental state's policy of choosing the winners based on verifiable performance. All who can perform are welcome. Earlier the *chaebol* were the undisputed performers, and now it's the innovation skills of the Korean people. The earlier social base of performers was limited, now it has been widened. This inner logic of the Korean state behind launching new techno-scientific initiatives is to enhance Korea's future technological competitiveness.[4] Nonetheless, some fear that the new tech-strategy may create a powerful elite – a techno-scientific class – that will lead Korea's march into the global knowledge economy while leaving the rest struggling for survival.

Ideas and arguments in this paper are organized in the following manner: Section "Korean Model of Technological Catch-Up" conceptualizes the distinctiveness of the Korean catch-up strategy under the rubric of the KMTC, based on the symbiotic alignment of statist and market perspectives aimed at using public money to meet market-based technological expectations. It discusses the rationale behind imitation-driven, reverse engineering led by the developmental state. By discussing the Korean national innovation system (KNIS), this section highlights two distinct but mutually reinforcing phases in the KMTC: first, the imitation-driven GRI System, and second, the *chaebol*-led, private sector funded innovation system. Section "Innovation, Entrepreneurship and Schumpeterian Logic" provides a societal perspective as a step to correct the anomalies inflicted by the symbiotic alignment of statist and market perspectives. This societal perspective is rooted in the idea of neo-Schumpeterian logic on innovation and entrepreneurship, which argues that cyclical behavior of the capitalist economy, has a single root cause: innovation which in turn is due to a privileged agent: the entrepreneur. It stresses the role of advances in basic sciences and social, managerial and organizational changes in forming clusters of basic innovation. It discusses how the financial crisis

[4]Budget allocation in different scientific areas indicates future technological orientation. By type of technologies, 33.8% of the total was invested in information technology while the investments in nanotechnology, biotechnology and environment technology accounted for 12.3%, 7.6% and 8.5%, respectively.

worked as a mechanism of Schumpeterian 'creative destruction,' which shook the foundation of 'crony capitalism' created to promote Korean national entrepreneurship, albeit in a narrow social setting. It highlights Korea's new-tech strategy based on the neo-Schumpeterian notion of innovation and entrepreneurship; however with the exclusive focus on expanding its societal base.

Section "State, Innovation and Korea's New Techno-Scientific Strategy" examines Korea's earlier era 'exclusive development' based on selecting winners among the chosen few *chaebol* responsible for executing the state objective to leapfrog into the 'techno-industrial sphere' by reverse engineering foreign technologies. It contrasts that with the post-crisis 'inclusive development' that supported a broad regime based on innovation in the new 'techno-scientific sphere.'[5] It argues that the 'developmental state' is not dead, but it has transformed itself into a new 'techno-scientific state,' which instead of intervening in the market to propel exports is facilitating growth of technological innovation and invention (Uttam 2006). It delves into earlier-era construction and contemporary-era reconstruction challenge to state, market and society relations (SMSRs) in Korea. It argues that the earlier-era construction of SMSRs was much easier as there were no powerful interest groups to penetrate in the state system. However, the post-crisis reconstruction of SMSRs faces various challenges as there are powerful interest groups to resist changes. The main argument in this section is that through the BK-21 and NURI projects, the Korean state has effectively shifted financial incentives away from *chaebol* towards an industry–university combine. This shift amounts to a fundamental transformation in the pre-established SMSRs. Section "BK-21/NURI and the Linking of State, Market and Society for Innovation" evaluates new SMSRs and links their objectives to the BK-21/NURI projects. It pinpoints the basic features associated with the operating structures of BK-21/NURI projects and analyzes their impact on transforming the nation's techno-scientific innovation system. By examining the reach, functioning and monitoring mechanisms of BK-21/NURI projects, it argues that these innovative initiatives have initiated Korea's transition from the *chaebol* power to knowledge power; from industrial to post-industrial era.

Section "Korea's Tech-Strategy in an International Context" provides empirical support by highlighting Korea's technological catch-up through a cautiously thought-out latecomer strategy used in the semiconductor industry. By competing head on with the established U.S. and Japanese semiconductor industries, Korea's leapfrogging into a technologically sensitive and highly competitive industry demonstrates the effectiveness of its catch-up strategy, however its persistent weakness in the non-memory sector of the semiconductor industry shows the limits of the old-era technological catch-up model. It points to the fact that now Korea must broaden the social base of science and technology to innovate and possibly invent new

---

[5]For the details regarding the 'techno-scientific sphere' and 'techno-industrial spheres,' see Hilpert U (ed) (1991) State policies and techno-industrial innovation. Routledge, London/New York.

endogenous technological paths, which can make Korea an internationally acclaimed technological leader, an innovator not merely an imitator in the semiconductor field.

Section "Policy Implications" derives policy lessons from the Korean state's experiment to widen the social-base of innovation. Korea's earlier focus on increasing '*chaebol* power' and its subsequent transition from the agrarian to industrial era was recognized globally and impacted various policies of late-late developing economies. Similarly, Korea's contemporary focus on enhancing the social-base of innovation and its intended objective to transform industrial Korea into a post-industrial society is being keenly observed with greater policy ramifications. The transparent, globally recognized performance regime of BK-21/NURI programs aimed at aligning state, market and society in a cooperative framework that imparts various policy lessons for many developing economies, enabling them to propel techno-scientific innovations. Section "Conclusion" concludes that BK-21/NURI is an attempt to push Korea from the imitation phase to the innovation/invention phase. Careful linking of the resources of the state, market and society through these initiatives has the potential to provide a firm base for Korea's socially-embedded innovation regime of techno-scientific excellence which can facilitate its endogenous technological regime capable of competing head on with other advanced economies.

Korean Model of Technological Catch-Up

Foreign Technology, Reverse Engineering and the Development State

The KMTC started to take shape in the wider framework of the Korean national innovation system (KNIS) [6], which came to prominence from the 1960s. The main function of the KMTC during the initial catch-up phase was to facilitate the technological capacity acquisition process in Korea through the dynamic interplay between foreign technology and indigenous R&D efforts.[7] The developmental state played the role of facilitator by investing hugely in the institutional infrastructure such as establishing GRIs responsible for training quality researchers and identifying appropriate technology inputs for domestic entrepreneurs. During Korea's initial catch-up phase, the KMTC was nearly entirely devoted to establishing basic

[6]The concept of the KNIS has been discussed frequently in S&T policy and innovation research since the late 1980s. For details, see Freeman (1987), Dosi (1988), Nelson (1993), Patel and Pavitt (1994), Chung (1999), Lundvall (1992), and Mahlich and Pascha (2007).

[7]For details see Eros and Park (1989).

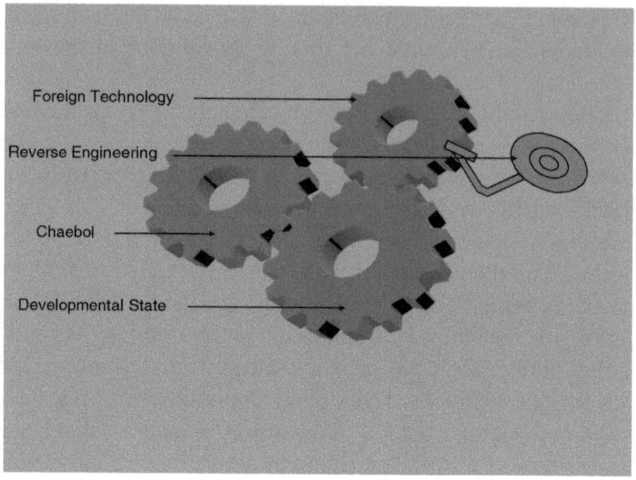

Fig. 3.1 Stake-holders of KMTC (Source: Author's conceptualization)

industrial infrastructure built upon the 'reverse engineering' of imported technology.[8] International competitiveness in this phase was secured by manufacturing products at reasonable quality and low cost. The stake-holders in the KMTC operated in a concerted manner to converge their interests in the national mission of technological catch-up. Figure 3.1 symbolizes key stake-holders and their mutual interdependence in propelling technological advances through the reverse engineering of foreign technological know-how.

As shown in the figure above, major stake-holders in the KMTC – the state, *chaebol* and foreign technology – demonstrate mutual coordination in their functioning, which leads to reverse engineering of foreign technology. Close coordination between the state and *chaebol* facilitated the inflow of foreign technology, often in the form of technology contracts and licensing agreements. This trend suggests that the KMTC facilitated an imitation-driven technological order which was further boosted by the establishment of GRIs, providing assistance to firms importing, acquiring, and absorbing foreign technology. Although Korea, a late-late industrializing country, had depended heavily on foreign technologies, it has made concerted efforts to accumulate indigenous technological capabilities. At the initial launch of its export-led industrialization plan, Korea was poorly endowed

[8]In the early years of industrialization, firms in the developing countries import/or imitate mature technologies, whose products have already been tested in the advanced countries. Large firms acquire production technologies mainly through turnkey plants or foreign licenses. In contrast, small firms take an imitative approach with reverse engineering of foreign products as the core strategy and evolve organically over a long period of time. R&D in a sense of generating new technologies is not needed during the initial phase of industrialization. Korea's first phase of industrialization was no exception to this general trend.

with factors necessary for industrialization except for a plentiful labor force. Furthermore, the technological competence of Korean firms was far below the world standard. Consequently, it was inevitable to look toward foreign sources of technology, and Korea witnessed a remarkable growth in the imports of foreign technology after the launch of the industrialization process in 1962. Table 3.1 shows a number of technology contracts and licensing agreements based on imported technology during the 1973–1989 period, when Korea relied heavily on reverse engineering.

Table 3.1 shows incremental growth in licensing agreements, engineering consultancy and the import of technology during the 1973~1989 phase, the heyday of technological catch-up. In 1989, the fees for licensing agreements exceeded USD900 million in 763 cases. Nevertheless, the process of technological capacity building in Korea has been characterized as a two-way process between imported technology and local innovation. The KMTC during the 'developmental state' phase can be broadly classified into statist and market perspectives, however with greater symbiotic relations: initially, the government-led, imitation-driven GRI system of the 1960–1970s being later complemented by *chaebol*-led private initiatives since the 1980s.

Table 3.1 Contracts and payments for Korea's imported technology

	Licensing agreements		Engineering consultancy		Imported technology	
	Number of contracts	(USD millions)	Number of contracts	(USD millions)	Number of contracts	(USD millions)
1973	67	11.5	28	5.3	95	16.8
1974	86	17.8	44	4.1	130	21.9
1975	99	26.5	57	2.2	156	28.7
1976	131	30.0	61	6.1	192	36.5
1977	168	58.1	94	9.2	262	67.3
1978	296	85.1	105	9.7	401	94.8
1979	288	93.9	131	14.3	419	108.2
1980	222	107.1	120	9.1	342	116.3
1981	247	107.1	101	12.4	348	119.5
1982	308	116.7	125	19.6	433	135.3
1983	360	149.5	190	73.3	550	222.8
Growth rates (in percentage)						
1973–1983	25.9%	29.2%	21.1%	30.0%	19.2%	29.5%
1984	437	213.2				
1985	454	295.5				
1986	517	411.0				
1987	637	523.7				
1988	751	676.3				
1989	763	930.3				

Source: Ministry of Science and Technology (Seoul, Korea), Ministry of Finance (Seoul, Korea), Science and Technology Annual, 1984; Imported Technology Annual, 1989

Imitation-Driven GRI System

The catch-phrase of this imitation era was 'reverse engineering' supported by the network of GRIs. Korea's per capita gross national product (GNP) in 1960 was USD79; less than that of Sudan and less than about one-third that of Mexico (Kim and Lee 1997). Facing acute problem of gross under development, the Korean government adopted import-substitution industrialization powered by imported foreign technology. In the early years of industrialization, the capacity of Korean industry to import and imitate foreign technology was severely limited. According to Kim (2001), private sector flattered in R&D investment, accounted for only 2% of the nation's total R&D expenditures in 1963. During this period, the research infrastructure in Korean universities was at best rudimentary, even the primary focus of top ranking universities was on undergraduate teaching undertaking little research (Kim and Lee 1997; Lim 2000). In the absence of research in industries and universities, the GRIs spearheaded technology sourcing, development, and facilitated its absorption (Choi 1984; Kim 1999, 2001; Shin and Kim 1994).

However, with the beginning of rapid industrialization in the 1960s, the demand for creating a domestic R&D base started to gain momentum. Given the increasing demand, the Korean government took a firm initiative to establish a government research institute – the Korea Institute of Science and Technology (KIST) – in 1966.[9] This and other GRIs however faced numerous obstacles in the 1970s. For example, they suffered from poor linkages with industry. Most Korean scientists and engineers recruited by KIST came from either academic institutions or R&D organizations that undertook advanced research. There was no demand from domestic industries for the kind of service that KIST could offer. From the domestic industry perspective, expertise was particularly lacking in manufacturing know-how and the development of various prototypes. Furthermore, KIST could not compete against foreign firms in supplying detailed blueprints and other manufacturing know-how or assist industries with solving teething problems in the crucial initial stages of production.

Despite the complex problems, KIST along with other GRIs made important contributions to industrial development in Korea.[10] One of the most important roles played by GRIs at this stage was to help industries strengthen their bargaining power in acquiring foreign technology. In an attempt to develop an effective linkage between GRIs and the private sector, the government coerced large firms

[9]KIST alone could not respond to the increasing demand from expanding industries. Thus, the Korean government established many GRIs under related ministries in order to meet immediate R&D demands from export-oriented strategic industries. As a result, 14 institutes in the S&T area had been founded by the end of the 1970s, mainly as spin-offs from KIST and other institutes. For details, see Lim (2000).

[10]Other government-funded research institutes include Gwangju Institute of Science and Technology (GIST) founded in 1993; Korea Institute for Advanced Study (KIAS), 1996; and Information and Communications University (ICU), 1998.

into joint research with them. Such joint research provided opportunities to some firms to acquire sufficient prior knowledge about technology to simplify importing it. This enabled them to identify prospective technology suppliers and to enhance their bargaining power in negotiating technology transfer arrangements. Once imported, such joint research provided a platform on which firms could assimilate and adapt technology rapidly. In other words, during the early years of industrialization, private firms entered into joint-research programs not so much to obtain significant research outcomes as to gain initial knowledge about technology, which they were interested in acquiring. GRIs also played a significant role in transferring technology to industry through reverse engineering of foreign technology – an activity that was beyond the capacity of Korean industry at the time.[11] Another important economic effect of local technology development by GRIs was a drastic cut in the prices of imported technologies and materials. The most important but unintended role of GRIs played during the early years of industrialization in Korea was the raising of experienced researchers. When the private sector was reluctant to invest in R&D, the GRIs generated a large number of experienced researchers who later left to join new GRIs and emerging corporate R&D centers in the 1980s.

Increasingly, the Korean government understood the need for ministry level government organizations to accelerate domestic S&T development and, as a result, established the Ministry of Science & Technology (MoST) in 1967 (Lim 2000). Since then, MoST has provided leadership and coordination in upgrading S&T infrastructure in the nation.[12] Also, in response to rising criticisms of the GRIs, the Korean government initiated a reorganization program aimed at expanding inter-institutional mobility of R&D resources and the sharing of equipment, etc. GRIs reduced their staff size by 20% in 1998, and their budgets for 1999 were cut by 20% over the previous year. GRIs have been regrouped into three research councils and were placed under the jurisdiction of the Prime Minister's Office. The three research councils were: (1) Korea Research Council for Fundamental Sciences, (2) Korea Research Council for Public Welfare Technology, and (3) Korea Research Council for Industrial Technology. The Chairman of the research council reports directly to the Prime Minister's Office.

---

[11]A case in point involves polyester film production for use in cassette tapes. When a Japanese company rejected Korea's request for technology transfer, for fear of losing its market in Korea, a Korean chemical firm in collaboration with KIST successfully undertook a reverse engineering task to invent around the production technology. At that time, Korea recognized process patents but not the product patents. Soon KIST had invented around the technology. When the Japanese company offered a transfer agreement, the Korean government rejected it in order to protect domestically developed technology. Korea now is the world's major supplier for audio and videocassettes tapes.

[12]Korea's Ministry of Science and Technology (MOST) was later renamed Ministry of Education, Science & Technology (MEST).

Chaebol-led, Privately-funded Innovation System

With the rapid market expansion and industrial widening, The KMTC faced dramatically increased demand for newer technologies, which the GRIs had difficulty meeting. Therefore, since the early-1980s, private enterprises in Korea rapidly increased their R&D spending with large business conglomerates taking the lead. As there was a wide gap between the domestic S&T base and the technological requirement of fast moving industrial and production activities, private enterprises had no option but to opt for in-house research. The internalization of the technology base by the *chaebol* has obvious advantages. Internal technological capabilities are apparently a basic requirement for business success as they enable companies to monitor market trends, to preempt competitors and to reap higher profits through economic rents. The problem is whether internalization is accompanied by increased learning or technological deepening; this is where Korea seems to face serious bottlenecks.

The technological catch-up spearheaded by fast expending *chaebol* established Korea's place in the various high-tech industries. In doing so, *chaebol* aggressively diversified their technology sourcing through adopting a number of technology strategies such as opening outposts in technological hubs overseas, including California's Silicon Valley, hiring Korean–American scientists and engineers with overseas experience, opening subsidiaries, collaborating with multi-national companies, and entering into extensive licensing ties with foreign technology firms.

The cost of excessive internalization in wide ranging technological activities has been apparent in many respects. In addition to the high financial burden of maintaining them, big research labs have not been so flexible; the fixed costs involve in dismantling the organizational structures in order to meet changing needs is often prohibitively high. The internalization of R&D activities by Korean *chaebol* has not come from specialization; rather it is the result of diversification of business activities, which require mostly quick product development. This system neglected learning and blocked further development of the KNIS (Suh 2000). Another limitation of excessive technological internalization is that it may weaken the need for closer cooperation with other innovation actors. For Korean *chaebol*, this was the case both domestically and internationally. This is quite contrary to the current trend for the increased externalization of R&D activities in most OECD countries. Strong internal ties between subsidiary companies weaken the incentive for cooperation with companies in other groups. A vicious cycle of self-propagating internal ties blocked further development of the KNIS (Suh 2000).

It is evident that both state-led and private sector-led R&D efforts mutually reinforced each other and paved the way for the remarkable success of KMTC. Korea's double digit economic expansion, popularly known as 'The Miracle on the Han River' is a live testimony of a successful catch-up model. However, the KMTC started to demonstrate its limitations as cutting-edge foreign technology was no longer available for transfer or outright sale.[13] On top of this, increasing cost pressures emanating from Southeast Asia and China started to cut market share, leading to the erosion of corporate profitability. The financial crisis of 1997 drove home the point that Korea had to rethink its core S&T policies and entire model based on reverse engineering.

Innovation, Entrepreneurship and Schumpeterian Logic

The KMTC fully served its purpose to propel Korea into the ranks of industrialized economies, accumulating wide-ranging cutting-edge manufacturing technologies.[14] However, the basic anomalies inflicted by the symbiotic alignment of state and big business based on the exclusion of large segments of society made the KMTC lose relevance in the face of ever increasing competition from advanced economies. The inability of the KMTC to push the nation in the new 'techno-scientific sphere' came as a major stumbling block. Indeed, the KMTC facilitated the *chaebol* R&D drive in the similar 'techno-industrial sphere.' However, the overall nature and direction of the KMTC could not allow Korea to shift to the new 'techno-scientific sphere.'

With the end of the Cold War, the KMTC started to demonstrate clear signs of fatigue. The financial crisis of 1997 provided a blunt reminder that the imitation and reverse engineering must give way to innovation and invention. It was felt that Korea should broaden the social base of its science to move beyond the old practice of technological catch-up. Many policy planners perceived the crisis as Schumpeterian 'creative destruction' and argued for a new technological path based on innovation and invention. Schumpeter's idea about innovation triggering business cycles attracted the attention of the Korean state. In Schumpeter's view, technological innovation is at the center of both cyclical instability and economic growth, with the direction of causality moving clearly from fluctuations in innovation to fluctuations in investment and from that to cycles in economic growth (Freeman 1986). Schumpeter's suggestion that large corporations with certain monopolistic abilities may be better equipped to develop and manage innovation as they have more resources and market power compared to their smaller peers suited Korea's diversified giants well.

Moreover, Schumpeter sees innovation as clustering around certain points in time-periods that he referred to as a "neighborhood of equilibrium" – when entrepreneurial perception of risks and returns warranted innovative commitments.

---

[13]For details see Choi (1988).

[14]Korea Institute of Industrial Technology (KITECH) launched the Korea Advanced Manufacturing Systems (KAMS) Project in 1992 with support from the Ministry of Trade, Industry, and Energy and the Ministry of Science and Technology. Consequently, Korea has gained a competitive edge in various high-tech industrial sectors such as semiconductors, consumer electronics, telecom equipment, etc.

These clusters, in turn, lead to long cycles by generating periods of acceleration (and eventual deceleration) in aggregate growth rates. The Neo-Schumpeterian logic stresses the role of advances in basic sciences, and social, managerial and organizational changes in triggering and facilitating clusters of basic invention and innovations (Freeman 1986).

Taking seriously the neo-Schumpeterian view of socially embedded innovation trajectories, the 'post-developmental state' in Korea initiated policies aimed at broadening the social base of S&T. The core logic behind the launching of BK-21/NURI was to bring more of society into the framework of techno-scientific innovation. Korea gave due consideration to entrepreneurs who can lead innovation-centered techno-economic culture. The main reason for Korea's success in catching-up with industrialized countries in some high-tech industries was due to entrepreneurs' risk-taking spirit who made commitments to unknown, high-risk technological frontiers such as semiconductors, automobiles, telecom equipment, etc.

By bringing larger segments of society into the framework of innovation, Korea's new techno-scientific strategy (KNTSS) can be viewed as a corrective mechanism to Korea's age-old time tested KMTC.

State, Innovation and Korea's New Techno-Scientific Strategy

Moving away from efforts to 'reverse engineer' economic success, the 'post-developmental state' in Korea seeks to carve a new techno-scientific sphere by initiating various monetary incentives under distinct policy initiatives and programs, the most prominent among these programs are BK-21 and NURI.[15] The Korean state's experience 'governing the market' by a set of innovative policies pushing big business to meet market expectations became rather useful to streamline support for a new techno-scientific strategy aimed at creating university–industry–government linkages based on shared competitive performance. The most important lesson that the developmental state has imparted to its successor 'post-developmental state' in Korea is to construct performance criteria based on internationally recognized, market-friendly transparent indicators. Indeed, corporate Korea has invested large sums into research and development under the guidance of the development state, however it remained in the 'techno-industrial sphere' with the core focus on imitation and innovation, not invention. It was prohibitively expensive for private sector companies to explore the new 'techno-scientific sphere' because of the huge investments needed in unpredictable frontiers of technology.

[15]The 'post-developmental state' in Korea has been referred to by some scholars as the 'new techno-economic state,' which contrary to the popular belief, did not transform into a 'neo-liberal state,' rather old interventionist policies shifted focus from promoting *chaebol* to encouraging universities to widen the social base for economic development. For details, see Uttam 2005.

By recognizing the fact that Korean conglomerates' R&D efforts were directed towards similar techno-industrial sectors which locked them into a number of low-profit making, mass-manufacturing industries, the state's proactive role in creating the KNTSS is reminiscent of its earlier era rôle promoting the KMTC. The state in Korea has committed large resources to promote key technological frontiers with specific list of products. However, in its quest to develop world-class scientific infrastructure, the government this time has shifted its focus from corporations to universities, albeit with clear incentives to link them to industries. The inception of the BK-21 and NURRI projects was a well-planned government response to the changing techno-scientific needs of the nation.

Developmental Vs. Post-developmental Korea

The state's role under the Park Chung-hee administration (1961–1979) as 'investor and planner' and its function as 'regulator' was consolidated under Chun Doo-hwan's rule (1980–1987). The successive governments of Roh Tae-woo (1988–1992) and Kim Young-sam (KYS) (1993–1997) consistently advanced technological catch-up, reinforcing a regulatory system to bolster R&D activities. The KYS government's S&T approach was aimed at building up autonomous S&T capabilities and at the same time withdrawing state intervention in the economy. However, the inertia of the developmental state built by Park Chung-hee was so powerful that piecemeal changes by the successive governments could not fundamentally alter it. The 'unbalanced growth strategy' employed by the 'developmental state,' which gave rise to inequality between the chosen and the deserted, continued unabated. This manifested in the unprecedented economic concentration led by *chaebol*[16]. They devoted large sums to enhance corporate R&D, but the focus remained on advancing the 'techno-industrial sphere,' not the 'techno-scientific sphere.' Large industrial capital, which grew in Korea in the 1970s and 1980s under a set of market barriers installed by governments, largely monopolized R&D activities. Lee (2005) argues that the 'developmental state' succeeded in creating a nationwide consensus that state-fostered business conglomerates were national enterprises, not individual property. This blurring of the divide between public and private was responsible for the cornering of major R&D resources by big business. He adds that any notion of collectivism, however, vanished because ownership and managerial rights were bequeathed exclusively to the owners' families, leading to the deepening of antagonism between the privileged group and labor. The relative deprivation felt by labor

[16]As of 1985, the five largest *chaebol* constituted 20.4% of Korea's total manufacturing sector assets. For further details, see Dong-song CHO, *Hanguk Chaebol Yongu* [A Study of the Korean Chaebol]. Seoul: Maeil Gyeongje Sinmunsa, 1991.

and even the middle class evolved into a significant social problem that challenged the basic intentions of developmental state and eroded public support for it.[17]

The basic nature and orientation of the Korean state takes a decisive turn with the onset of the 1997 financial crisis and the inauguration of the Kim Dae-jung administration (1998–2002). Kim Dae-jung was instrumental in launching a new political center away from the traditional support-base of earlier administrations. The break from the past was so decisive that scholars referred to the post-crisis administration as the 'post-developmental state' since it initiated new state, market and social relations aimed at social cohesion based on voluntary cooperation. In order to achieve this, the post-developmental state started to foster civic organizations as a third sector in order to accumulate social capital. The post-developmental state refrained from 'governing the market' in an authoritarian manner. This prime mission of the state became the creation of an environment in which society and market cooperate in an effort to tackle social fragmentation. While, the developmental state's approach posited a dichotomy between state, market and society relations, the post-developmental state's mass-participatory democracy approach emphasizes symbiotic relations between state, market and society. The Post-developmental state tried to broaden science and technology's social base in Korea. Under the Roh Moo-hyun regime (2003–2007), policies and programs initiated by the Kim Dae-jung administration continued as polity in Korea remained firmly under the 'progressives'. The inauguration of the Lee Myung-bak administration (2007) led to the review of various revisionist policies aimed at augmenting the market role in the promotion of scientific excellence. Table 3.2 summarizes government-led programs and policies under successive regimes aimed at upgrading the nation's S&T infrastructure and capabilities.

Construction–Reconstruction of SMSRs for New Techno-Scientific Regime

The three major disjunctions in the modern history of Korea – Japanese colonization, the Korean War and the division of the peninsula – deconstructed all types of interest groups and provided the state with an opportunity to construct state, market and societal relations (SMSRs) from scratch. Park Chung-hee's powerful 'developmental state' retained financial control as a means to construct a government-big business alliance, where the state acted as a senior partner. Under the ambitious 'big push' for heavy and chemical industries in the early 1970s, the state channeled large resources to select *chaebol* and provided a firm grounding for domestic entrepreneurs to participate in the national mission of technological catch-up.

[17]The sharp division in Korea between 'conservatives' supporting 'developmental state' politics and policies and 'progressives' arguing for a 'post-developmental state' and a 'mass-participatory' economy polarized public opinion in the post-1997 crisis period.

Table 3.2 Characteristics of government-led R&D programs

Government administration	Characteristics of R&D programs related to the TH indicators
Park Chung-hee (1961–1979)	Government's strong push to run governmental research institutes and the joint research between universities and public organizations
Chun Doo-hwan (1980–1987)	Restructuring of government-sponsored research institutes, e.g., the integration of KAIS (Korea Advanced Institute of Science) and the KIST (Korea Institute of Science and Technology) into KAIST (Korea Advanced Institute of Science and Technology)
Roh Tae-hoo (1988–1992)	The gradual opening of research organizations in both private and public sectors; e.g., KIST became independent from KAIST in 1989
Kim Young-sam (1993–1997)	Dominance of governmental agencies from early 1990 to 1997 when Korea started to be subjected to International Monetary Fund conditions
Kim Dae-jung (1998–2002)	BK21 project started in 1999 to increase the research capacity of universities through large central government subsidies, thus increasing research autonomy at universities
Roh Moo-hyun (2003–2007)	Continual promotion of BK21 and internationalization of R&D, particularly in the academic sector. PBS was introduced in the governmental sector
Lee Myung-bak (2007~present)	Corporatiztion and internationalization of R&D

Source: Compiled from various Korean government documents

During this period, though the Korean state compromised 'external autonomy,' it gained 'internal autonomy' and bolstered its capacity to construct SMSRs. Under these SMSRs, the state vigorously promoted its junior partner, big business. However, things started to change in the early-1990s as the *chaebol* became gigantic and the power balance began to shift in their favor. The narrow social base of 'developmental alliance' started to tarnish the country's 'economic miracle,' and the crisis of 1997 gave the Korean people an opportunity to openly question the rationale behind the old SMSRs.

The inauguration of the Kim Dae-jung administration provided political consent to challenge the old SMSRs. 'DJnomics' based on 'participatory democracy' faulted the old SMSRs.[18] The devastating financial crisis shattered the myth of *chaebol* invincibility, and the Korean state was ready for a major readjustment. BK-21 was the post-developmental state's program of choice for widening the social base of the old SMSRs, while effectively retaining financial control. The two-phase BK-21 is funding a 'people-centered, institution-led' innovation drive – a new convergence of interests needed to provide the crucial techno-scientific

[18]The idea of participatory democracy was already developed before Kim Dae-jung took office. He asserted the need for democratization to tackle economic problems related to Korea's unbalanced concentration of industrial output, labor distribution, and regional growth in his book *Daejung Gyeongjeron* [Mass-participatory Economy] (Seoul: Chongsa 1986).

excellence needed. BK-21 provides a detailed action plan and associated monetary incentives to restructure Korea's research infrastructure with the explicit aim of decentralizing resources. In other words, BK-21 is playing a similar role in regard to reshaping SMSRs as Park's HCI program did in the 1970s .

BK-21/NURI and the Linking of State, Market and Society for Innovation

The BK-21 and NURI programs' intended nurturing of globally competitive research universities and graduate programs has attracted a great deal of attention – more, in fact, than its relative share of resources might at first seem to warrant.[19] Phase-1 of BK-21 accounted for only 5% of government R&D expenditures, 10% of university R&D funding from all sources, and just 1% of gross R&D expenditure in Korea. The New University for Regional Innovation (NURI) Project, which ran through 2004~2008, had USD1.4 billion in funding. Nonetheless, the importance of BK-21/NURI is found in the programs' ability to pull in more of society, though in a highly competitive format.

Structure, Operation and Significance of the BK-21/NURI

BK-21 aims to nurture high-quality scientific manpower by recreating research universities and graduate programs in Korea. Based on the proposal submitted by research group (*sa-up-dan*), BK-21 allocates funds to universities with a firm commitment to achieve promised results. In order to qualify for BK-21 funding, an institution/department/research group must satisfy several conditions. It must have a doctorate program with enrolled PhD candidates. The number of faculty members participating in the research group must be at least seven for liberal arts and social science groups, ten for basic science groups, and 10–25 for applied science groups. It is mandatory for the participating professors to have produced or exceeded the average number of publications by their peers for the prior 3 years. All participating research groups must secure matching funds from their universities equal to at least 5% of the BK-21 funding they seek. Applied science and interdisciplinary science research groups must secure funds from industry amounting to at least 10% of total funding sought. However, research groups from regional universities must secure only 3–5% of the total funding. BK-21 recipient research groups are selected at the beginning of a 7-year period based on the merit of their respective departments and

[19]In Phase-I, from 1999 to 2005, BK-21 allocated about USD1.4 billion in funding. Phase-II, which began in 2006 and is scheduled to run through 2012, allocates and additional of USD2.1 billion.

universities. Underperforming institutions and research groups face the danger of elimination from the list of BK-21 financial support programs.

The funding is allocated after carefully considering academic discipline, geographic location, and the scale of the research. By academic discipline, science and engineering receive 85% of the Phase-II funding for the large scale research projects. The remainder is split between liberal arts, social science and inter-disciplinary fields (75%) and basic science (25%). By location, about 25% of the total program budget is allocated for universities located outside of Seoul. Regional universities compete in separate rounds in each academic field. Top universities such as KIST, POSTECH, although located outside Seoul, are restricted from applying for regional university support. In order to further support for regional universities, there is the New University for Regional Innovation (NURI) Project, which is in line with the post-developmental state's national policy of 'Balanced Development of the Nation.'

The weak linkages between industry, university and government have been perceived as a major constraining factor regarding the effectiveness of national innovation systems since they limit the utilization of existing knowledge as well as the formation of new knowledge through the combination of complementary knowledge from different sectors (OECD 1999). Research collaboration between industry and university is increasing due to recent administrative measures, such as establishment of technology transfer offices within universities, but most likely due to the BK-21 funding to improve Korean universities' R&D capabilities. As a result, the number of patents which are co-invented by members of different organizations has been sharply increasing since the inception of BK-21. The operational aspects of BK-21 links government–university–industry into a symbiotic relationship which carries wider ramifications for restructuring inter-linkages between state, market and society. Here, the incorporation of universities equals the incorporation of more of society since people from all walks of life come to study in higher education. No doubt, BK-21 funds have cemented government–university–industry relationships; however the program has also created hierarchies with exclusionary policies benefiting top institutions. The enforcement mechanism instituted by BK-21 has already invited resentment from various stake holders. Many professors have registered protests about blatant discrimination in the selection process.

Korea's Tech-Strategy in an International Context

The Semiconductor Industry as a Case Study

Korea's technological catch-up as a cautiously thought out 'follower strategy' was explicitly demonstrated in its leapfrogging into various industrial sectors: chemicals, shipbuilding, automobiles, etc. However, the nation's successful jump into the highly competitive semiconductor industry, primarily dominated by the firms of advanced countries, such as Japan, the U.S. and EU states, has attracted attention worldwide. Therefore, the semiconductor industry provides a relevant case to study how late entrants overtake early movers. Japan had overtaken the still earlier mover the United States by implementing its own follower strategy. Until the early 1980s, the Korean semiconductor industry only consisted of factories assembling discrete devices for foreign companies. However, the Korean semiconductor industry registered an annual growth rate of 25.5% between 1984 and 1992, more than two and half times the growth rate (at 10.6%) of the world semiconductor industry. Currently, Korea's Samsung Electronics leads the memory chip market with 30.3% of world market share, followed by Hynix, another Korean firm, with 19.1% share. A latecomer in the semiconductor industry, Korea successfully managed to leapfrog into the top ranks of dynamic random access memory (DRAM) producers. Semiconductor exports brought in USD32.7 billion in 2008, racking up 7.7% of Korea's total exports.

Korea's 'follower strategy,' for semiconductors adopted innovative policies similar to those learned from shipbuilding and its other previously conquered industries. Based on the logical rationale put forth by one study (Cho et al. 1998), the following are the most important policy innovations responsible for Korea's international competitiveness in the semiconductor industry:

Entrepreneurial Leadership

Entrepreneurial leadership has been one of the important factors which catapulted Korea into the league of advanced semiconductor manufacturing nations. Tested in the global market place, Korean entrepreneurs who successfully established diversified business empires virtually 'bet their companies' on entering the DRAM market. With no prior individual or organizational experience related to the semiconductor business, no significant demand for semiconductors in the domestic market, and a huge internal investment – Korean companies' full-scale entry into the semiconductor industry was seen as sheer entrepreneurial courage and risk-taking capacity. The country's investment in the semiconductor industry increased at an annual average rate of about 4.5% from 1984 to 1991 (Chu 1992). Korean semiconductor manufacturers' risk-apatite was such that in 1985 when the industry underwent a severe worldwide recession, they continued to invest. This odd investment was based on their strategic insight into the act of catching up with leading industry players.

Resource Leveraging

Resource leveraging became possible as Korea's business conglomerates accumulated huge capital during the remarkable success of its export-oriented industrialization. Samsung, Hyundai and LG competed fiercely to invest in the semiconductor industry by channelling resources from other subsidiaries.

Economy of Speed

Korea's experience with the economy of speed in other industrial sectors provided companies relevant experience to build plants quickly. For example, Samsung completed conclusion of its first production line in just 6 months, about half of the industry average time (Cho et al. 1998). Korean manufacturers' ability to 'compress time' for production was an important factor in their success. Time compression is a manifestation of a focused approach driven by entrepreneurial leadership.

Human-Embodied Technology Transfer

Korea undertook strategic alliances; licensing selectively for essential technology transfers as evident by alliances between Samsung-Texas Instruments, LG-Hitachi and Hyundai-Texas Instruments. However, more than formal technology transfer, Korea followed a unique strategy of human-embodied technology transfer. Korean companies recruited U.S. educated Koreans with necessary industrial experience in the U.S. semiconductor industry. Samsung's entry into the semiconductor business was made possible by a team of Korean engineers who had been recruited from a variety of semiconductor firms in the United States (Samsung Electronics 1989). Human-embodied technology transfer has been regarded as the most effective and fastest way to transfer and absorb technologies.

Simmering Domestic Rivalry

Simmering rivalry among *chaebol* to leapfrog into any lucrative business domain was so intense that sensing semiconductors as a ripe opportunity, all *chaebol*-linked semiconductor companies chose DRAM as the main target. This led to focused investment in the DRAM area. Also, competitive expansion and upgrading of DRAM facilities by Samsung, LG and Hyundai created economies of scale that became the core of Korea's catch-up in this industry.

Role of the State

The state came to support the technologically sensitive semiconductor industry. Tariff reductions, preferential interest rates, subsidies for R&D, and other incentives were implemented to facilitate the industry's growth. In 1986, the Korean government launched the VLSI project, a decade after Japan's collaboration involving domestic chip makers and government institutions.

Selective Benchmarking

Korean companies selectively benchmarked the best performers in the business by imitating and learning from them. The Korean semiconductor industry learned a great deal from its Japanese counterpart. Carefully studying and analyzing preceding examples, Samsung set Toshiba as its role model, which demonstrated how a latecomer can succeed with a clear target of product segment and aggressive investment in the manufacturing process. The follower trend became indigenous as Hyundai and LG began to imitate Samsung.

Export Orientation

Export orientation played an important role as Korean companies entered the semiconductor business without a local market for a base but with the strategic intent of capturing a share of overseas markets.

The above mentioned factors worked well to operationalize Korea's 'follower strategy' based on the KMTC, however to construct a viable 'pioneer strategy' capable of competing with advanced economies, such as those of Japan, the U.S. and EU, Korea needs to move the semiconductor industry from the catch-up trap. In order to operationalize 'pioneer strategy,' Korea urgently needs to build technological capability in the non-memory semiconductor business field, which requires new SMSRs that can unleash fresh talent to push the new techno-scientific sphere and put an end to Korea's follower strategy in this field.

Policy Implications

The structure, nature and reach of the KNTSS have important policy implications for Korea and other economies. Most of the BK-21/NURI policies that are aimed at achieving high performance are rooted in the state-engineered high-growth era, which created clear performance measurements based on corporate Korea's export performance in the global market place. Subsidized credit facilitated by the state was linked with the company's performance in the competitive market place. Inception, implementation and innovation of Korea's S&T policies incorporate the same zeal, commitment and efforts of the HCI program after the Nixon Shock in 1971. The enforcement mechanism devised by the NTSS resembles the way the earlier era state pushed firms to either perform in the export sector or perish.

Phase-I of the KNTSS's flagship program BK-21 helped push the number of published SCI-level Korean-originated papers from 3,765 in 1998, the year before it began, to 7,947 in 2005, the year it ended. This remarkable jump imparts an important policy lesson – monetary incentives work best when it comes to motivating researchers to produce quality findings, university departments to increase research activities, and all related parties to collaborate with one another.

The other important policy lesson that one can derive from the KNTSS is in the socio-economic realm, namely upgrading the national S&T level requires the involvement of more of society. This type of social empowerment, virtually mandated by the BK-21/NURI programs, has the potential to engage a wider segment of society and transform state, market and society relations, creating a firm base for long-term techno-scientific advancement. Another related policy lesson: that state should not disassociate itself from the innovation process as the resources required for rising in the techno-scientific sphere cannot be generated by the corporate sector alone. To encourage corporate research to move beyond the techno-industrial sphere, state-funded R&D still has an important role to play.

Conclusion

This paper concludes that the KMTC worked well as a mechanism to imitate/reverse engineer foreign technologies, but its success pushed R&D exclusively towards a similar techno-industrial sphere and left the crucial techno-scientific sphere neglected. The end of the Cold War and crippling financial crisis signaled that the KMTC had lost its relevance in the fast-changing post-Cold War reality. By recognizing the problems in the symbiotic relationship between the state and big business that excluded labor, the study concludes that there is an urgent need to widen the social-base of scientific innovation.

It confirms that government-based monetary incentives, part of the KNTSS through BK-21/NURI, do not distort market-based incentives but rather act as a catalyst for mutual interest and state, market and societal convergence that broadens the social base and spurs the innovation needed for a techno-scientific revolution in Korea. Similar to the developmental state's 'big push' to promote HCI, current BK-21/NURI programs have the potential to recast SMSRs in a post-developmental state. By facilitating a convergence of government–university–industry interests, the KNTSS has initiated the restructuring of government, market and societal relations. However, this time, the Korean state faces stiff resistance from entrenched and powerful interest groups. To overcome this resistance, the post-developmental state has fostered a 'third sector' powered by civic groups to carefully monitor the old nexus between various interest groups.

The study also confirms that the enforcement mechanism based on the competitive selection of research funding recipients may create a rigid hierarchy among the institutions and researchers, which in turn might limit wider social participation. Careful monitoring is required to prevent elitism from creeping into BK-21 and similar programs. However, as seen from the socio-economic and political perspectives, the role of the Korean state is not only crucial for techno-scientific innovation, it also has important ramifications for relations between the state, society and market.

After a long search for a viable technological path, the post-developmental state in Korea has initiated programs that broaden social participation in the process of upgrading S&T in Korea. As success in the semiconductor industry suggests, even though these new initiatives have the power to restructure old SMSRs, acquiring and maintaining a competitive edge requires upgrading the entire regime for innovation. Korea's success in making the transition from the KMTC to KNTSS will test the resolve of its new technological strategy.

References

Cho D (1991) Hanguk chaebol yongu [A study of the Korean chaebol]. Maeil Kyongje Sinmunsa, Seoul

Cho D, Kim D, Rhee D (1998) Latecomer strategies: evidence from the semiconductor industry in Japan and Korea. Organ Sci 9(4):489–505

Choi HS (1984) Industrial research in less developed countries. ESCAP, Regional Centre for Technology Transfer, Bangalore

Choi HS (1988) Direction for technological self reliance: Korean approach. Technol Forecast Soc Change 33:23–32

Chu DY (1992) Bondoche Sanup eui Keupkyukhan Byunwha wa Urieui Daeung [Rapid changes in the semiconductor industry and our response]. Korea Maekyung Press, Seoul

Chung S (1999) Unification of National Innovation Systems: How Should it be Different between Germany and Korea?, Science and Technology Policy Institute, Seoul, Korea

Dosi G (ed) (1988) Technical change and economic theory. Printer, London

Eros JL, Park WH (1989) The adoption and diffusion of imported technology. Croom Helm, New York

Freeman C (1987) Technology and Economic Performance: Lessons from Japan, London: Pinter

Johnson C (1982) MITI and the Japanese miracle: the growth of industrial policy 1925-1975. Stanford University Press, Stanford

Kim D (1986) Daejung Gyeongjeron [Mass-participatory economy]. Chongsa, Seoul

Kim L-S (1997) Imitation to Innovation: The Dynamics of Korea's Technological Learning, Boston, Massachusetts, Harvard Business School Press

Kim L, Lee J (1997) Korea's entry into the computer industry and its acquisition of technological capability. Technovation 6:277–293

Kim L (1999) Learning and innovation in economic development. Edward Elgar, Cheltenham

Kim, Linsu (2003) "Crisis, Reform, and national Innovation in south Korea," (pp.86–107), in William W. Keller and Richard J. Samuels (ed.), Crisis and Innovation in Asian Technology, Cambridge Online Books

Lee C (2005) Korea's national innovation system. Proceedings of the ASIALICS 2nd international conference, Jeju, Korea, 17–20 Apr 2005

Lim Y (2000) Development of the public sector in the Korean innovation system. Int J Technol Manage 20(5/6/7/8):684–701

Lundvall B (1992) National systems of innovation: towards a theory of innovation and interactive learning. Pinter, London

Mahlich J, Pascha W (2007) Innovation and technology in Korea: challenges of a newly advanced economy. Physica, Heidelberg

Nelson R (1993) National innovation systems: a comparative analysis. Oxford University Press, Oxford

OECD (1999) Managing national innovation systems. OECD, Paris

Patel P, Pavitt K (1994) The nature and economic importance of the national innovation system, STI review 14. OECD, Paris

Samsung Electronics (1989) Samsung Jeonja Ishipnyunsa [Samsung Electronics twenty-years]. Seoul, Korea

Shin T, Kim H (1994) Research foresight activities and technological development in Korea: science and technology policies in national R&D programs. Technol Forecast Soc Change 9 (1):31–45

Song SH (2009) R&D spending jumped 10 percent last year, Korea Herald, 16 Sept 2009

Suh JH (2000) Korea's innovation system: challenges and new policy agenda. Paper presented at INTECK, United Nations University, 2000-4, Maastricht, 7–10 Nov 2000

Uttam, Jitendra (2006) Korea's New Techno-Scientific State: Mapping a Strategic Change in the 'Developmental State', China Report (42)3:257–268

Chapter 4
Innovation & Technology in Korea an International Perspective

Ji Hong Kim

Introduction

In the early stage of Korea's economic development in the 1960s, Korea started out with a large pool of unemployed or under-employed workers and little capital accumulation or technology. Despite these adverse conditions, Korea has achieved remarkable economic development over the past three decades. In 2005, Korea's per capita GNP reached $16,400 while its total trade volume expanded to approximately $289 billion. The manufacturing sector's share of GNP also increased from 14% in 1961 to 28% in 2005, while the agricultural, forestry and fisheries sector's share decreased from 37% to 3.5% during the same period. As a result of this rapid development, Korea today is a recognized OECD member (Fig. 4.1).

Figure 4.2 presents the growth of nominal GDP per capita and the decomposition of Korea's economic growth over the past four decades. It shows that the contribution of knowledge, presented as total factor productivity (TFP), was a key factor in Korea's rapid economic growth.

The structural transformation of the Korean economy has been successful both in the width and depth of its industrial structure. Korea has changed from an investment-driven economy to an innovation-driven economy. In particular, the Korean strategy to utilize outside capital and technologies while accumulating indigenous capabilities is still valid for many economic sectors.

The Korean development process offers valuable lessons for other developing economies since the industrialization process was not always smooth. There were times of high inflation and severe economic setbacks. During the early stage of economic development, the shortage of foreign exchange was the most decisive constraint. As Korea pursued an export-led growth strategy from an agrarian economy to earn foreign exchange, it became more vulnerable to the vicissitudes of the international environment. The Korean government has been preoccupied with adjusting to the international environment and concerned with the balance of payments. The financial crisis of 1997 manifested the limitations of the input-driven or investment-driven growth paradigm.

J. Mahlich and W. Pascha (eds.), *Korean Science and Technology in an International Perspective*, DOI 10.1007/978-3-7908-2753-8_4,

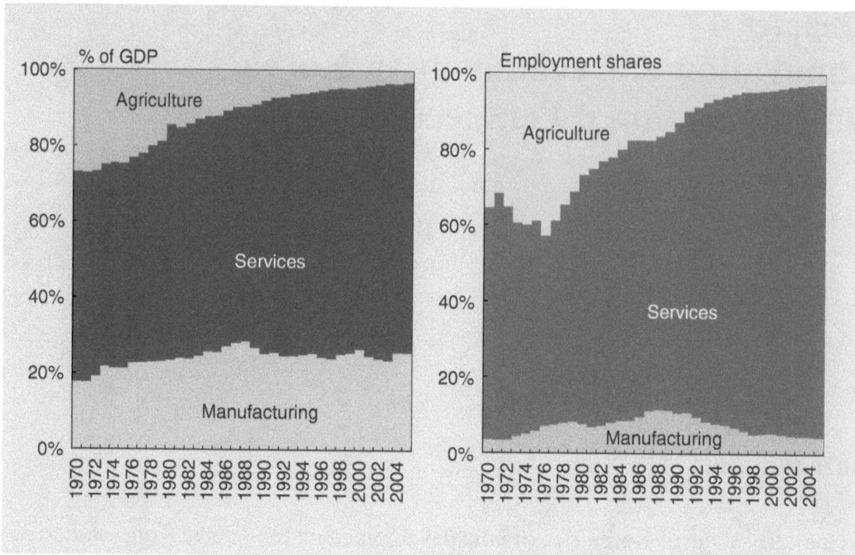

Fig. 4.1 Change in industrial structure (Source: National Statistics Office, Korea 2009)

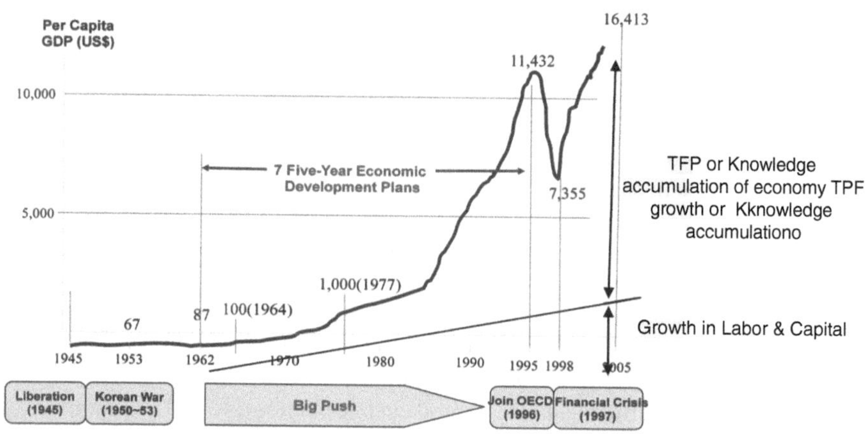

Fig. 4.2 Growth of GDP per capita in Korea (Source: Korea Development Institute (KDI) and World Bank Staff Estimate)

In the aftermath of the financial crisis, policy efforts shifted toward making the Korean economy knowledge-based; an economy where innovation could thrive, enhance overall productivity and thereby sustain economic growth.[1] Korea's new

[1]A knowledge economy is one that uses knowledge as the key driving force. It is an economy in which knowledge is acquired, created, disseminated and used effectively to enhance economic development.

growth strategy is built upon the four pillars of the knowledge economy: macroeconomic framework, information infrastructure, human resource development, and innovation systems.[2] The successful transition to an innovation-driven economy involves elements such as long term investment in education, developing innovation capabilities, modernizing the information infrastructure, and having an economic environment that is conducive to market transactions.

The main purpose of this paper is to describe the Korean economy's transition and draw possible lessons for developing economies. First, the rapid economic growth and structural transformation is reviewed. Second, the results and transition efforts are briefly explained. Third, major Korean policies on innovation and technology are discussed. Lastly, several lessons from Korea's experience are summarized.

Overview of Korea's Development Process

The industrialization of Korea since the 1960s is not just a process of capital accumulation but numerous successes and failures by risk-taking entrepreneurs, painstaking learning efforts of the workforce, and trial-and-error government policy to create the right business environment[3].

Korean industrial transformation and increases in income levels have been achieved through the intensive learning process where technological capability building and human resource development have played a decisive role. Another distinctive aspect of the Korean industrialization process is the active role that government played by intervening in the market to initiate the transformation. The Korean development process can be divided into three stages: (1) Factor Driven (1960s through 1970s), (2) Investment Driven (1970s through 1990s), (3) Innovation Driven (1990s through 2000s) (Fig. 4.3).

Development Strategies and Policies: Chronological Review

Before Industrialization

Poorly endowed with natural resources and devastated by the Korean War (1950–1953), Korea, until the end of the 1950s, had remained an agrarian society and her industrial activities were mostly confined to light industries such as simple

[2]The four pillars are elaborated in various articles. This section is borrowed from "Korea as a Knowledge Economy," published as Chap. 2 in the World Bank Institute Learning Resources Series' *Knowledge Economy.*

[3]Ibid.

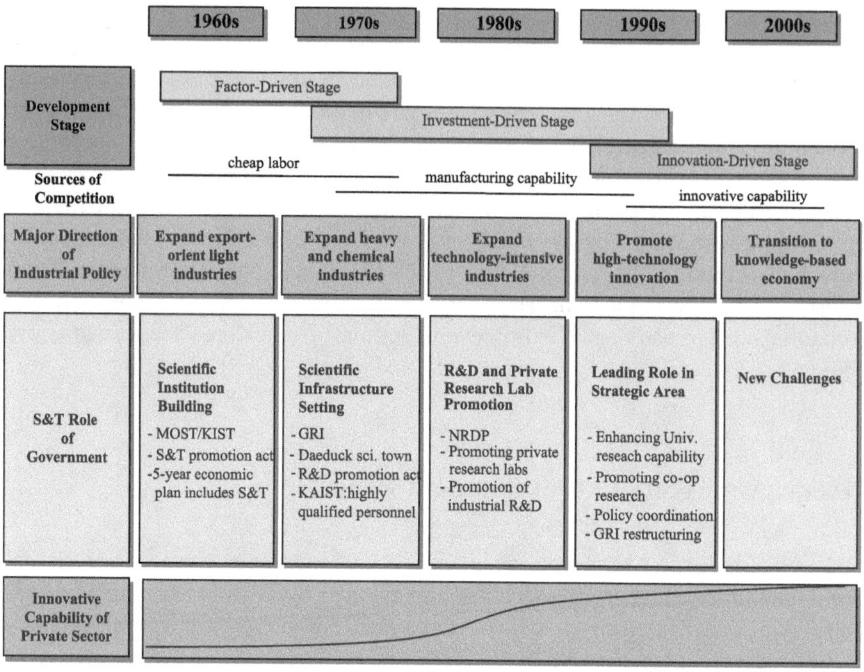

Fig. 4.3 Three stages of economic growth in Korea (Source: Korea Development Institute)

assembly and processing of raw materials[4]. During this period, however, the government had implemented two important policies – compulsory education and land reform. The Constitution of 1949 declared that it was the duty of the people to educate and the right of the people to be educated, and the government introduced compulsory education for primary schools. The introduction of compulsory education helped create an abundant pool of knowledgeable people that would be instrumental for industrialization in later years. The land reforms of 1947 and 1949 laid another foundation for later industrialization in that more equitable wealth distribution enabled greater opportunities for more people. But, the efforts of building institutional bases for the newly born country were abruptly stopped by war in 1950.

Launching Industrialization: 1960s

The development strategies of the 1960s aimed at terminating the vicious cycle of low savings, low investment and low growth, through policies designed to promote an increase in government savings and a rise in foreign capital inflow. Priority was given to an export-led industrialization. The development strategy of the 1960s

[4]Ibid.

was based on the promotion of both export and import-substitution industries. Korea had limited raw materials, a non-integrated industrial structure and a skilled labor force that received low real wages. Thus, at this stage of development, it was inevitable that export should mainly consist of labor-intensive processing of imported raw materials and intermediate goods. The capital accumulation which had been attained through this process was used both for the development of the agricultural sector and for the promotion of the heavy and chemical industries in order to accelerate industrialization. In addition, while the government assumed leadership in development efforts, private creativity and initiative were encouraged. Market principles were accepted as the basic premise of the economic order in the 1960s. Such measures as normalization of interest rates, adoption of flexible exchange rates and liberalization of trade were adopted to improve the operation of the price mechanism.

Upgrading Industrial Structure: 1970s

In the mid-1970s, faced with sudden wage hikes, the government adopted a new set of development strategies, shifting away from the promotion of labor-intensive export industries of the 1960s to the development of heavy and chemical industries (HCI). In order to finance the massive investment requirements, the government established the National Investment Fund, which provided long-term subsidized loans to the strategically selected industries to develop their capital formation in an effective and orderly manner. In addition, government intervention was visible in other ways such as high tariff barriers to protect targeted HCI projects. Entry barriers were also erected to certain industries to limit the rise of domestic and foreign competition.

The development of heavy and chemical industries is seen as a means for the economy to adapt itself to changes in its international and domestic environments. The targeted industries, including machinery, metallurgy, chemicals and shipbuilding, set a foundation for the development of other industries. Alongside the industrial targeting, the plan explicitly stated the importance of technological and human resources development and included pertinent strategies. In particular, the education system for training technicians was overhauled to raise quality and produce a greater diversity of skills

This policy shift significantly deepened the industrial structure, but also accelerated inflation and wage hikes and increased economic inefficiencies. The premature and excessive investments in HCIs caused severe distortions in resource allocation especially at the expense of upgrading labor-intensive industries. The adverse effects were mainly due to the overly ambitious investments, which outgrew the technological and financial capacities of the economy. Large enterprises were crucial in the process of heavy-chemical industrialization because of the economies of scale, but the importance of large enterprises led to the concentration of economic power in the hands of a few big Korean businesses or "chaebols." Moreover, the development of assembly industries without the concomitant development of parts and materials industries deepened the dependency of the economy on foreign imports.

Rationalizing Industrial Structure: 1980s

To cope with the new challenges, promotion of economic autonomy became one of the key policy goals of the 1980s. The government simplified various approval and authorization procedures. Economic deregulation reduced government intervention and allowed more individual freedom and was actively pursued. Steps were also taken toward the internationalization and liberalization of the economy. The accelerated import liberalization policies resulted in an almost complete liberalization of manufactured imports and tariff rates were also substantially lowered. These initiatives were also intended to promote the international competitiveness of domestic industries through greater exposure to foreign competition. Fair trade and competition policies were strengthened to reduce the inefficiencies of domestic industrial structures and curtail abusive monopolistic practices.

These stabilization, deregulation and liberalization policies helped check inflationary pressures and substantially improved the international competitiveness of the export industries. Still, problems from the big gap between large and small firms, imbalances between urban and rural sectors, and unequal income distribution remained. In the 1980s, therefore, the government exerted major efforts to improve income distribution and to enhance social equity. It is worth noting that social equity and welfare significantly improved in the late 1980s. The government enacted the minimum wage law in 1988 and introduced the national pension system in 1988 and nationwide medical insurance system in 1989.

Liberalizing the Economy: 1990s

Policy measures introduced during the 1980s seemed to be successful: for instance, the economy regained its high growth path, and 1986 marked the first current account surplus since the launch of the export-led industrialization. But the trade balances returned to red in 1989, and a new set of problems surfaced. Manufacturing wages rose rapidly, surpassing productivity growth. Combined with the appreciation of the Korean won, the competitiveness of the economy deteriorated, with chronic current account deficits. Excessive expansion of domestic demand and rapid wage hikes during the period of trade surplus led to the return of balance of payment deficits. However, the new democratic order that brought more individual freedom and capital liberalization – but not effective and prudential regulation, financial sector reform or restructuring – also hastened the 1997 financial crisis. Many analyses trace the root cause of this crisis to deeply embedded but long-neglected structural weaknesses in the economy. They range from the so-called "crony capitalism" and structural vulnerabilities in the corporate and financial sectors to large current account deficits. Mismanagement of the country's external liabilities is also cited. Others attribute the cause of the crisis to uncontrollable outside shocks. External elements often cited are contagion from the crisis that erupted in Thailand in the summer of 1997, a sudden reversal of the flow of foreign capital by the herd behavior of overseas lenders, sharp price drops in semiconductors and other major export items, and the depreciation of the Japanese yen.

Due to domestic wage hikes and the appreciation of the Korean won, Korea continued to pursue high-value added manufacturing in the 1990s by promoting high-tech innovation. Together with the build-up of a modern and accessible information infrastructure, there was continued expansion of R&D capability and a skilled labor force, the result of a government-backed and aggressive expansion of higher education.

The proportion of high- and medium-high-technology products among Korea's exports has grown sharply since the early 1990s, owing to the development of substantial technological capabilities in a number of high-technology areas. Korean firms now have the largest world market share in DRAM semiconductors, TET-LCD and other high-tech industries.

Knowledge Economy: 2000s

In 1998, the government launched major reforms in four areas; the public sector, financial sector, chaebol, and labor market. In particular, it committed to the transition into an innovation-driven economy to enhance overall productivity. After the financial crisis, Korea performed well in its efforts to catch up with the leading economies and instigated waves of industrial upgrading to become a leader in targeted high-tech industries. This period is marked by a substantial increase in R&D spending by both the public and private sectors and by attempts to improve knowledge flows and technology transfers across the system. The Korean innovation policy sought to accelerate the shift from a catch-up to a creative innovation system.

The catch-up model centered upon large–scale strategic technology development with government–affiliated research institutes and large global conglomerates taking the leading role. Korea has a strong comparative advantage in high-tech manufacturing, enjoying a 2.1% surplus in related manufacturing trade in 2005. Korea also showed a 0.9% trade surplus in medium-tech trade. China's manufacturing trade surplus is strongly dependent on low-tech industries; her high-tech manufacturing still shows a comparative disadvantage. However, China has continuously improved her international competitiveness and is expected to show a comparative high-tech advantage in the near future (Fig. 4.4).

Overall Assessment

Despite the drive for high growth, the technological base of Korean firms was well below that of the developed countries and thus exposed a serious structural weakness for the economy. In Korea's industrial development, the low-wage workforce

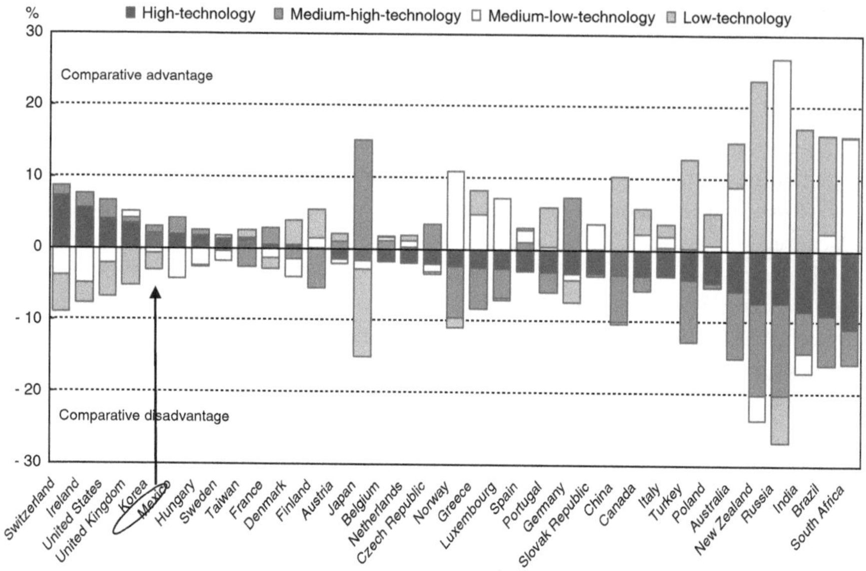

Fig. 4.4 Contributions to the manufacturing trade balance, 2005. As a percentage of manufacturing trade (Source: OECD (2008), OECD science, technology and industry scoreboard 2007)

in the 1960s made up for an acute shortage of capital and technology and laid the foundation for light industry development.

In the late 1970s, heavy and chemical industries emerged, followed by semiconductor and information-technology development in the 1980s. Throughout this period, however, Korean firms continued to rely heavily on imports of key components and capital goods. This was in part a result of the focus of Korean firms on developing component assembling technologies that yielded quick investment returns. Korea achieved a globally competitive position in processing technologies. But when it came to development technologies for core components and capital goods, Korea continued to depend heavily on technologies from developed countries.

The problem of dependence on technologies from developed countries was exacerbated by the rapid advances of the Chinese and Southeast Asian economies. Thus, Korea faced the dilemma of lagging behind developed countries but at the same time being chased closely by emerging countries.

With new waves of technological advances continually sweeping across the globe, knowledge (information) is increasingly being developed, stored, and transferred in digital forms, and many advanced economies are making significant strides with the transition to a truly knowledge-based economy. Indeed, it is estimated that nearly half of the GDP of leading OECD countries is already produced by knowledge-based industries. For Korea, the challenge has been to catch up with advanced economies and positively accelerate the transition to a truly knowledge-based economy. In addition, Korean innovation policy has not been conducive to the creation of innovative start–ups, to technology transfer, or to building-up basic research capabilities, all of which are increasingly important as Korea moves toward knowledge frontiers.

This process required the transformation of techniques, organization, and composition of production in the direction of higher productivity. Major development strategies and policy goals are summarized in Table 4.1:

The initiation of catch-up growth impacted new investments, and the successful maintenance of high savings propensity depended on the ability to sustain high rates of return on investments. Export-oriented growth through international price-quality competition is indeed conducive to achieving this objective, and Korea is an exemplary case for the role of export promotion in achieving industrialization in latecomers. According to NAM Duck-Woo, the following factors contributed to successful economic development: economic – (1) outward-looking strategy, (2) good use of foreign resources, (3) favorable international environment, (4) education, (5) faith in free enterprise system, (6) activist role of government/ non-economic – (1) ethnic and cultural homogeneity and a strong Confucian tradition that places a high value on education, achievement, and loyalty to the nation; (2) security threat; and (3) political leadership. (Nam 1997)

The World Bank also pointed to the following features of Korea's economic development process: first, at the macroeconomic dimension, the high rates of investment and savings imply that Korea's economic transformation has been achieved by massive capital investment. Korea's capital accumulation has been possible through increases in domestic savings, and the earlier shortages in domestic savings were rectified through foreign capital inducement. Second, industrial and labor compositions have been changed in the direction of higher productivity, as the share of manufacturing has steadily increased. Trade structures have also fundamentally changed from a primary goods exporter to manufactured products. Capital goods exports accounted for more than 40% of total exports as of 2002. Third, the changes in the structures of industry, employment and trade have been preceded with great improvements in human resources and technology, two of the most important factors for sustaining economic growth that enables efficiency gains. Despite the debates on the nature of East Asian growth performances, it is apparent that Korea has poured tremendous efforts into upgrading her knowledge and human resource bases. Korea's industrialization process is not only the process of capital accumulation; but it is also the learning process, a key concept of the knowledge economy.

Science & Technology Policy in Korea

Korea has recognized the importance of developing indigenous capabilities in science and technology for successful industrialization, and therefore has made building such capabilities one of the key policies over the past four decades.

Table 4.1 Development strategies and major policies for the Korean economy

	Development goals	Major policy directions	Macroeconomic policy framework	Human resource development	Science and technology
1960s	• Building production base for export-oriented industrialization	• Expanding export-oriented light industries •Mobilizing domestic and foreign capital	• Preparing legal and institutional bases to support industrialization	• Decreasing illiteracy • Establishing national infrastructure • Vocational training • Improving teaching quality	• Scientific institution building: Legal & administrative framework
1970s	• Building self-reliant growth base	• Industrial targeting: Promoting HCI and upgrading industrial structure • Building social overhead capital • Industrial rationalization	• Maximizing growth: expanding policy loans •Governmentintervention in the market	• Increasing college graduates with engineering degrees • Expanding higher education system	• Scientific infrastructure setting: Specialized S&T institutions, Daeduck Science Town
1980s	• Expanding technology-intensive industries	• Decreasing export subsidies and expanding import liberalization • Supporting technology development	•Stabilization • Enhancing private autonomy and competition	• Developing semi-skilled HR	• R&D and private research center promotion • NRDP
1990s	• Promoting high-technology innovation	• Building information infrastructure	• Liberalization • Reform and restructuring	• High skilled HR in strategic fields-IT, BT, etc. • Developing a life-long learning system • Increasing research productivity	• Leading role in strategic areas: HAN
2000s	• Transitioning to knowledge-based economy	• Promoting venture business and SME	• Globalization • Balanced national development	• Improving quality of university education • Regional development	• Building national and regional innovation systems

Source: Korea as the Knowledge Economy, WBI

Table 4.2 Historical perspectives: S&T policy measures to build up indigenous R&D

Imitation 1960s and 1970s	• Foundation of KIST, MoST • The S&T Promotion Act (1967) • Establishment of GRIs (1970s) in the areas of machinery, shipbuilding, chemicals, marine science, electronics • Tax credit for R&D investment (1974) • Development of human resources for R&D (KAIST)
Transformation 1980s	• National R&D program (NRDP, 1982) • Establishment of Daedeok Science Town • Promotion of private research: financial and tax incentives to stimulate R&D investments (reduction of tax for technology-based start-ups (1982); tax credit for technology and manpower development expense)
Innovation 1990s	• Promotion of university-based research Science Research Center • Five-year plan for innovation (1997) • Establishment of the national Science and Technology Council (1999) • S&T vision 2025 (1999) • Promotion of university research: SRC, ERC, NRL, etc. • HAN projects, Frontier R&D programs, etc. • Establishment of inter-ministerial coordination body: NSTC
Innovation 2000s to present	• First national Technology Roadmap (2001) • E-government vision 2006 (2002) • Broadband and IT Korea Vision (2007) • New organization of MOST (2004) – Deputy prime minister, establishment of the office of the Ministry of Science, Technology and Innovation (OSTI) • Innovation of overall coordination system: Ministerial office of S&T innovation in October 2004 • Launch of the Ministry of Education, Science and Technology (MEST) (2008)

Source: Various Korean government publications

Science & technology policies can be divided into three periods: (1) imitation, (2) transformation, (3) innovation (Table 4.2).

Catching-Up Period: Imitation

The technological base of Korean firms was well below that of the developed countries and thus showed a serious structural weakness for the economy. Since the First Five Year Economic Development Plan, science and technology areas got special attention. Creating institutional frameworks for mobilizing resources toward targeted areas was among the government's primary tasks, whereas the assimilation of technologies was among the main tasks of industry. Industrialization is the result of both accumulation and assimilation through concerted government and industry efforts.

When Korea launched its industrialization drive in the early 1960s, it had to rely on imported foreign technologies. By doing so, Korea pursued two key objectives:

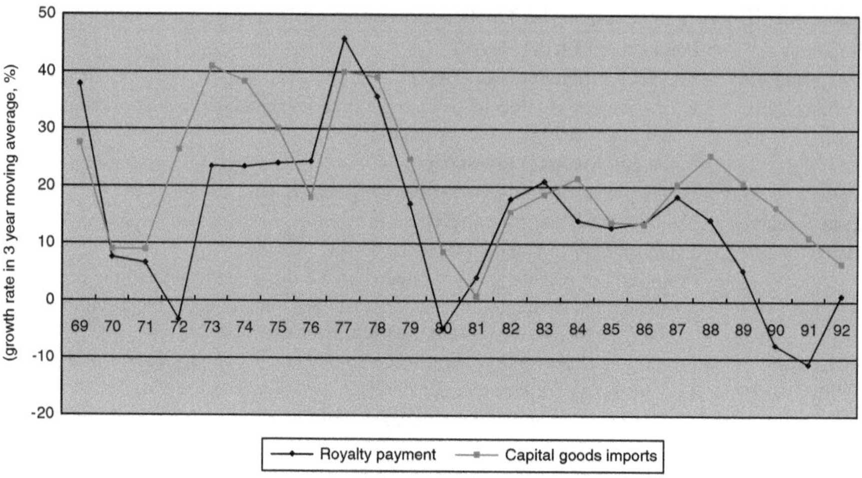

Fig. 4.5 Royalty payment and capital goods imports (Source: National Statistical Office, Korea, 2009)

First was to promote the inward transfer of foreign technology, and second was to develop domestic absorptive capacity to digest, assimilate, and improve upon the transferred technologies and to adapt them to domestic production. The second objective required a relatively skilled labor force, which was fortunately abundant because of the compulsory and aggressive education policies.

Rapid increase in foreign capital inflow has contributed to economic growth by not only financing the expansion of production capacity, but also increasing productivity through the concomitant transfer of advanced technologies. As shown in Fig. 4.5, there is a close co-movement between capital goods imports and royalty payments for licensed foreign technologies. This implies that Korea has been making great efforts to industrialize.

The technology assimilation strategy used various channels, such as original equipment manufacturing, foreign licensing-based production, reverse engineering of imported capital goods, and learning from the building of turnkey plants. These channels of informal technology assimilation enabled Korea to minimize its dependence on FDI, which had become more prominent since the 1997 financial crisis, and to maintain independence from multinational corporations. The strategy proved to be a success, and Korean firms were able to assimilate technologies rapidly enough to undertake subsequent expansion and improvement with little assistance from foreign suppliers (Table 4.3).

In the early years of launching full-scale economic development plans, the government recognized that science and technology would play important roles in the process. In the 1960s, two noteworthy institutions were established: Korea Institute of Science and Technology and Ministry of Science and Technology. These two, together with KAIS established in 1971, have exerted powerful

Table 4.3 Inflows of foreign capital (Unit: (current) USD millions)

	First plan (1962–1966)	Second plan (1967–1971)	Third plan (1972–1976)	Fourth plan (1977–1980)
Loans	291	2,166	5,432	10,256
- Public	116	811	2,389	4,084
- Commercial	175	1,355	3,043	6,172
Direct Foreign Investment	17	96	557	425
Total	308	2,262	5,989	10,681

Short-term capital and bank loans are excluded
Source: Korean Government, The Fifth Five-year Economic and Social Development Plan, 1982

influences over the S&T community in Korea.[5] Despite government efforts to build S&T institutions in the 1960s and establish an S&T infrastructure, such as KIST and specialized government research institutes (GRIs) in the 1970s, S&T policies played a limited role in those years.

Building Technological Capabilities

In accordance with the stages of economic development, the government has successively reoriented S&T policy. In the earlier years, more emphasis was put on building the infrastructure for technological development, whereas in later years the emphasis shifted toward more specific targeted technological development. Especially, changes in the economic environments in the early 1980s induced Korea to embark on serious investments in indigenous R&D. On the one hand, Korean industrial development had reached the stage at which domestic industries found it more difficult to be competitive in the international market because they were reliant on imported technologies and employed domestic labor that was becoming more and more expensive. On the other hand, Korean industries had grown to become potential competitors in the international market, making foreign companies increasingly reluctant to transfer technologies to Korea; thus, it was inevitable that Korea would have to develop an indigenous base for research and innovation. Meeting the challenge required highly trained scientists and engineers as well as financial resources to support R&D activities.

According to the Ministry of Science and Technology, Korea's gross expenditure on research and development (GERD) has tremendously grown, both in the size and intensity, especially in the 1980s and 1990s: the percentage share of GERD increased from 0.25% in 1963 to 2.64% in 2003. The number of researchers has increased 100 times over four decades (Fig. 4.6).

[5]MOST has been the main designer of Korea's overall S&T policy; KIST has played the role of technological functionary in responding to industrial demands for rapid economic growth; and KAIS (later KAIST) first implemented the concept of the research-oriented university into the Korean higher education system.

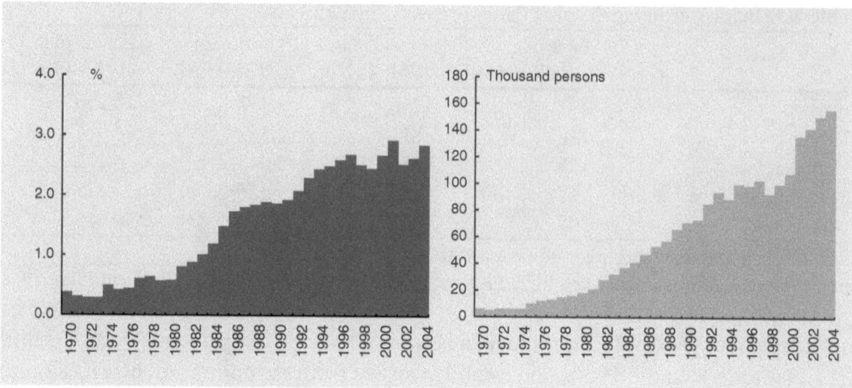

Fig. 4.6 GERD = GDP ratios and R&D personnel (Source: National Statistics Office, Korea 2009)

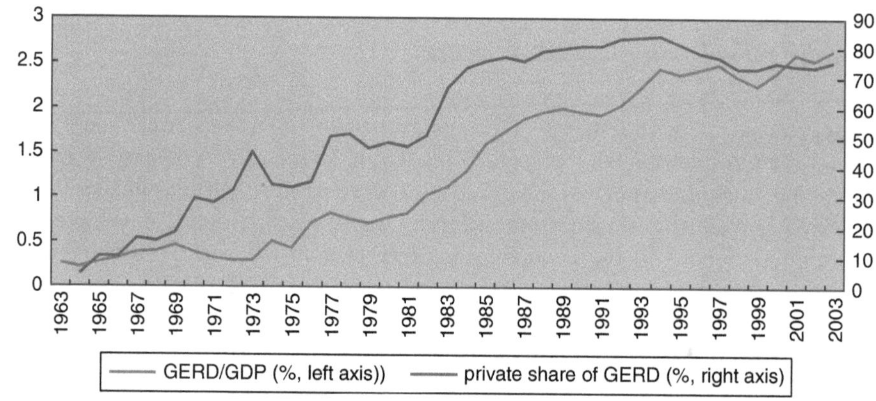

Fig. 4.7 R&D trend, 1963–2003 (Source: Ministry of Science and Technology, internal report)

The rapid increase in R&D has been possible through active expansion of private sector investment. During the earlier years of industrialization, private sector R&D spending was negligible; but as the rapid economic growth has called for commensurate investment in technology development, private enterprises have continuously increased R&D. Consequently, the funding sources have also greatly changed: government share of GERD has continuously decreased, and in recent years, only one-fourth of GERD came from the government (Fig. 4.7).

According to Woo's report, the process of technological capability building in Korea can be characterized as a dynamic process of the interplay between imported technologies and indigenous R&D efforts. Figure 4.8 plots the trend of the ratio of royalty payment over business R&D expenditures (BERD) on the horizontal axis and the trend of R&D intensity noted in terms of BERD over sales from 1976 to 2002. The ratio of royalty payment to BERD had substantially decreased until the early 1980s, which implies that the growth of BERD exceeded that of royalty

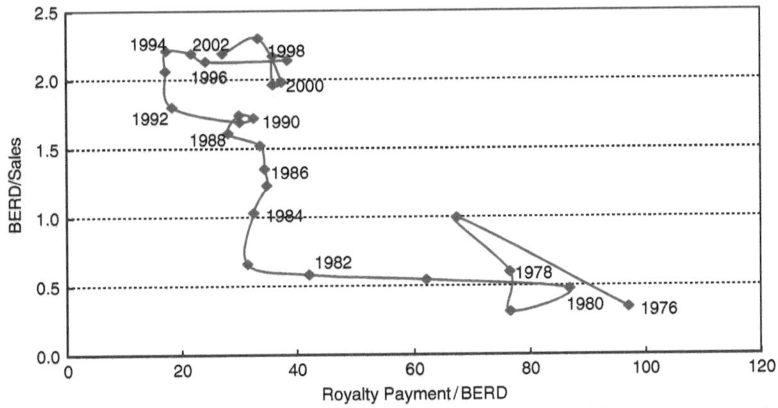

Fig. 4.8 Changing relationship between royalty payments and R&D (1976–2002) (Source: Chun-Sik WOO)

payments. The R&D intensity, however, had not risen and remained at the level of 0.5%. There is a clear change in the trends since the early 1980s. While the ratio of royalty payment to BERD had remained at the level of 30%, R&D intensity had started to increase.

The rapid increase in R&D investments has led to a corresponding increase in indigenous innovation and adoption of foreign technologies. This can be seen from Korea's improved performance in various international indicators of innovation and technology adoption, such as the number of scientific and technical journal articles written and published by Korean authors, the number of patents granted to Korean inventors by the United States Patent and Trademark Office (USPTO), and the amount of royalties and license fees paid and received by Korea (Table 4.4, Fig. 4.9).

Within a very short period of time, Korea has accumulated tremendous technology development capabilities – in particular, the private sector's willingness and ability to spend on R&D and the presence of a large number of relatively well-educated researchers. The ratio of patents to private R&D increased in the 2000s, and Korea's share of privately funded R&D in the national total for research has surpassed the OECD average. It has been found that the successful transition to the knowledge economy typically involves elements such as long-term investments in education, developing innovation capability, modernizing the information infrastructure, and having an economic environment that is conducive to innovation (Fig. 4.10).

Building Knowledge Economy

After having fixed the monetary and foreign exchange turmoil and its underlying causes during the 1997–1998 Asian crisis, Korea launched a major plan to make the

Table 4.4 Patents (triadic patent families)

	Total number			Percentage of total		Per million population	
	2006	1996	Growth (%)	2006	1996	2006	1996
United States	15,942	12,931	2.1	31.1	33.2	53.3	47.4
EU-25	14,787	13,106	1.2	28.9	33.6	31.8	29.2
Japan	14,187	10,608	2.9	27.7	27.2	111.1	84.2
Germany	6,171	5,508	1.1	12.0	14.1	74.9	67.1
Korea	2,785	322	21.6	5.4	0.8	57.7	7.0
France	2,499	2,147	1.5	4.9	5.5	39.5	35.9
United Kingdom	1,663	1,659	0.0	3.2	4.3	27.4	28.4
Netherlands	1,005	805	2.2	2.0	2.1	61.5	51.6
Switzerland	856	816	0.5	1.7	2.1	114.8	115.2
Sweden	847	914	−0.8	1.7	2.3	93.3	103.3
Canada	767	427	5.9	1.5	1.1	23.5	14.3
Italy	767	701	0.9	1.5	1.8	13.0	12.3
China	484	23	30.5	0.9	0.1	0.4	0.0
Israel	477	209	8.2	0.9	0.5	67.6	35.9
Australia	390	233	5.2	0.8	0.6	18.7	12.5
Austria	388	213	6.0	0.8	0.5	46.8	26.8
Belgium	372	362	0.3	0.7	0.9	35.3	35.6
Finland	341	356	0	0.7	0.9	64.7	69.3
Denmark	277	226	2.0	0.5	0.6	51.0	42.8
Spain	236	91	9.5	0.5	0.2	5.4	2.3
Norway	131	75	5.6	0.3	0.2	28.2	17.1
Chinese Taipei	119	38	11.4	0.2	0.1	5.2	1.8
Singapore	113	34	12.0	0.2	0.1	25.6	8.9
Ireland	70	28	9.3	0.1	0.1	16.4	7.5
Russian Fed.	63	58	0.8	0.1	0.1	0.4	0.4
New Zealand	54	32	5.2	0.1	0.1	13.1	8.6
Hungary	41	25	5.0	0.1	0.1	4.1	2.4
South Africa	30	29	0.6	0.1	0.1	0.6	0.7
Turkey	22	4	17.4	0.0	0.0	0.3	0.1
Mexico	17	10	5.3	0.0	0.0	0.2	0.1
Total	51,226	38,964	2.7	100.0	100.0	18.4	14.9

Source: OECD, STI, 2008/2

country a full-fledged knowledge economy. This led notably to major investments in information and communication technology (ICT), making the country one of the world's most digitalized, and in higher education, pushing enrolment rates to all time highs. This effort, coordinated at the highest level of the government, helped the country to recover quickly from the Asian crisis. In mid 2000s, Korea was again leading the world in terms of GERD as a percentage of GDP as well as researchers per 1,000 employees (Figs. 4.11 and 4.12).

The Korean government has been facilitating R&D, early market development industry clusters, and other sources of innovation. The role of government is supplemented by private industries in areas such as product development,

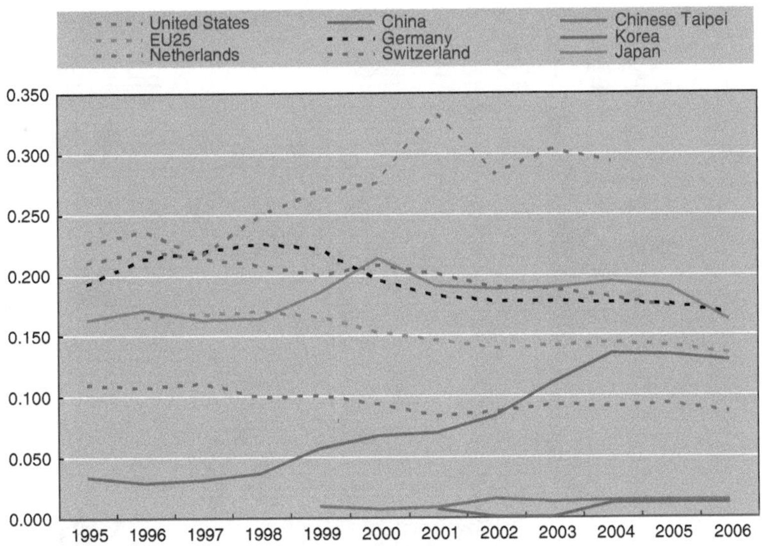

Fig. 4.9 Ratio of triadic patent families to industry-financed R&D (selected, 1995–2006). Note: Gross domestic expenditure on R&D (GERD) financed by industry, millions of US$ get rid of (2000) using purchasing power parities, lagged by 1 year (Source: OECD, STI, 2008/2)

applications and solution development. The coordination of the public and private sectors is illustrated as follows (Fig. 4.13):

For example, the rapid development of the Korean information infrastructure hinged on key government organizations that were responsible for the informatization strategy. These organizations were restructured in the 1990s and included the Informatization Promotion Committee, chaired by the prime minister; the Informatization Strategy Meeting, chaired by the president; and the Ministry of Information and Communication. Concurrently, the government established three master plans for the development of the information society: in 1995, the Informatization Promotion Act was enacted, and the first master plan for promoting informatization was formulated a year later. In 1998, during the second master plan, Cyber Korea 21 was established to cope with the changing environment that resulted from the Asian financial crisis. And in 2002, when most of the policy goals set up by Cyber Korea 21 had been achieved ahead of the original schedule, the third blueprint, e- Korea Vision 2007, was laid out.

Information Infrastructure

In the early 1970s, Korea's information infrastructure was inadequate and the provision of ICT services insufficient to meet the telecommunication demands associated with rapid economic growth. In 1975, only 3% of Koreans had a telephone. To improve efficiency in the provision of telecommunication services,

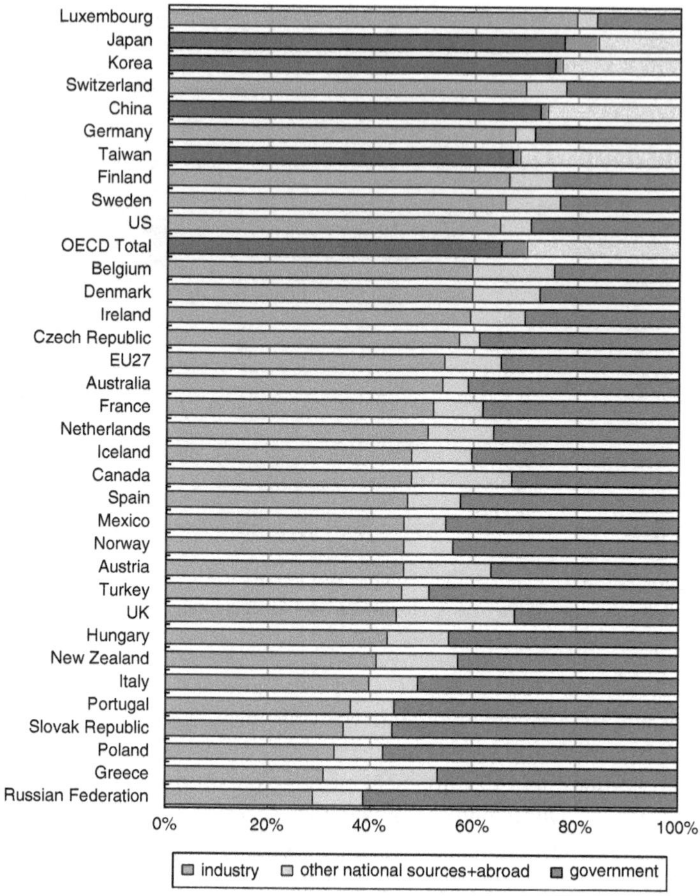

Fig. 4.10 R&D expenditure by source of financing, 2006. As a percentage of the national total (business enterprises, government, other). OECD + China, Taiwan, Russian Federation, Brazil. India. Highlights: Korea, Japan, Taiwan, China, OECD average, Source: OECD, STI, 2008/2

the government decided to rely on the invisible hands of the price mechanism and thus focused on introducing competition into the ICT infrastructure sector. A series of sequential but rapid policy measures were implemented for the deregulation and liberalization of the ICT services sector, along with privatizing the government-owned telecom operators (Fig. 4.14).

The reform of the ICT infrastructure sector resulted in tremendous improvements in terms of ICT penetration rates. Most noteworthy are the recent penetration rates of cell phones and the Internet. From 1995 to 2003, the proportion of Koreans with cell phones increased by nearly 20 times, to 70%, while the proportion of Internet users increased by a whopping 75 times, to 60%. Similarly, Korea is currently among the leading countries in the world in terms of the proportion of broadband Internet subscribers.

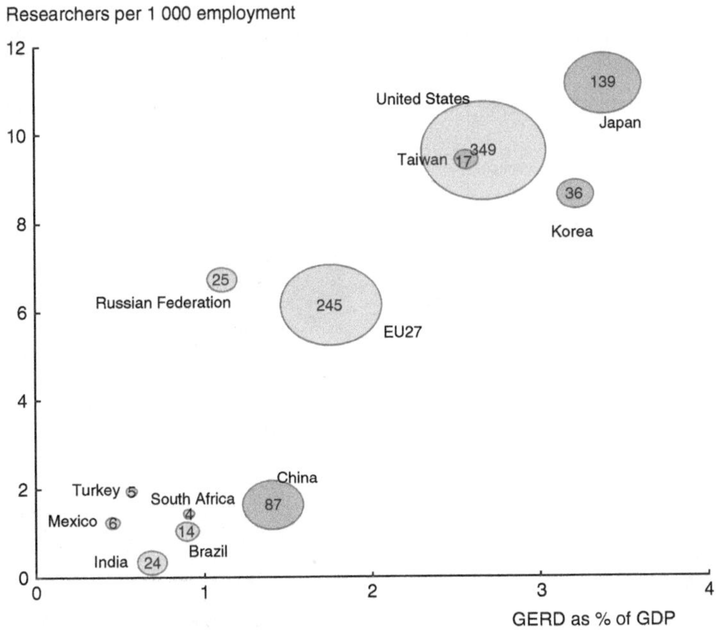

Fig. 4.11 R&D Inputs: GERD as a percentage of GDP, in billions of current USD PPP, and researchers per 1,000 persons employed (2006). Note: The size of the bubble and the number therein represents R&D expenditures in billions of USD in purchasing power parity (Source: OECD, STI, 2008/2, Jean Eric Aubert)

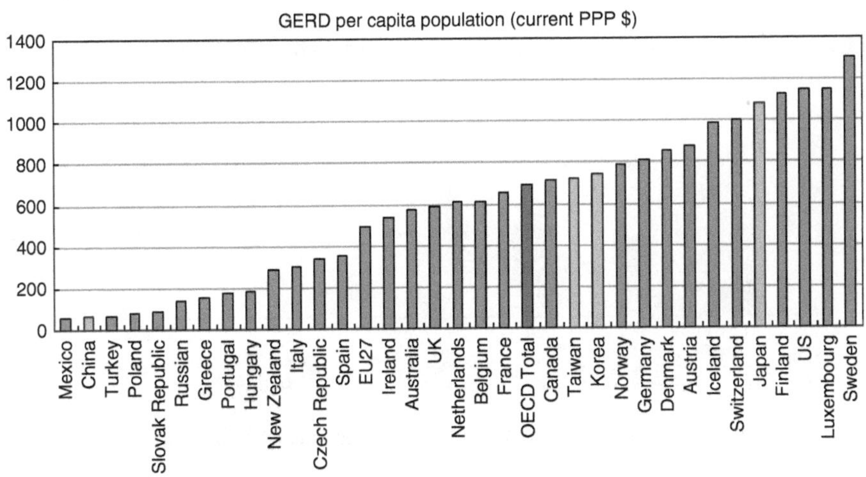

Fig. 4.12 GERD per capita population (current PPP$) (Source: OECD, STI, Scorecard, 2008/2)

To build infrastructure efficiently and economically, while actively responding to technological development and changing demands, diverse implementation methods have been used. Networks in commercial and densely populated areas

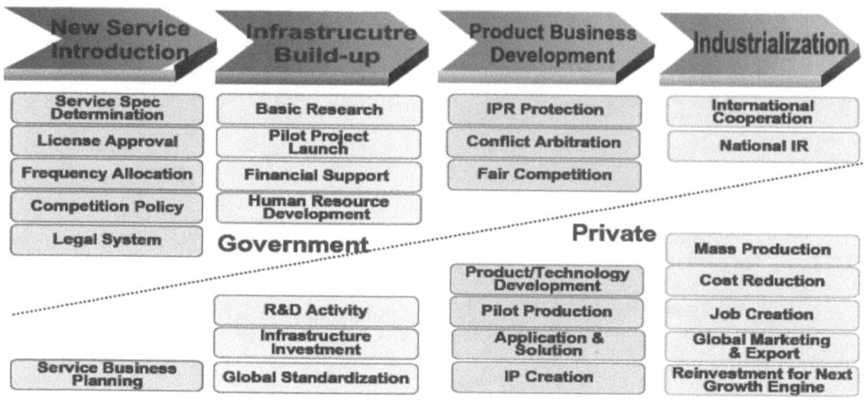

Fig. 4.13 Government and private-sector R&D support (Source : DaeJe JIN)

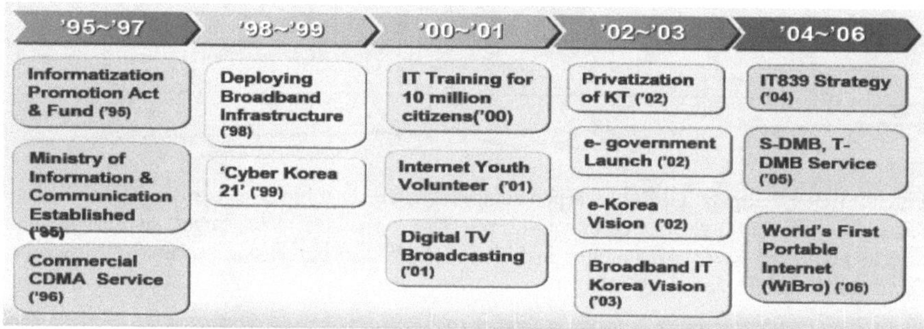

Fig. 4.14 Major government policies in ICT (Source: various Korean government publications)

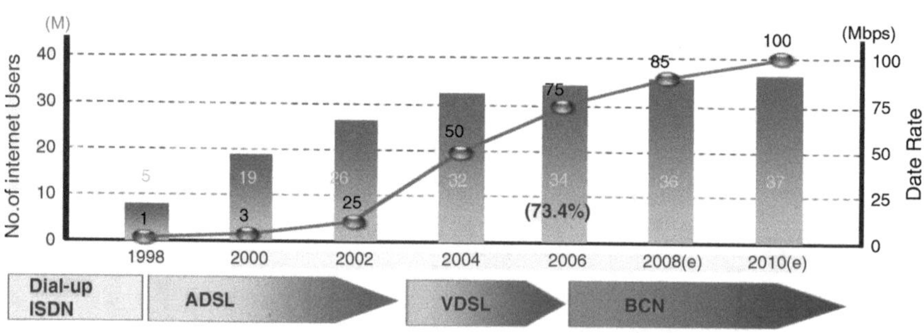

Fig. 4.15 Internet penetration (Source: Korean government internal report)

have been built with optical cables; networks that extend to subscriber premises have been built partly with optical cables and partly by digitizing and enhancing the speed of existing telephone lines or CATV networks or by building new wireless local loops (Fig. 4.15).

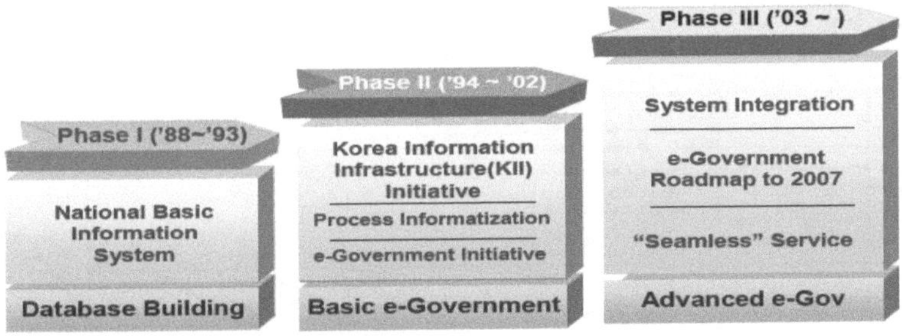

Fig. 4.16 e-Government plan (Source: Korean government internal report)

In order to finance the rapid development of and investment in Korea's information infrastructure, the Informatization Promotion Fund was established as a special vehicle to overcome short-term budgetary constraints. From 1993 to 2002, the fund reached a total of US$7.78 billion. About 40% of that was from government budgetary contributions, 46% came from private enterprises (of which licensing fees for new communication services composed the major portion), and the remaining 14% came from miscellaneous profits and interests. Korea has constructed ICT networks connecting all areas of the country, and ICTs are used extensively in numerous economic and social activities. The number of individuals using ICT-related services is also constantly rising: the number of subscribers to Internet banking services reached 22.58 million as of March 2005, and e-commerce has rapidly increased from 50 billion won in 1998 to 314 billion won in 2004, which is equivalent to 40% of GDP.

Led by an e-government initiative, the public sector is also extensively using ICTs. The entire process of managing government documents (i.e. production, distribution, transfer and retention) will be digitized. Since 2003, central and local government organizations have adopted standardized e-Document systems and have begun to distribute documents electronically.[6] The number of organizations using the systems expanded to the private sector in 2007, along with security improvements leading to the establishment of an e-document tracking system. The process of archive management (collection, retention, and destruction) has been digitized and standardized (Fig. 4.16).

The e-government window enables citizens to obtain necessary documents and information online, and submit forms without visiting government offices. Since its launch in 2002, this central government portal has provided general information via the Internet for an estimated 4,900 civil services. For example, the Administration Information System (AIS) enables the electronic processing of all administrative aspects related to architecture, housing and construction management. Those seeking approval for building designs do not need to submit thousands of hand-drawn blueprints anymore or visit related offices. As a result, the number of online service users exploded from 2002 to2005 and continues to increase (Figs. 4.17 and 4.18).

[6]In 2004, about 97% of documents were dealt with through the e-approval system in the government agencies, compared with only 21% in 1998.

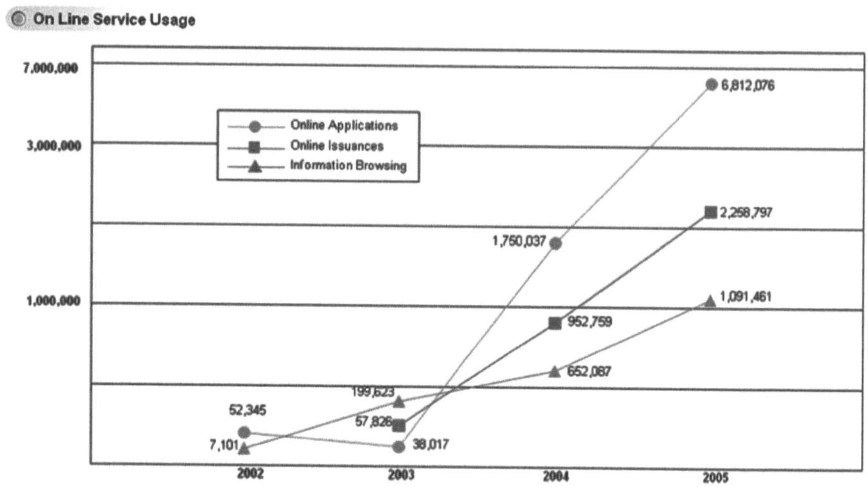

Fig. 4.17 Online service users (Source: Korean government)

The ultimate goal of the government is unified e-government services available from a single portal. In addition, the service areas will be extended to advanced areas such as e-healthcare, e-welfare, e-transport, etc. (Fig. 4.19).

Government Role and Innovation in Korea

The role of the government in the efforts of developing indigenous innovation capabilities evolved together with the phases of industrial development For example, to enable firms to finance the massive importing of capital goods and the building of turnkey plants in line with its technological assimilation strategy in the 1960s, the Korean government brought in large-scale, long-term foreign loans and allocated them to selected industries. In the 1970s, as the economy had developed and was moving into heavy industries, the government created GRIs in the fields of heavy machinery and chemicals to compensate domestic industries for their technological weakness. These GRIs, such as the Korea Institute of Machinery and Metals (KIMM) and the Electronics and Telecommunications Research Institute (ETRI), worked with private industry to enhance technological capabilities for further industrial development.

In the past, S&T policy had been supply-oriented in order to provide foundations for adopting foreign technologies. During the 1980s, the direction of S&T policy has turned toward encouraging domestic R&D activities. The Ministry of Science and Technology launched the National R&D Program in 1982, and mission-oriented R&D programs by other ministries subsequently started afterwards. The following figure highlights the policy tools used during the last four decades, such as tax deductions, financial support, and R&D subsidies (Fig. 4.20):

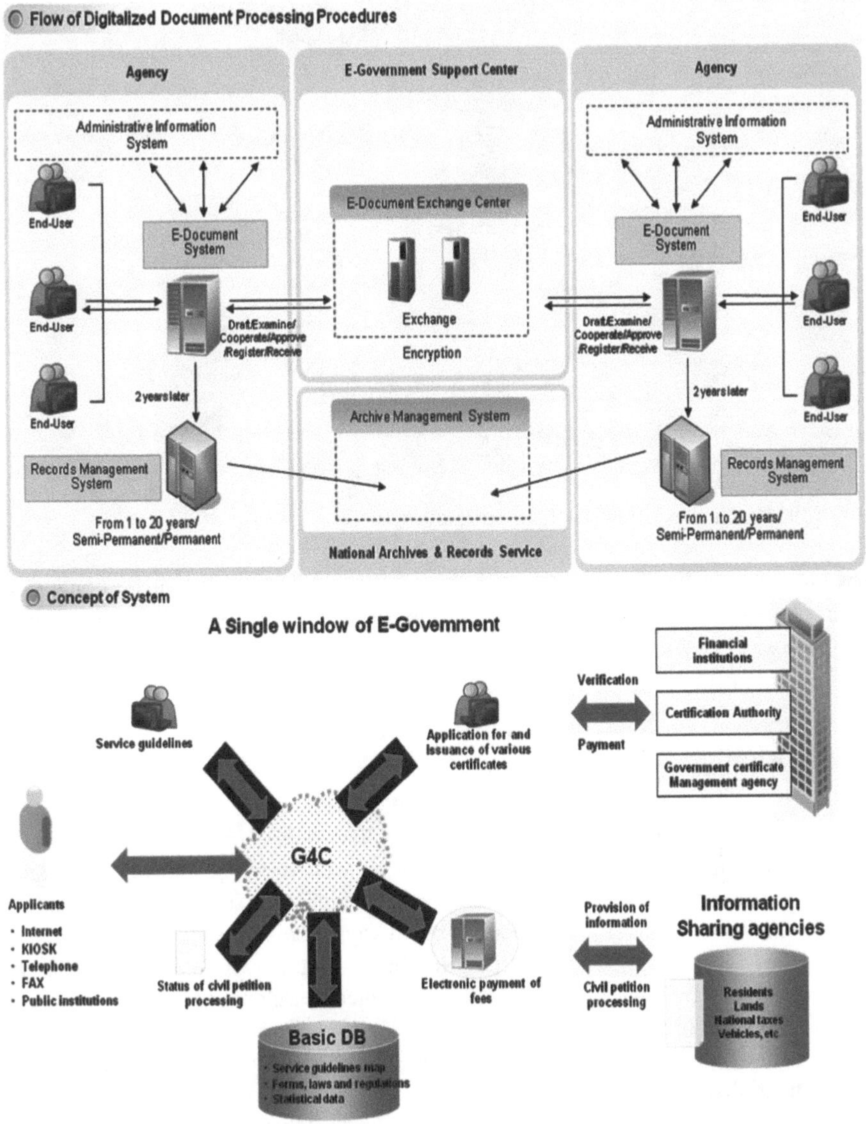

Fig. 4.18 Flow of digitalized document processing procedures (Source: Korean government)

Underlying the changing relationships, both the private sector and the government have made concerted efforts to develop technological capabilities. First, there was a fundamental shift in business strategy. In earlier years, international competitiveness relied mostly on such cost factors as low wages and scale economies based on mass production. And as imported technologies required simple assimilation and adaptation, there was no need to organize R&D activities. In later years, however, as the cost advantage of cheap skilled labor was exhausted and the economic structure

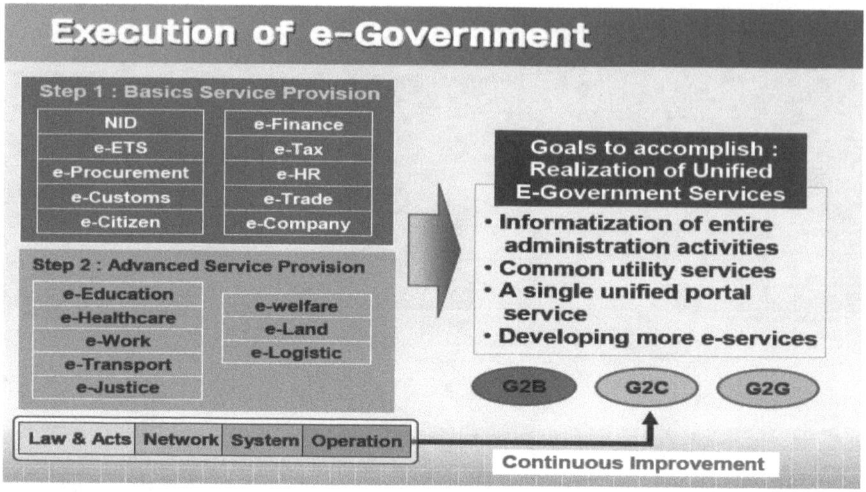

Fig. 4.19 Vision of e-government (Source: Dae Je CHIN)

became more technology-intensive, there was a pressing need for institutionalized R&D activities. The private sector met this need by establishing in-house R&D laboratories (Fig. 4.21).

Conclusion

Over the last 40 years, the Korean economy has moved from resources-based development to innovation-based growth. The transformation shows a gradual transition to a more sophisticated and advanced industrial structure. The Korean economy has faced numerous challenges but always seems able to properly respond and successfully adapt. For example, in the 1960s, Korea took advantage of expanded trade by following an outward-oriented development strategy. In the 1970s, however, the Korean economy experienced economic deterioration mainly due to excessive government intervention and the over-ambitious heavy and chemical industrialization drive. The 1980s have been a period of structural adjustment aimed at the promotion of continued economic growth with price stability. The 1990s are characterized as the period of liberalization of the economy, especially in the areas of financial reform and interest rates. The successful transformations have been possible because of government policy frameworks and active industry engagement. Government sets the development goals that clearly indicate where the country should move. The development goals and major policy directions have changed according to the development stages and changes in domestic and international economic conditions. In the course of economic development, S&T policy and its framework have evolved in response to the changes in the industrial structure. Key characteristics of Korean S&T policy can be summarized as follows:

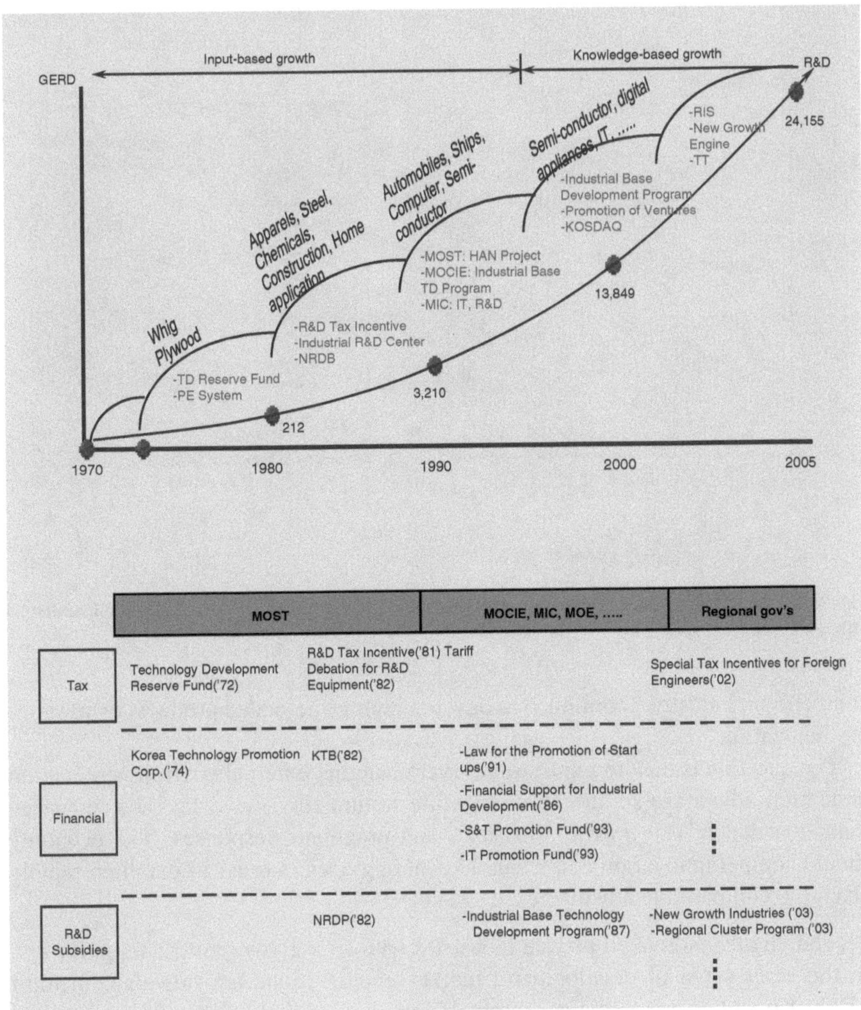

Fig. 4.20 S&T policy tools (Source: Tae Young SHIN)

First, export-oriented economy and competition created strong pressure for R&D investment

With a small domestic market, Korea had no choice but to adopt an export-oriented development strategy in the early stage of economic development. It enabled Korea to take advantage of the economy of scale and improve productivity rapidly. Most less developed countries (LDCs), including Korea, have little influence over the unpredictable international environment, and have to formulate economic development strategies taking external environments as constraints. Korea remained an open economy, and some structural developments made it vulnerable to external shocks. Economic dynamism and successful globalization were goals to achieve with an integrated, multi-tasked approach. Faced with fierce domestic and global

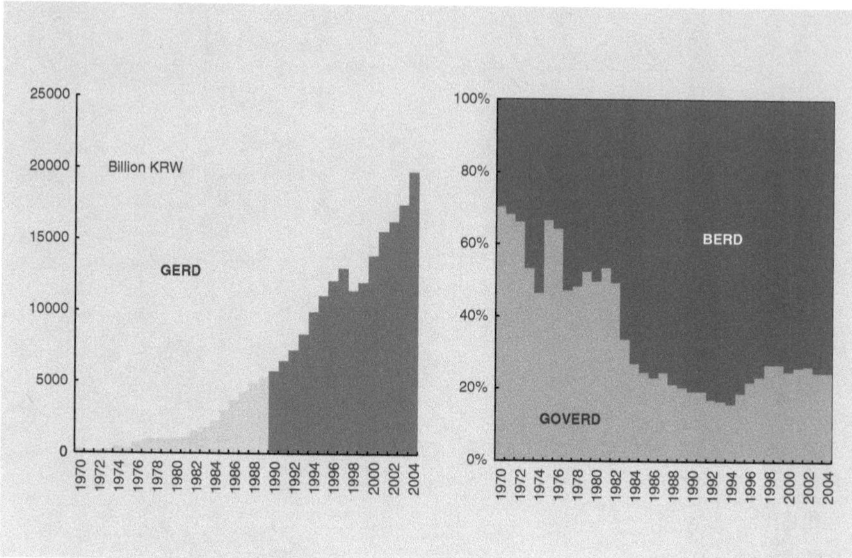

Fig. 4.21 GERD growth and share of government & private R&D (Source: National Statistics Office, Korea)

competition, industries continued to develop high value-added products by promoting innovation.

The question is how to adjust to the ever changing external environment, taking maximum advantage of the changes while minimizing the risks. The industrial transition depends on swift, consistent, and pragmatic responses. The economy should compel innovation and technological upgrades in order to maintain rapidly changing comparative advantages.

Second, highly-motivated private industries with strong government support
In the early stage of development, market imperfections are prevalent in most LDCs. Korea had adopted a good mix of free-play market forces and government guidance. From the beginning, the government sought to identify economic objectives and priorities and facilitate the achievement of these goals. It tried to ensure that the private sector possessed the wherewithal for this purpose.[7] However, one cannot draw up a totally planned economy. The essential characteristic of the Korean government-business relationship is that the business community and the various government departments have been in close communication with each other. The result is a style of industrial development which has allowed Korean business considerable initiative even when subject to administrative guidance facilitated by a variety of government aids and incentives.

[7]Throughout the 1960s and 1970s, the Korean government was deeply involved in the economic development process, not only in drawing up development plans, but also in encouraging the private sector to achieve the goals set in the plans.

Economic plans are continuously revised in response to changing economic conditions. The Korean government is not a substitute for the market; rather it has complemented the functioning of the market, especially in the earlier years when the markets and institutions were still at an inceptive stage. This approach led to industrial targeting and Korea has performed exceptionally well over the last few decades in its efforts to catch up with the world's leading economies. Korea has instigated waves of industrial upgrading to become a world leader in some of the most high-tech industries.

Third, High absorptive capacity: rich pool of human resources

With proper governmental support and guidance, innovative Korean entrepreneurs and diligent laborers have been able to maximize the numerous business opportunities open to them, and thus help advance the economy. Entrepreneurship is essential in implementing government policies during all stages of industrialization. Therefore, it is critical to foster a cadre of entrepreneurs who can display business leadership. In addition, a disciplined workforce and capable bureaucrats are fundamental for moving economic development forward. There is no doubt that continuous training for entrepreneurs and bureaucrats is a precondition for development, and education builds a nation's ability to absorb new knowledge and technology. The Korean government assumed full responsibility for taking measures necessary to promote human resource development. Education and industrialization help each other sustain and accelerate mutual development. Education made technological absorption and therefore industrialization possible, while industrialization enhanced the return on investment in education.

Fourth, government led the development of S&T, where GRIs played important roles

It has been noted, that one of Korea's strategies is to build R&D capability through expansion of GRIs rather than strengthening the R&D capability of universities. Government intervention in general has brought great positive effects on human resource development and science and technology. Investment in innovation and technology is always advocated as one of government's priority areas, and Korea's case shows that practical, incremental approaches are effective in line with the overall development stages. Expanding S&T infrastructure was the primary policy goal in the earlier years, whereas to enhance the quality of education and innovation systems was the main focus in later years. Investment in these areas is not only important for their own sake; it is also important to enhance the productivity of workers, firms and industry as a whole. In addition to education and innovation systems, as the information and communications technologies become critical for propelling national economic growth, Korea has made great efforts to expand investment in these areas. Korea's drive to build an information society came rather late, gaining movement after the financial crisis in 1997.

Fifth, imbalance in several dimensions

Korea's efforts to build an indigenous innovative base are not without challenges. The Korean economy experienced distortion in several dimensions of S&T.

1. Industrial Technology vs. Basic Sciences
   There exists still an over-emphasis on short-term, industrially oriented research at the expense of longer-term, general research; a weakly developed research capacity in the universities; lagging productivity in services; relatively weak internationalization of the domestic research system; and under-utilization of labor resources. None of these problems are easy to resolve quickly, particularly because many are legacies of past successes. The government needs to increase efforts to shift the existing innovation system into a more creative mode.

   Bold actions are needed to extend and activate interactions between the science base and industry, and more generally the broader economic and social communities. The current situation may help to implement the reorganization of the GRI system. Drastic measures are needed to intensify and modify the existing links between the industrial sector and the broader community and include striking an appropriate balance between precarious and guaranteed funding and new promotion and evaluation systems. These measures should be accompanied by a strong push to stimulate university-industry collaboration through appropriate incentives, including matching funds. The development of research capacities in universities, which have traditionally focused on teaching, needs to be further encouraged. The development of Korea's innovative capacities requires action on the fundamental structure of the education system to promote the value of entrepreneurship, autonomy and creativity.
2. Large firms vs. SMEs
   Most of the private R&D is being performed by the chaebols, and little is being conducted by small and medium enterprises (SMEs) or universities. This disproportionate distribution of R&D investment is not sustainable for the economy as a whole; insufficient resources are being devoted to R&D in basic sciences, which is necessary for the long-term development of science and technology.

   The dynamism of the small-scale sector is weak in Korea. This is particularly true for the segment of new technology based firms, which is typically a source of economic and industrial renewal. Efforts were made to stimulate the venture capital industry facilitating the emergence of NTBFs, but they quickly revealed their limits. The way the business sector is being solicited must change first, and should be reconfigured through measures concerning the development of infrastructure (procurement) and the reshaping of the industrial basis (sub-supplier chains).
3. Regional imbalances – high concentration of economic activity, wealth and human resources in metropolitan Seoul

   Korean aims to accelerate the shift from a "catch-up" to a "creative" innovation system, as articulated in the 2004 Implementation Plan for the National Innovation System. The catch-up model centered upon large-scale strategic technology development with government-affiliated research institutes and large global conglomerates taking the leading role. It has not been conducive to the creation of innovative start-ups, to technology transfer, or to building basic research capabilities, all of which are increasingly important as Korea moves towards knowledge frontiers. The creative model instead relies upon greatly increased spending on R&D – by both the public and private sectors – and improving knowledge flows and technology transfer across the system. Under this model, innovation policy also seeks to address regional imbalances in the national economy, with a rich portfolio of programs designed to develop innovation capacities outside of the Seoul metropolitan region.

The Korean economy has entered a new and critical phase in its development, and there are few guarantees of continuing success. With the growth of labor and capital inputs already slowing, continued movement toward OECD averages, through input-based growth, will become more difficult. Furthermore, with the lowest fertility rates in the OECD area and increasing competition from newly industrializing countries, especially China, Korea faces new challenges. More than ever, sustaining economic growth will depend upon boosting innovation. Korea's transition to a knowledge economy is not completed. To sustain economic growth in the future, it is essential to refurbish the economic system, so creativity and entrepreneurship can thrive.

Appendix: Innovation

In the competitive environment and globalized era, speed and innovation are essential. As such, it has become crucial to derive additional value added from various means of product differentiation via innovative designs, effective marketing, efficient distribution, reputable brand names, etc. Thus, it is critical to be able to contribute productively to global value chains as well as generate entirely new ones. The key is not necessarily production, but innovation and high-value services.

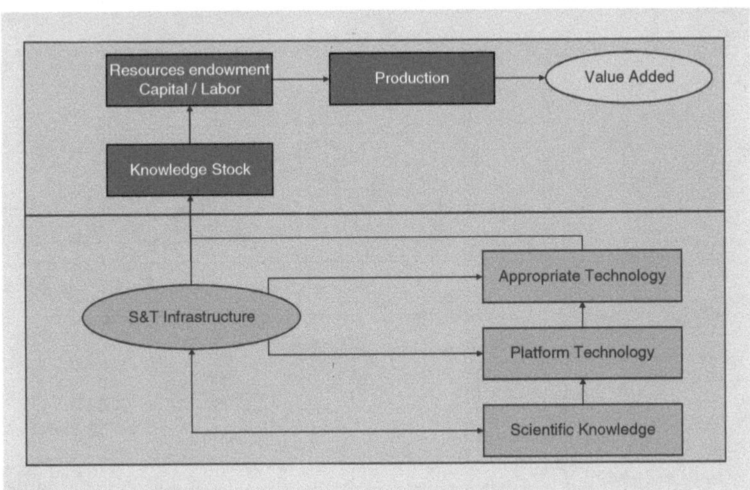

Source: Tae Young SHIN

In light of the above, sustainable economic growth depends on developing successful strategies that involve the continual use and creation of knowledge at the core of the development process. In the pursuit of S&T capability, knowledge strategies also hinge critically on the domestic innovative effort and underlie the move to produce products and services with greater value-added to ensure high wages. The following two concept charts are useful for understanding the connection between innovation and the rest of the economy:

Concept of innovation by OECD

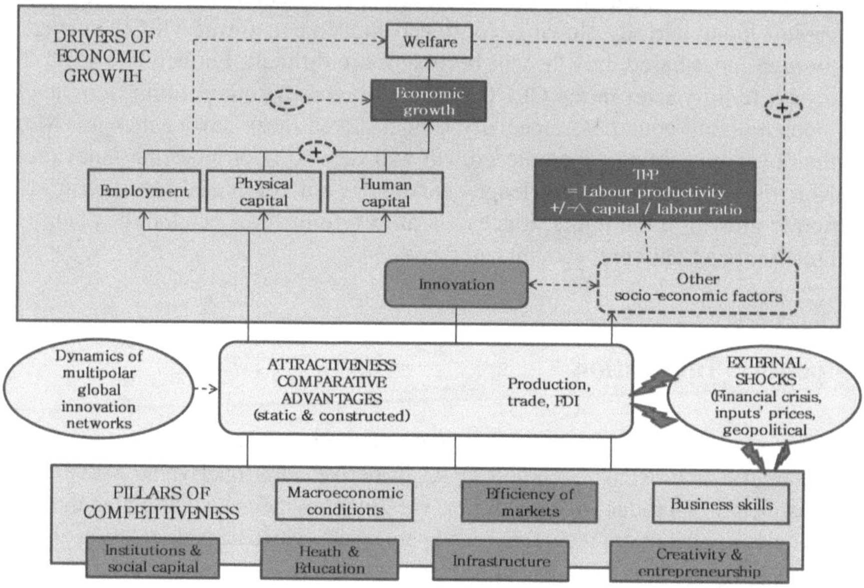

Concept of innovation by macro model. Source: Tae Young SHIN

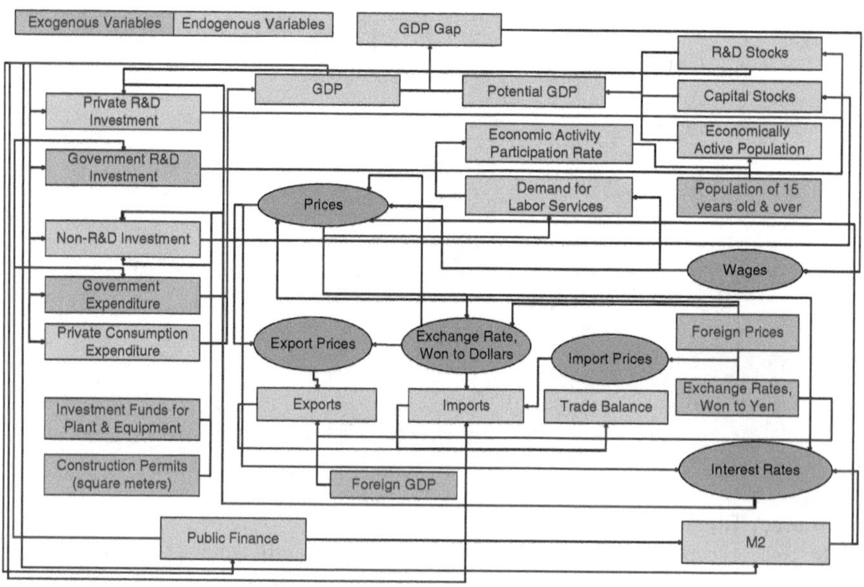

References

Adams JD (1990) Fundamental stocks of knowledge and productivity growth. J Polit Econ 98(4):673–703

Amsden A (1989) Asia's next giant: South Korea and late industrialization. Oxford University Press, Oxford

Aoki M, Hyung-Ki K, Okuno-Fujiwara M (eds) (1997) The role of government in East Asian economic development: comparative institutional analysis. Oxford University Press, Oxford/ New York

Aubert J-E (2005a) Knowledge economies: a global perspective. In: Intellectual capital for communities – nations, regions, and cities. Elsevier, New York

Aubert J-E (2005b) Promoting innovation in developing countries – a conceptual framework. World Bank Policy Research Working Paper 3554. World Bank, Washingon, DC

Bureau of statistics (2009) National Statistical office, Korea

Cha D-S, Kim KS, Perkins DH (1997) The Korean economy 1945–1995: performance and vision for the 21st century. Korea Development Institute, Seoul

Chen DHC, Dahlman CJ (2004) Knowledge and development: a cross-section approach. World Bank Policy Research Working Paper 3366, Washington, DC

Chin D (2008) Korean experience in informatization. Unpublished Paper

Chung U-C (2004) The Korean economy before and after the crisis. In: Chung D-K, Eichengreen B (eds) The Korean economy beyond the crisis. Edward Elgar, Cheltenham, UK

Dahlman C, Andersson T (eds) (2000) Korea and the knowledge-based economy: making the transition. World Bank, Washington, DC

Dornbusch R, Park YC (1987) Korean growth policy. Brookings Paper on Economic Activity

Government of the Republic of Korea (1962) The first five-year economic development plan 1962–1966. Government of the Republic of Korea, Seoul

Government of the Republic of Korea (1982) The fifth five-year economic development plan 1982–1986. Government of the Republic of Korea, Seoul

Government of the Republic of Korea (1992) The seventh five-year economic and social development plan 1992–1996. Government of the Republic of Korea, Seoul

Government of the Republic of Korea (1999) DJnomics: a new foundation for the Korean economy. Published for the Ministry of Finance and Economy. Korea Development Institute, Seoul

Government of the Republic of Korea (2004) Dynamic Korea: a nation on the move. Ministry of Finance and Economy, Kwachon, Republic of Korea

KITA (Korea Industrial Technology Association) (1997/2004) Major indicators of industrial technology. KITA, Seoul

Lederman D, Saenz L (2005) Innovation and development around the world: 1960–2000. World Bank Policy Research Working Paper 3774, Washington, DC

Lee S-C (1991) The heavy and chemical industries promotion plan (1973–1979). In: Cho L-J, Kim YH (eds) Economic development in the Republic of Korea: a policy perspective. University of Hawaii Press, Honolulu

Lee YK (2002) Analysis of government funds to higher education institutions and efficiency measures. KDI, Seoul

MOFE (Ministry of Finance and Economy) (1999) Korea: an economy transformed. MOFE, Kwachon, Republic of Korea

MOFE (Ministry of Finance and Economy) (2004) Economic surveys: Korea. MOFE, Kwachon, Republic of Korea

Mytelka LK (2000) Local systems of innovation in a globalized world economy. Indus Inn 7(1):15–32

Nam D-W (1997) Korea's economic growth in a changing world. Samsung Economic Research Institute, Seoul

OECD (2008) Science, technology and industry scoreboard 2007 STI, 2008(2)

Park HJ (2004) Political economy of economic development. In: Chung C-S, Choi K (eds) Economic development and economic crisis management in Korea. KDI School of Public Policy and Management, Seoul
Presidential Commission on Education (1995) The new education system towards globalization and information society. Ministry of Education, Seoul
Shin TY (2007) S&T development in Korea. Unpublished Paper
Wade LL, Kim BS (1978) Economic development of South Korea: the political economy of success. Praeger, New York/London
WBI (World Bank Institute) and KDI (Korea Development Institute) (2003) Technology, skills and internet services in Korea: moving towards a knowledge-based economy. Report 23905-KO, Washington, DC
WBI (World Bank Institute) and KDI (Korea Development Institute) (2004) Innovation systems: world bank support of science and technology development. World Bank Working Paper 32, Washington, DC
WBI (The World Bank) and KDI (Korea Development Institute) (2006) Global integration and technology transfer. Washington, DC
WBI (World Bank Institute) and KDI (Korea Development Institute) (2007) Korea as a knowledge economy: evolutionary process and lessons learned. World Bank, Washington, DC/Seoul; (1993) The East Asian Miracle: economic growth and public policy. Oxford University Press, New York (for the World Bank)
Woo CS (2008) Korea Development Institute, Unpublished Paper
Yusuf S (2003) Innovative East Asia: the future of growth. Oxford University Press, New York

Chapter 5
Comment on JH Kim Paper

Iain Paterson

Korea's transition from being a poor, low resource economy, developing steadily throughout the second half of the twentieth century through the stages of industrialisation and manufacturing, to becoming an important player among the knowledge economies of this century is truly a remarkable story, as traced in Ji Hong Kim's paper.[1]

Kim's paper reaffirms much of the findings of the World Bank and OECD, in particular regarding the strengths and deficits of Korea as a competitor in the knowledge (driven) economy. The World Bank has developed a Knowledge Assessment Methodology that can be used to assess the extent of countries' participation and relative standing in the knowledge economy. Around 150 countries are evaluated in the international comparison. It may be instructive to take a closer look at this "scoreboard" to see where Korea is placed, in particular with respect to other developed countries and especially in comparison with other emerging Asian economies.

The World Bank's Knowledge Assessment Methodology distinguishes three broad Knowledge Index categories. These assessment categories are themselves generated from data on three specific factors. The Knowledge Index (KI) thus combines information relating to the state of play in the areas of

- *Education and Human Resources*
- *The Innovation System*
- *Information and Communication Technology (ICT)*

The World Bank considers that, in addition, a Knowledge Economy must also operate under an *Economic and Institutional Regime* that provides incentives for the efficient use of existing and new knowledge in order for entrepreneurship to flourish. Thus, this fourth factor is also taken into account in an aggregation that generates an overall Knowledge Economy Index (KEI).

[1]Kim JH (2009) Innovation & technology in Korea in international perspective. WKO, Vienna.

J. Mahlich and W. Pascha (eds.), *Korean Science and Technology in an International Perspective*, DOI 10.1007/978-3-7908-2753-8_5,

We now take a more detailed look at the recent position of Korea in terms of Knowledge Assessment Methodology. For purposes of comparison, data is standardized on a scale of 0 (lowest) to 10.

Figure 5.1 shows that overall, in terms of the Knowledge Economy Index (KEI), Korea is ranked 29th among world economies. The vertical arrow marks the corresponding data column in decreasing order of KEI score. The ranking of Korea is indicated by the horizontal arrow. Rated at the top are Denmark, Sweden, Finland, the Netherlands and Norway. Other major Asian economies – Taiwan, Singapore, Japan and Hong Kong – are also above Korea, located at or around the 20th position.

Rank		Country	KEI	KI	Economic Incentive Regime	Innovation	Education	ICT
1	2	Denmark	9.52	9.49	9.61	9.49	9.78	9.21
2	-1	Sweden	9.51	9.57	9.33	9.76	9.29	9.66
3	-1	Finland	9.37	9.39	9.31	9.67	9.77	8.73
4	0	Netherlands	9.35	9.39	9.22	9.45	9.21	9.52
5	2	Norway	9.31	9.25	9.47	9.06	9.6	9.1
6	6	Canada	9.17	9.08	9.45	9.44	9.26	8.54
7	2	United Kingdom	9.1	9.06	9.24	9.24	8.49	9.45
8	6	Ireland	9.05	8.98	9.26	9.08	9.14	8.71
9	-3	United States	9.02	9.02	9.04	9.47	8.74	8.83
10	-5	Switzerland	9.01	9.09	8.79	9.9	7.68	9.68
11	-3	Australia	8.97	9.08	8.66	8.88	9.69	8.67
12	3	Germany	8.96	8.92	9.06	8.94	8.36	9.47
13	4	Iceland	8.95	8.76	9.54	8.07	9.41	8.8
14	-3	New Zealand	8.92	8.97	8.79	8.66	9.78	8.46
15	-5	Austria	8.91	8.78	9.31	9	8.48	8.85
16	-3	Belgium	8.8	8.77	8.87	8.93	9.14	8.25
17	2	Luxembourg	8.64	8.37	9.45	9	6.61	9.51
18	3	Taiwan, China	8.45	8.79	7.42	9.27	7.97	9.13
19	1	Singapore	8.44	8.03	9.68	9.58	5.29	9.22
20	-4	Japan	8.42	8.63	7.81	9.22	8.67	8
21	7	Estonia	8.42	8.31	8.76	7.56	8.32	9.05
22	-4	France	8.4	8.64	7.67	8.66	9.02	8.26
23	4	Hong Kong, China	8.32	7.92	9.54	9.04	5.37	9.33
24	-1	Spain	8.28	8.18	8.6	8.14	8.33	8.07
25	1	Slovenia	8.15	8.17	8.1	8.31	8.31	7.88
26	-4	Israel	8.01	7.93	8.24	9.4	6.86	7.54
27	3	Hungary	8	7.88	8.35	8.21	7.73	7.7
28	4	Czech Republic	7.97	7.9	8.17	7.78	8.23	7.7
29	-5	Korea, Rep.	7.82	8.43	6	8.6	8.09	8.6
30	-5	Italy	7.79	8.18	6.62	8	7.96	8.59
31	3	Lithuania	7.77	7.7	7.98	6.7	8.4	7.99
32	4	Latvia	7.65	7.52	8.03	6.63	8.35	7.58
33	-4	Portugal	7.61	7.34	8.42	7.41	6.95	7.66
34	6	Malta	7.58	7.18	8.78	7.95	5.86	7.74
35	-2	Cyprus	7.5	7.47	7.6	7.81	6.65	7.95
36	3	Slovak Republic	7.47	7.37	7.78	6.89	7.26	7.95
37	-2	Poland	7.41	7.38	7.48	7.03	8.02	7.09

Taiwan 18, Singapore 19, Japan 20, HK 23.

Fig. 5.1 Knowledge economy index (Source: World Bank, http://info.worldbank.org/etools/kam2/KAM_page5.asp (weighting by population))

In fact, Korea's success in knowledge-intensive sectors is underlined by the comparative assessment of its standing in terms of innovation, education and ICT, respectively.[2]

These respective rankings are:

- *Innovation* – 23rd place (see Fig. 5.2), whereas Switzerland, Sweden and Finland are 1st, 2nd and 3rd, respectively. Singapore at 4th, Taiwan 10th, Japan 12th and Hong Kong 15th are all rated higher than Korea.
- *Education* – 28th place (see Fig. 5.3), whereas Denmark, New Zealand and Finland take the top three places and Japan is in 14th position. This ranking has Korea above Taiwan (31st), Hong Kong (68th), and Singapore (70th).
- *ICT* – 19th place, with Switzerland, Sweden and Netherlands at the top. Singapore (8th), Hong Kong (7th), and Taiwan (10th) are rated higher than Korea, but not Japan (28th) (Fig. 5.4).

The ratings for the above three indicators – innovation, education and ICT – together make up the basis for the Knowledge Index (KI). Figure 5.5 shows Korea ranked as high as 20th position by KI, somewhat behind Taiwan in 14th and Japan in 19th, but ahead of Singapore in 26th and Hong Kong in 28th position.

The difference in Korea's Knowledge Index (20th) and its overall performance as a knowledge economy (29th in KEI, Fig. 5.1) is explained by a relatively low position with respect to the category *Economic Incentives and Institutional Regime*. As Fig. 5.6 shows, Korea ranks only 56th there, many places behind Singapore (ranked 1st), Hong Kong (3rd), Japan (32nd) or Taiwan (38th).

Korea's Economic Incentive and Institutional Regime performance is an exception to its otherwise higher rankings. It is the only area where many Eastern European, Middle Eastern and Latin American countries are rated higher than Korea.

This deficit is markedly visible in comparison to the highest rated economies in the East Asia and Pacific region, namely Australia, New Zealand, Taiwan, Singapore and Hong Kong and is illustrated by the relative shortfall in the leftmost bar in the chart (as shown in Fig. 5.7), which also shows it regressing since 1995.

The question arises as to which particular areas of policy contribute most to the deficit in the assessment of Korea's economic regime. Figure 5.8 illustrates those areas in comparison with Taiwan and Hong Kong:

- As for *Economic Regime* Korea scores considerably less than both of these countries[3] in terms of
  - Trade as a percentage of GDP
  - Tariff and Non-tariff Barriers
  - Exports of Goods and Services as a percentage of GDP
  - Intensity of Local Competition
  - Cost to Register a Business in relative terms

[2]These three factors are aggregated to produce the Knowledge Index.

[3]Data for 2007.

Rank		Country	KEI	KI	Economic Incentive Regime	Innovation	Education	ICT
1	-5	Switzerland	9.01	9.09	8.79	9.9	7.68	9.68
2	-1	Sweden	9.51	9.57	9.33	9.76	9.29	9.66
3	-1	Finland	9.37	9.39	9.31	9.67	9.77	8.73
4	1	Singapore	8.44	8.03	9.68	9.58	5.29	9.22
5	2	Denmark	9.52	9.49	9.61	9.49	9.78	9.21
6	-3	United States	9.02	9.02	9.04	9.47	8.74	8.83
7	0	Netherlands	9.35	9.39	9.22	9.45	9.21	9.52
8	6	Canada	9.17	9.08	9.45	9.44	9.26	8.54
9	-4	Israel	8.01	7.93	8.24	9.4	6.86	7.54
10	3	Taiwan, China	8.45	8.79	7.42	9.27	7.97	9.13
11	2	United Kingdom	9.1	9.06	9.24	9.24	8.49	9.45
12	-4	Japan	8.42	8.63	7.81	9.22	8.67	8
13	6	Ireland	9.05	8.98	9.26	9.08	9.14	8.71
14	2	Norway	9.31	9.25	9.47	9.06	9.6	9.1
15	4	Hong Kong, China	8.32	7.92	9.54	9.04	5.37	9.33
16	-5	Austria	8.91	8.78	9.31	9	8.48	8.85
17	2	Luxembourg	8.64	8.37	9.45	9	6.61	9.51
18	3	Germany	8.96	8.92	9.06	8.94	8.36	9.47
19	-3	Belgium	8.8	8.77	8.87	8.93	9.14	8.25
20	-3	Australia	8.97	9.08	8.66	8.88	9.69	8.67
21	-3	New Zealand	8.92	8.97	8.79	8.66	9.78	8.46
22	-4	France	8.4	8.64	7.67	8.66	9.02	8.26
23	-5	Korea, Rep.	7.82	8.43	6	8.6	8.09	8.6
24	1	Slovenia	8.15	8.17	8.1	8.31	8.31	7.88
25	3	Hungary	8	7.88	8.35	8.21	7.73	7.7
26	-1	Spain	8.28	8.18	8.6	8.14	8.33	8.07
27	4	Iceland	8.95	8.76	9.54	8.07	9.41	8.8
28	-5	Italy	7.79	8.18	6.62	8	7.96	8.59
29	6	Malta	7.58	7.18	8.78	7.95	5.86	7.74
30	-2	Cyprus	7.5	7.47	7.6	7.81	6.65	7.95
31	4	Czech Republic	7.97	7.9	8.17	7.78	8.23	7.7
32	107	Aruba	7.38	7.26	7.74	7.73	7.03	7.01
33	4	Croatia	7.28	7.28	7.26	7.67	6.56	7.62
34	-4	Barbados	7.16	7.58	5.92	7.63	8.09	7
35	-7	Greece	7.39	7.58	6.82	7.57	8.21	6.94
36	7	Estonia	8.42	8.31	8.76	7.56	8.32	9.05

Taiwan 10, Singapore 4, Japan 12, HK 15

Fig. 5.2 Innovation (Source: World Bank, http://info.worldbank.org/etools/kam2/KAM_page5.asp (weighting by population))

- In the field of *Governance*, Korea is rated somewhat behind Taiwan and Hong Kong regarding
  - Regulatory Quality
  - Political Stability
  - Control of Corruption

So deficits in the above mentioned policy areas identify a notable lag for Korea, not only behind Taiwan and Hong Kong, but in general, with respect to most other High Income countries.

Rank		Country	KEI	KI	Economic Incentive Regime	Innovation	Education	ICT
1	2	Denmark	9.52	9.49	9.61	9.49	9.78	9.21
2	-3	New Zealand	8.92	8.97	8.79	8.66	9.78	8.46
3	-1	Finland	9.37	9.39	9.31	9.67	9.77	8.73
4	-3	Australia	8.97	9.08	8.66	8.88	9.69	8.67
5	2	Norway	9.31	9.25	9.47	9.06	9.6	9.1
6	4	Iceland	8.95	8.76	9.54	8.07	9.41	8.8
7	-1	Sweden	9.51	9.57	9.33	9.76	9.29	9.66
8	6	Canada	9.17	9.08	9.45	9.44	9.26	8.54
9	0	Netherlands	9.35	9.39	9.22	9.45	9.21	9.52
10	-3	Belgium	8.8	8.77	8.87	8.93	9.14	8.25
11	6	Ireland	9.05	8.98	9.26	9.08	9.14	8.71
12	-4	France	8.4	8.64	7.67	8.66	9.02	8.26
13	-3	United States	9.02	9.02	9.04	9.47	8.74	8.83
14	-4	Japan	8.42	8.63	7.81	9.22	8.67	8
15	2	United Kingdom	9.1	9.06	9.24	9.24	8.49	9.45
16	-5	Austria	8.91	8.78	9.31	9	8.48	8.85
17	3	Lithuania	7.77	7.7	7.98	6.7	8.4	7.99
18	4	Cuba	4.36	5.37	1.31	5.14	8.36	2.61
19	3	Germany	8.96	8.92	9.06	8.94	8.36	9.47
20	4	Latvia	7.65	7.52	8.03	6.63	8.35	7.58
21	-1	Spain	8.28	8.18	8.6	8.14	8.33	8.07
22	7	Estonia	8.42	8.31	8.76	7.56	8.32	9.05
23	1	Slovenia	8.15	8.17	8.1	8.31	8.31	7.88
24	4	Czech Republic	7.97	7.9	8.17	7.78	8.23	7.7
25	-7	Greece	7.39	7.58	6.82	7.57	8.21	6.94
26	4	Ukraine	6	6.58	4.27	5.83	8.15	5.77
27	-4	Barbados	7.16	7.58	5.92	7.63	8.09	7
28	-5	Korea, Rep.	7.82	8.43	6	8.6	8.09	8.6
29	-2	Poland	7.41	7.38	7.48	7.03	8.02	7.09
30	-2	Belarus	4.93	6.19	1.15	5.79	8.02	4.74
31	3	Taiwan, China	8.45	8.79	7.42	9.27	7.97	9.13
32	-5	Italy	7.79	8.18	6.62	8	7.96	8.59
33	-5	Uruguay	6.49	6.54	6.35	5.37	7.79	6.45
34	3	Hungary	8	7.88	8.35	8.21	7.73	7.7
35	-5	Switzerland	9.01	9.09	8.79	9.9	7.68	9.68
36	8	Bulgaria	6.99	6.94	7.14	6.43	7.65	6.74
37	3	Slovak Republic	7.47	7.37	7.78	6.89	7.26	7.95

Taiwan 31, Singapore 70, Japan 14, HK 68

Fig. 5.3 Education (Source: World Bank, http://info.worldbank.org/etools/kam2/KAM_page5.asp (weighting by population))

Indeed it would be feasible for Korea to achieve around 20th place as a knowledge economy – in line with other East Asian countries – if its economic regime were rated as favourably as its knowledge Indicators. This will require a distinct improvement in the aspects concerning economic incentives and the institutional regime, as indicated above.

JH Kim drew attention in his paper to some of the specific challenges confronting the innovation system in Korea, in particular due to the imbalances between

Rank		Country	KEI	KI	Economic Incentive Regime	Innovation	Education	ICT
1	-5	Switzerland	9.01	9.09	8.79	9.9	7.68	9.68
2	-1	Sweden	9.51	9.57	9.33	9.76	9.29	9.66
3	0	Netherlands	9.35	9.39	9.22	9.45	9.21	9.52
4	2	Luxembourg	8.64	8.37	9.45	9	6.61	9.51
5	3	Germany	8.96	8.92	9.06	8.94	8.36	9.47
6	2	United Kingdom	9.1	9.06	9.24	9.24	8.49	9.45
7	4	Hong Kong, China	8.32	7.92	9.54	9.04	5.37	9.33
8	1	Singapore	8.44	8.03	9.68	9.58	5.29	9.22
9	2	Denmark	9.52	9.49	9.61	9.49	9.78	9.21
10	3	Taiwan, China	8.45	8.79	7.42	9.27	7.97	9.13
11	2	Norway	9.31	9.25	9.47	9.06	9.6	9.1
12	7	Estonia	8.42	8.31	8.76	7.56	8.32	9.05
13	-5	Austria	8.91	8.78	9.31	9	8.48	8.85
14	-3	United States	9.02	9.02	9.04	9.47	8.74	8.83
15	4	Iceland	8.95	8.76	9.54	8.07	9.41	8.8
16	-1	Finland	9.37	9.39	9.31	9.67	9.77	8.73
17	6	Ireland	9.05	8.98	9.26	9.08	9.14	8.71
18	-3	Australia	8.97	9.08	8.66	8.88	9.69	8.67
19	-5	Korea, Rep.	7.82	8.43	6	8.6	8.09	8.6
20	-5	Italy	7.79	8.18	6.62	8	7.96	8.59
21	5	United Arab Emirates	6.73	6.72	6.75	6.69	4.9	8.59
22	6	Canada	9.17	9.08	9.45	9.44	9.26	8.54
23	-3	New Zealand	8.92	8.97	8.79	8.66	9.78	8.46
24	-4	France	8.4	8.64	7.67	8.66	9.02	8.26
25	-3	Belgium	8.8	8.77	8.87	8.93	9.14	8.25
26	-1	Spain	8.28	8.18	8.6	8.14	8.33	8.07
27	5	Qatar	6.73	6.63	7.05	6.45	5.37	8.06
28	-4	Japan	8.42	8.63	7.81	9.22	8.67	8
29	3	Lithuania	7.77	7.7	7.98	6.7	8.4	7.99
30	3	Slovak Republic	7.47	7.37	7.78	6.89	7.26	7.95
31	-2	Cyprus	7.5	7.47	7.6	7.81	6.65	7.95
32	1	Slovenia	8.15	8.17	8.1	8.31	8.31	7.88
33	6	Malta	7.58	7.18	8.78	7.95	5.86	7.74
34	4	Czech Republic	7.97	7.9	8.17	7.78	8.23	7.7
35	3	Hungary	8	7.88	8.35	8.21	7.73	7.7
36	-4	Portugal	7.61	7.34	8.42	7.41	6.95	7.66
37	4	Croatia	7.28	7.28	7.26	7.67	6.56	7.62

Taiwan 10, Singapore 8, Japan 28, HK 7

Fig. 5.4 ICT (Source: World Bank, http://info.worldbank.org/etools/kam2/KAM_page5.asp (weighting by population))

- Large Firms/and SME's (four chaebols account for 40% of Korea's R&D)
- Metropolitan Seoul/and the rest of the country
- Government research Institutes/and universities

The World Bank views the relatively slow development of US patent registration as evidence of weakness in Korea's innovation system. Whereas the rate of growth of patents in Korea (as measured by the USPTO patent count) is similar to that of Japan and Finland, the actual level of activity in Korea, in terms of the number of patents registered annually, still lags behind these countries by about 20 and 10 years, respectively. Taiwan, on the other hand, had caught up with Japan by 2000, from a starting point similar to that of Korea in the 1980s.

Rank		Country	KEI	KI	Economic Incentive Regime	Innovation	Education	ICT
1	-1	Sweden	9.51	9.57	9.33	9.76	9.29	9.66
2	2	Denmark	9.52	9.49	9.61	9.49	9.78	9.21
3	0	Netherlands	9.35	9.39	9.22	9.45	9.21	9.52
4	-1	Finland	9.37	9.39	9.31	9.67	9.77	8.73
5	2	Norway	9.31	9.25	9.47	9.06	9.6	9.1
6	-5	Switzerland	9.01	9.09	8.79	9.9	7.68	9.68
7	-3	Australia	8.97	9.08	8.66	8.88	9.69	8.67
8	6	Canada	9.17	9.08	9.45	9.44	9.26	8.54
9	2	United Kingdom	9.1	9.06	9.24	9.24	8.49	9.45
10	-3	United States	9.02	9.02	9.04	9.47	8.74	8.83
11	6	Ireland	9.05	8.98	9.26	9.08	9.14	8.71
12	-3	New Zealand	8.92	8.97	8.79	8.66	9.78	8.46
13	3	Germany	8.96	8.92	9.06	8.94	8.36	9.47
14	3	Taiwan, China	8.45	8.79	7.42	9.27	7.97	9.13
15	-5	Austria	8.91	8.78	9.31	9	8.48	8.85
16	-3	Belgium	8.8	8.77	8.87	8.93	9.14	8.25
17	4	Iceland	8.95	8.76	9.54	8.07	9.41	8.8
18	-4	France	8.4	8.64	7.67	8.66	9.02	8.26
19	-4	Japan	8.42	8.63	7.81	9.22	8.67	8
20	-5	Korea, Rep.	7.82	8.43	6	8.6	8.09	8.6
21	2	Luxembourg	8.64	8.37	9.45	9	6.61	9.51
22	7	Estonia	8.42	8.31	8.76	7.56	8.32	9.05
23	-5	Italy	7.79	8.18	6.62	8	7.96	8.59
24	-1	Spain	8.28	8.18	8.6	8.14	8.33	8.07
25	1	Slovenia	8.15	8.17	8.1	8.31	8.31	7.88
26	1	Singapore	8.44	8.03	9.68	9.58	5.29	9.22
27	-4	Israel	8.01	7.93	8.24	9.4	6.86	7.54
28	4	Hong Kong, China	8.32	7.92	9.54	9.04	5.37	9.33
29	4	Czech Republic	7.97	7.9	8.17	7.78	8.23	7.7
30	3	Hungary	8	7.88	8.35	8.21	7.73	7.7
31	3	Lithuania	7.77	7.7	7.98	6.7	8.4	7.99
32	-4	Barbados	7.16	7.58	5.92	7.63	8.09	7
33	-7	Greece	7.39	7.58	6.82	7.57	8.21	6.94
34	4	Latvia	7.65	7.52	8.03	6.63	8.35	7.58
35	-2	Cyprus	7.5	7.47	7.6	7.81	6.65	7.95
36	-2	Poland	7.41	7.38	7.48	7.03	8.02	7.09
37	3	Slovak Republic	7.47	7.37	7.78	6.89	7.26	7.95

Taiwan 14, Singapore 26, Japan 19, HK 28.

Fig. 5.5 Knowledge index (Source: World Bank, http://info.worldbank.org/etools/kam2/KAM_page5.asp (weighting by population))

Recommendations for improvements have been forthcoming from both the World Bank and, in policy detail, from the OECD. Key suggestions include

- Emphasizing creative forms of learning, and less rote
- Enhancing collaboration between government, universities and industries
- Moving to reduce income inequality between sectors of the economy. (Nevertheless, it may be noted that the GINI coefficient for Korea, which measures inequality of incomes, is fairly low for the economy as a whole)

Rank		Country	KEI	KI	Economic Incentive Regime	Innovation	Education	ICT
1	1	Singapore	8.44	8.03	9.68	9.58	5.29	9.22
2	2	Denmark	9.52	9.49	9.61	9.49	9.78	9.21
3	4	Hong Kong, China	8.32	7.92	9.54	9.04	5.37	9.33
4	4	Iceland	8.95	8.76	9.54	8.07	9.41	8.8
5	2	Norway	9.31	9.25	9.47	9.06	9.6	9.1
6	6	Canada	9.17	9.08	9.45	9.44	9.26	8.54
7	2	Luxembourg	8.64	8.37	9.45	9	6.61	9.51
8	-1	Sweden	9.51	9.57	9.33	9.76	9.29	9.66
9	-5	Austria	8.91	8.78	9.31	9	8.48	8.85
10	-1	Finland	9.37	9.39	9.31	9.67	9.77	8.73
11	6	Ireland	9.05	8.98	9.26	9.08	9.14	8.71
12	2	United Kingdom	9.1	9.06	9.24	9.24	8.49	9.45
13	0	Netherlands	9.35	9.39	9.22	9.45	9.21	9.52
14	3	Germany	8.96	8.92	9.06	8.94	8.36	9.47
15	-3	United States	9.02	9.02	9.04	9.47	8.74	8.83
16	-3	Belgium	8.8	8.77	8.87	8.93	9.14	8.25
17	-5	Switzerland	9.01	9.09	8.79	9.9	7.68	9.68
18	-3	New Zealand	8.92	8.97	8.79	8.66	9.78	8.46
19	6	Malta	7.58	7.18	8.78	7.95	5.86	7.74
20	7	Estonia	8.42	8.31	8.76	7.56	8.32	9.05
21	7	Chile	7.09	6.53	8.76	6.85	6.48	6.27
22	-3	Australia	8.97	9.08	8.66	8.88	9.69	8.67
23	-1	Spain	8.28	8.18	8.6	8.14	8.33	8.07
24	-4	Portugal	7.61	7.34	8.42	7.41	6.95	7.66
25	3	Hungary	8	7.88	8.35	8.21	7.73	7.7
26	-4	Israel	8.01	7.93	8.24	9.4	6.86	7.54
27	4	Czech Republic	7.97	7.9	8.17	7.78	8.23	7.7
28	1	Slovenia	8.15	8.17	8.1	8.31	8.31	7.88
29	4	Latvia	7.65	7.52	8.03	6.63	8.35	7.58
30	1	Mauritius	5.48	4.63	8.01	3.63	4.03	6.23
31	3	Lithuania	7.77	7.7	7.98	6.7	8.4	7.99
32	-4	Japan	8.42	8.63	7.81	9.22	8.67	8
33	3	Slovak Republic	7.47	7.37	7.78	6.89	7.26	7.95
34	107	Aruba	7.38	7.26	7.74	7.73	7.03	7.01
35	-4	France	8.4	8.64	7.67	8.66	9.02	8.26
36	-2	Cyprus	7.5	7.47	7.6	7.81	6.65	7.95
37	-2	Poland	7.41	7.38	7.48	7.03	8.02	7.09
38	3	Taiwan, China	8.45	8.79	7.42	9.27	7.97	9.13
39	4	Croatia	7.28	7.28	7.26	7.67	6.56	7.62
40	0	Oman	5.36	4.77	7.15	4.94	4.47	4.9
41	8	Bulgaria	6.99	6.94	7.14	6.43	7.65	6.74
42	5	Qatar	6.73	6.63	7.05	6.45	5.37	8.06
43	0	Namibia	4.28	3.37	7.01	3.14	2.65	4.34
44	-4	Turkey	5.55	5.07	6.98	5.83	4.46	4.92
45	13	Romania	6.43	6.25	6.98	5.74	6.47	6.55
46	-7	Greece	7.39	7.58	6.82	7.57	8.21	6.94
47	5	United Arab Emirates	6.73	6.72	6.75	6.69	4.9	8.59
48	-6	Bahrain	6.04	5.8	6.75	4.29	5.82	7.3

Fig. 5.6 (continued)

Rank		Country	KEI	KI	Economic Incentive Regime	Innovation	Education	ICT
49	-5	Italy	7.79	8.18	6.62	8	7.96	8.59
50	-3	Costa Rica	6.03	5.84	6.6	6.25	5.19	6.07
51	-6	Kuwait	5.85	5.63	6.5	4.98	4.93	6.96
52	5	Armenia	5.65	5.37	6.48	6.25	6.36	3.52
53	-5	Uruguay	6.49	6.54	6.35	5.37	7.79	6.45
54	-10	Dominica	5.65	5.47	6.19	3.67	6.4	6.34
55	0	Malaysia	6.07	6.06	6.11	6.82	4.21	7.14
56	-5	Korea, Rep.	7.82	8.43	6	8.6	8.09	8.6
57	-6	Jordan	5.54	5.39	5.99	5.59	5.62	4.95
58	13	Saudi Arabia	5.31	5.1	5.94	3.97	4.89	6.43
59	-4	Barbados	7.16	7.58	5.92	7.63	8.09	7
60	2	Trinidad and Tobago	5.59	5.49	5.88	6.1	4.43	5.95
61	-12	South Africa	5.38	5.33	5.55	6.85	4.68	4.45
62	-18	Botswana	3.88	3.37	5.38	4.06	2.65	3.41
63	-1	Georgia	5.21	5.15	5.36	5.22	6.46	3.78

Taiwan 38, Singapore 1, Japan 32, HK 3

Fig. 5.6 Economic incentive and institutional regime (Source: World Bank, http://info.worldbank.org/etools/kam2/KAM_page5.asp (weighting by population))

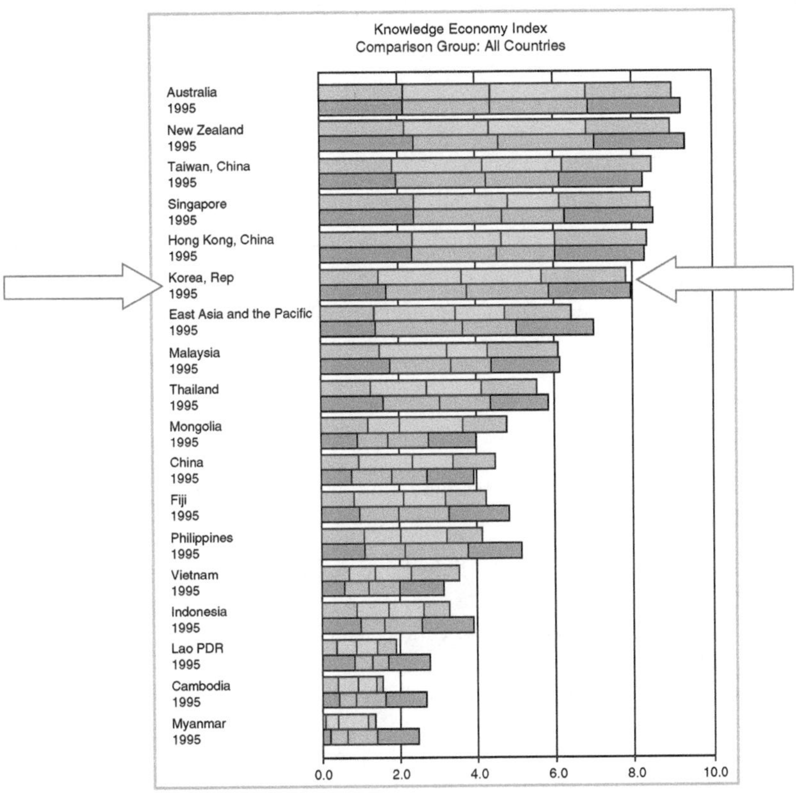

Fig. 5.7 Cross country comparison of contributions to the knowledge economy index 2009 and 1995 (Source: World Bank, Knowledge assessment methodology, 2009 (weighted by population). Note: *Bars* represent scores for (*left* to *right*): economic regime, innovation, education, ICT. http://info.worldbank.org/etools/kam2/KAM_page9.asp)

Economic Ragime	Korea, Rep.		Taiwan, China		Hong Kong, China	
	actual	normalized	actual	normalized	actual	normalized
✓ Gr. Capital Formation as % of GDP, 2003-2007	29.80	7.50	20.80	3.13	21.60	3.75
✓ Trade as % of GDP, 2007	90.00	3.75	140.00	7.50	404.00	9.38
✓ Tariff & Nontariff Barriers, 2009	70.20	3.13	85.20	8.75	95.00	10.00
✓ Intellectual Property Protection (1-7), 2008	5.00	7.14	4.90	6.43	5.40	7.86
✓ Soundness of Banks (1-7), 2008	5.50	5.71	4.60	2.86	6.60	9.29
✓ Exports of Goods and Services as % of GDP, 2007	46.00	4.38	74.00	7.50	207.00	9.38
✓ Interest Rate Spread, 2007	1.00	10.00	n/a	n/a	4.00	8.13
✓ Intensity of Local Competition (1-7), 2008	5.20	4.29	6.10	10.00	6.00	9.29
✓ Domestic Credit to Private Sector as % of GDP, 2007	108.00	7.50	n/a	n/a	140.00	9.38
✓ Cost to Register a Business as % of GNI per Capita, 2009	16.90	3.13	4.10	6.88	2.00	8.13
✓ Days to Start a Business, 2009	17.00	6.25	42.00	4.38	11.00	8.13
✓ Cost to Enforce a Contract (% of Debt), 2009	10.30	10.00	17.70	7.50	14.50	8.13
Select All Variables of Economic Regime						

Governance	Korea, Rep.		Taiwan, China		Hong Kong, China	
	actual	normalized	actual	normalized	actual	normalized
✓ Regulatory Quality, 2007	0.88	7.06	0.94	7.65	1.89	10.00
✓ Rule of Law, 2007	0.82	7.65	0.67	7.06	1.40	8.24
✓ Government Effectiveness, 2007	1.26	7.65	1.05	6.47	1.80	8.24
✓ Voice and Accountability, 2007	0.66	8.24	0.74	8.82	0.59	7.65
✓ Political Stability, 2007	0.45	6.47	0.47	7.06	1.05	8.82
✓ Control of Corruption, 2007	0.36	7.06	0.41	7.65	1.61	8.24
✓ Press Freedom (1-100), 2008	30.00	8.24	20.00	9.41	30.00	8.24
Select All Variables of Governance						

Key: ⇨ Lower scores (normalised) than Taiwan and HK

Fig. 5.8 Aspects of relatively weaker performance in economic regime and governance in comparison with Taiwan and Hong Kong (Source: World Bank, Knowledge assessment methodology, 2009. http://info.worldbank.org/etools/kam2/KAM_page3.asp)

Other criticisms of the Korean innovation system have been raised. Hemmert (see Mahlich and Pascha 2007)[4] also points out that

- A sizable 0.6% of GDP of Publicly Funded R&D remains under government control
- There are six responsible ministries and agencies with overlapping and sometimes conflicting objectives, and that also
- International links frequently are underdeveloped.

[4]Mahlich J, Pascha W (eds) (2007) Innovation and technology in Korea. Physica-Verlag, New York.

In summary, the critical points raised should be put into perspective. Korea has already surpassed the goal of 3% of GDP on research and development spending, which has clearly eluded the ambitions of the European Union and most of its member states thus far. Further, three quarters of Gross R&D expenditures comes from the business sector.

Likewise, the achievement of one of the highest levels of broadband penetration by Korea bodes well not only for future ICT development but for the innovation system as a whole.

An afterthought: One area of Korea's economy which could potentially stand to gain from innovation is renewable energy. According to the International Energy Agency, Korea's share of renewable sources in its energy supply (0.6%) has decreased below its already low 1990 level. (In comparison, the figure for the OECD as a whole is 6.5%, up from 5.8%). Successful innovation in this field offers the prospect for Korea to loosen the constraints due to its comparative lack of natural resources.

Chapter 6
Korean Innovation Governance Under Lee Myung-Bak – A Critical Analysis of Governmental Actors' New Division of Labour

Margot Schüller, Marcus Conlé, and David Shim

Introduction

One of the major challenges in the development of South Korea's innovation system is the search for the institutional structure that most suitably supports the country's ambitious plans to become one of the leading innovation-driven economies. In particular, this involves the search for a state structure that provides the most effective stimulus for the development of the indigenous innovative capacity within the national innovation system (NIS). In order to align policies and public services to the requirements of an effective system, the structure from which these policies and services emanate has to be adapted accordingly.

As was the case in other *latecomer* countries in East Asia, the Korean government has been playing a crucial role in Korea's race for global technological leadership (Kim 1993; Lim 2000; OECD 1996). In recent years, however, several observers have pointed to the growing maladaptation of public institutions, which are arguably still based on the catch-up model (Suh 2000). As Korea is striving to be an innovation leader, the catch-up model has to give way to public institutions that focus on the support of knowledge creation and commercialization instead of restricting their emphasis to the transfer and absorption of foreign knowledge. The public domain's lack of effectiveness has contributed to several major deficiencies and imbalances (Hong 2005):

Firstly, although the business sector accounts for about three-quarters of gross expenditure for research and development (R&D), only a handful of the largest conglomerates (chaebol), such as Hyundai, account for virtually all of this expenditure. In contrast, Korean small and medium-sized enterprises (SMEs) play only a minor role with regard to innovation. Moreover, even most of the large firms still rely on foreign technology for their high-tech exports. Secondly, Korea has a well-developed infrastructure of public research institutes. However, their research efforts appear to be increasingly irrelevant to the industry. On the one hand, Korea's SMEs conduct little R&D and their absorptive capacity is still too low for them to really be able to utilize the research results. On the other hand, the public institutes'

J. Mahlich and W. Pascha (eds.), *Korean Science and Technology in an International Perspective*, DOI 10.1007/978-3-7908-2753-8_6,

R&D is not aligned with the R&D conducted by the large firms. Thirdly, Korea's universities are predominately focused on education. As a consequence, basic research is vastly neglected and the opportunities for academic spin-off activities are rather low. Altogether, these features imply weak linkages within as well as between all sub-sectors of Korea's NIS; thus, coordination is not well-developed (Hemmert 2007). The innovation system's efficiency is considered to be problematic, as high R&D expenditures do not correspond to significantly higher outputs in terms of patents, scientific results, and new products (Schüller and Shim 2010).

Recent policy efforts on the part of Korea's government have sought to address the coordination problem by instigating public-sector reforms. In 2004, the institutional structure of the state was reorganized in order to improve innovation governance. This reform was welcomed by both Korean and OECD scholars as its direction was linked to the international MONIT (Monitoring and Implementing National Innovation Policies) project, an initiative to develop recommendations for the reform of innovation governance in OECD countries (OECD 2005). Yet surprisingly, a new round of reforms with a different focus was started in 2008, when Lee Myung-bak took office following his landslide victory in the national elections. Once again, the declared aim of the reforms was to improve coordination between the government agencies in charge of the formulation and implementation of innovation policies. The recent changes were founded neither on different attitudes to the importance of innovation policy nor on an already manifest failure of past reform efforts. Evidently, divergent ideologies and political strategies played a role in the different thrust of the various reforms. In particular, Lee Myung-bak was responding to his campaign pledge to strive for a small government and to increase the efficiency of the public sector. Consequently, we concentrate on the concrete differences in the rationales of both approaches to public governance reform rather than look into the reasons behind the initiation of a new round of reforms.

Ideologies provide mental models for evaluating given reform options in the face of uncertainty about the effectiveness of particular efforts (Denzau and North 1994). That is, policy makers from both the old and the new administrations have identified the same three major problems described above, but have assessed different institutional structures in the state sector as being more problematic. Nonetheless, both administrations have availed themselves of the two large "shopping baskets" of reform options – the New Public Management (NPM) and the Whole-of-Government (WOG) approaches – currently available on the market for policy advice (Christensen and Lægreid 2007a). Thus, although differences in ideology have accounted for the relative salience of particular reform "packages", both reforms have been grounded in the social sciences and cannot be reduced to (possibly misguided) ideological considerations per se; rather, they have to be treated as reasonable alternatives. In this contribution, we seek to trace and to discuss the differences in the characteristics of the reforms. Specifically, we ask: (1) What are the main structural features of both reforms? (2) How do they address the problems of Korea's NIS? (3) What are the strengths and weaknesses of each reform approach?

In order to address these questions, we proceed as follows: In section "Public Governance," we provide a short overview of the recent literature on public governance. We then detail the structural changes of the 2004 and 2008 reforms, highlighting the different reform foci (section "Korea's Reforms of Public Innovation Governance"). Section "Discussion" compares the reforms in light of the deficiencies and imbalances pointed out above. In doing so, we profit from insightful interviews with representatives from governmental agencies, research institutes and universities conducted in Korea during summer 2009. Section "Conclusion" provides some conclusions.

Public Governance

The concept of the *national innovation system* (NIS) has recently received significant attention in policy circles as it promises to provide hints for increasing the international competitiveness of a country's economy (Lundvall 2007). Above all, it is attractive to policy makers because it awards the state a greater role in the country's economic performance.[1] Yet in order to make a contribution to economic development, the state has to institutionalize effective organization. Until only recently, however, the literature on NIS was focused on discussing the comparative effectiveness of particular policy choices, without concern for the design of the structure within which policies are selected, implemented and evaluated (OECD 2005). Policy makers who wanted to undertake reforms to increase the state's effectiveness therefore referred to solutions that have emerged in other research fields.

The organizational structure of the state can be described according to its vertical and horizontal dimensions (Roness 2007). While the vertical dimension concerns the allocation of tasks and responsibilities between different levels of the state hierarchy, the horizontal dimension pertains to the allocation of rights at the same administrative level. Reforms, in turn, affect the allocation of tasks and responsibilities within both dimensions. They involve the centralization or devolution of authority (vertical dimension) and specialization or integration (horizontal dimension). In principle, several forms of organization are possible, ranging from a large all-task institution at the highest level, for example, an all-encompassing superministry, to lowest-level institutions specialized in a single task and bestowed with all-encompassing autonomy. The key public governance question concerns how to allocate tasks and authority in order to realize the desired objectives. Public sector reforms around the world are currently informed by two kinds of approaches.

One approach, which gained momentum during the 1980s, is New Public Management. NPM consists of a set of elements that are based on the economics

[1]As Freeman (1995) points out, the main idea dates back to Friedrich List, who proposed policies to accelerate the catch-up of Germany's economy in the nineteenth century.

of organization.[2] Its main idea is to introduce markets into a bureaucratic context (Christensen and Lægreid 2007a). In order to do so, tasks (such as the provision of public services, public procurement, regulation, etc.) are conceived of as a bundle of "products" offered to customers (i.e., other public institutions or societal actors) (Hood 1991). Insofar as public institutions can be treated as (service) production entities, the problem of how to organize the production process arises. NPM favours specialization within both the vertical and horizontal dimensions (Roness 2007). That is, tasks are supposed to be unbundled and allocated to specified formal entities, preferably new legal institutions such as "executive agencies", "responsibility centres", "state-owned enterprises", etc. (Pollitt 1995). The process of creating these institutions is known as "agencification". Since agencification leads to the separation of ownership, represented by the government, and control, exercised by the management of the agency, contractual arrangements have to be established, as specified by agency theory (i.e., the centrepiece of the economics of organization). Accordingly, the discretion given to the agency in improving the implementation of the particular task counteracts continuous monitoring. In turn, the emerging information asymmetry between the owner and the agency is remedied by aligning the incentives through *ex post* performance evaluation of the targets, which are specified in a performance (or incentive) contract.

Altogether, the main features of NPM therefore boil down to an increased market orientation, the devolution of authority, managerialism and the use of contracts (Christensen and Lægreid 2007a; Hood and Peters 2004). Given NPM's theoretical foundation, these features promise to increase (static) efficiency and, hence, lessen the fiscal problems governments have to cope with. Specific reform elements include the following (cf. Pollitt 1995):

- Separating the various tasks of the state, e.g., public service provision and the purchasing of those services
- Devolving authority to single-task agencies (or lower levels within public institutions)
- Instituting strong management in those public institutions which previously operated according to other principles (e.g., universities)
- Capping the agencies' budgets and obligating them to access third-party funds (from other agencies or societal actors)
- Introducing performance targets, indicators and output objectives to evaluate the performance of an agency, and in particular its management
- Shifting the basis of public employment from permanency and pure wage schemes towards fixed-term contracts and performance-related pay schemes
- Emphasizing service quality and "customer" satisfaction

While NPM relates to both the vertical and horizontal dimensions, the focus is clearly on the former. In the course of reforms in several countries, however, the

[2]See Milgrom and Roberts (1992) for the main features of the economics of organization.

relative neglect of the horizontal dimension has had considerable side effects.[3] Most of all, the fragmented state organization induced by agencification aggravates the problem of horizontal coordination (Peters 1998) and, hence, a more holistic response to several policy issues, so that dynamic efficiency may even decline (Christensen and Lægreid 2007b). This problem is particularly evident not only with regard to innovation policy (cf. OECD 2005) but also with regard to other policy areas such as environmental policy. Policy makers are increasingly aware of the systemic nature of these policy areas. As a consequence, policy makers feel that they are unable to secure the institutional support and cooperation necessary to respond to these future-related issues in a satisfactory manner. Moreover, the "cross-cutting" nature of these policy areas has effectively blurred the spheres of competence within which particular state institutions intervene. Consequently, a considerable duplication of policies may result. Even where policies do not overlap, the interconnectedness of the targeted system renders it likely that the policies of one agency will have significant negative effects on those devised by other agencies. To overcome all of these problems, closer collaboration between institutions at the same level, in order to make better use of resources and create synergies, is required (Pollitt 2003; Christensen and Lægreid 2007b).

As a reaction to these problems, some governments have strengthened horizontal coordination. The Whole-of-Government (WOG)[4] approach seeks to carry out a more holistic strategy which is based on a rather sociologically informed organization theory.[5] It establishes strategies with a stronger emphasis on long-term effectiveness or (qualitative) outcome orientation, in contrast to the myopic "purchaser" efficiency and (quantitative) output focus of NPM (Christensen and Lægreid 2007b). Instead of incentive alignment, the main focus of the proposed strategies is the alignment of the cultures of the various institutions involved. The basic idea is that institutions have formed their own cultures, which help their members make sense of their environment and provide them with routines to solve the usual problems that arise during task implementation. As a consequence,

[3]The paradoxes of NPM include the problem that well-specified output targets, which are introduced to provide more leeway for innovation in task fulfilment, can unintentionally promote isomorphism, particularly if these targets are coupled with benchmarking and "best practice" comparisons. Instead of causing agencies to use their discretion to innovate (doing something differently), indicator competition leads to mimetic institutions. Furthermore, agencification was originally thought to "depoliticize" public management. But since political executives continue to be blamed when difficulties arise at the agency level, the hiring and firing of agency managers has become a frequently used method to retain control. As a consequence, the politicization of public management may even have increased. On these and other paradoxes see Hood and Peters (2004).

[4]There are different concepts which are, however, identical in content. Initially, the concept "joined-up government" was used more often (Christensen and Lægreid 2007b).

[5]Organizational sociology comprises diverse theories such as contingency theory, resource dependence theory, network theory, organizational ecology, and the sociological institutionalist approaches. The economist Oliver Williamson's transaction-cost economics is usually also considered to have contributed to this field of study. For an overview of the broad spectrum of organizational sociology see Scott (2004).

institutions have developed their own stock of knowledge that does not necessarily conform to that of other institutions. These differences in knowledge express themselves most visibly in different policy communities with distinct preferences, rationales, and educational backgrounds (OECD 2005). In order to achieve cooperation between public institutions, inter-organizational learning has to take place. Accordingly, public sector reforms have to focus on the establishment of common values, trust building, and teamwork (Christensen and Laegreid 2007b).

Altogether, WOG stresses reform measures that allow for a coherent policy approach towards specific targets, including certain civilian groups (e.g., the aged), regions (e.g., cluster policies), or policy sectors (e.g., innovation policy) (Pollitt 2003). Reforms involve the establishment of cooperation between institutions through more or less formal means, depending on the intensity of cooperation aimed for. Specific elements include the following (cf. OECD 2005):

- Collaboratively authoring policy frameworks, white papers, sector strategies or action plans with the aim of guiding policy making and agenda setting
- Introducing an auditor with the responsibility of assessing policy coherence and coordinating policy learning
- Establishing intergovernmental policy councils, government committees and other institutional means to foster the shared identification of problems and solutions
- Founding task forces or creating other frameworks (e.g., informal learning groups) to foster inter-organizational cooperation
- Centralizing tasks by promoting a "lead agency" (prime minister's office, a ministry) to a higher status vis-à-vis other institutions and assigning it the task of coordinating policies and possibly allocating budgets
- Merging institutions (e.g., super-ministries)

While the formal integration of institutions ("hierarchy" in Oliver Williamson's terms) through mergers reduces the need for bargaining and developing understandings (Peters 1998), the proponents of WOG appear to favour hybrid modes of governance, that is, informal networks and formal exchange and learning platforms, which allow for both long-term cooperation and flexibility. As with NPM, the underlying principle of WOG is taken from economic organization. It is based on the phenomenon of production networks, which are comprised of formally independent firms engaged in cooperation and competition (or "coopetition", Brandenburger and Nalebuff 1996). But the transfer of this idea into the public sector adds complexity to the coordination problem, since public institutions are less concerned with realizing a profit than with meeting their objectives, as spelt out by idiosyncratic performance targets or instructions. Whereas cooperation in the business sector does have a direct connection to the profitability of a firm, there may not exist such a straightforward relation in the public sector. Instead, horizontal objectives may be incompatible with the performance targets that guide the work of the agencies. Unless "cross-cutting" targets receive a status similar to the organization-specific ones, horizontal coherence will thus be very difficult to realize through network types of organization (Pollitt 2003).

But if such a change in emphasis is actually introduced, another problem – lack of accountability – is bound to arise (Peters 1998; Pollitt 2003). The effectiveness of performance management is obviously compromised if there are multiple tasks without a clear priority ranking and/or if multiple institutions are involved in the implementation of a particular task.[6] Moreover, horizontal coordination is much more time- and resource-consuming than vertical coordination (Christensen and Lægreid 2007b). As mentioned above, cooperation requires investment in knowledge and skill development in order to effect cultural change. In particular, cooperation cannot be dictated from above but has to grow in a bottom-up fashion. Hence, although policy makers face the demand for quick results, they need to find the patience to let efforts in horizontal coordination blossom.

As this discussion shows, both approaches focus on different types of coordination problems faced by policy makers who want to reform public governance. As we can see, a trade-off exists between the specialization and flexibility advocated by NPM and the political capacity and policy coherence emphasized by WOG, and this trade-off is not appropriately addressed theoretically by either of the two approaches. The approaches are useful as a means for thinking about reforms, but they do not provide ready-made solutions. Policy makers therefore have to adapt their public governance systems through a trial-and-error process that includes the idiosyncratic features of the national systems. In fact, the diffusion of NPM elements around the world has occurred rather selectively depending on a mixture of environmental and historical factors (Hood 1995; Christensen and Lægreid 2007a). The same will likely apply to the more recent WOG reform elements. Hence, we concur with Pollitt (1995) and view these approaches as "shopping baskets" from which policy makers can draw to shape their own reform "packages".

Korea's Reforms of Public Innovation Governance

Despite Korea's impressive growth in inputs – in particular, gross expenditure on R&D (GERD)[7] – and in some output indicators such as high-tech exports, as well as major advancements in the IT and electronics sectors (Yim 2005), major deficiencies and imbalances have cropped up in recent years. Significantly, critics have pointed to the manifest inefficiency of the Korean NIS in terms of scientific outcomes such as patents and research findings compared to the OECD average (Hong 2005). As already mentioned above, the NIS approach envisions a strong role for the government. Thus, it appears obvious that the state needs to get involved and effect some systemic changes in order to advance the NIS. However, since

[6]For multi-task agency models, which discuss the trade-offs for the agent as a result of allocating effort to several tasks, see Holmström and Milgrom (1991).

[7]According to the latest government figures, South Korea spent 3.37% of its GDP on R&D in 2009, making it the fourth largest R&D investor among OECD member countries (Lee 2010).

Korea already has a history of strong government involvement in economic development, it is not the commitment of the state which appears to be problematic but rather the tasks it fulfils. Two reforms, conducted in 2004 and 2008, sought to improve the quality of state activity by reorganizing public governance, that is, the structure within which tasks are devised and implemented. In the following, we highlight the major characteristics of both reforms. As we illustrate, the 2004 reform was oriented towards horizontal coherence, whereas the 2008 reform was linked to increasing flexibility.

The 2004 Reform

When the 2004 reform programme was designed, the Korean government was particularly concerned with the lack of effectiveness of government spending in the science sector. The administration attributed this problem to the lack of a coherent innovation policy. Too many ministries and research institutes appeared to be involved in the design and implementation of innovation policies and programmes, leading to a perceived overlap of functions and programmes and to the duplication of research activities. Before 2004, each ministry pursued its own projects without any visible consultations with other institutions. In addition to the Ministry of Science & Technology (MOST), 13 other ministries were involved in S&T policies and programmes. Among those, the most important ministries in terms of their share of the R&D budget were the Ministry of Commerce, Industry and Energy (MOCIE, for industrial policies), the Ministry of Information and Communication (MIC, for policies concerned with the development of information and communication technologies), the Ministry of Education and Human Resource Development (MOEHRD, for human resource policies) and the Ministry of National Defence (MND, for defence policies) (see Fig. 6.1).

Before the reform, (horizontal) coordination between ministries was addressed through the establishment of the National Science & Technology Council (NSTC) and regular ministerial meetings. The NSTC had been made responsible for the evaluation and coordination of R&D programmes by the Science and Technology Framework Law of 2001. However, no effective coordination had been achieved, for the following reasons (Hong 2005, pp. 69–72): (1) The coordinating units lacked power vis-à-vis other ministries. (2) There was insufficient implementation of policies due to weak linkages between the policy-design and budget processes. (3) There was a lack of understanding of S&T on the part of some government officials (for example, on the part of the Ministry of Planning and Budget). (4) The information required for effective policy coordination was lacking. These issues reinforced the impression that effective coordination required a more powerful agency. A crucial element of the 2004 reform was therefore a "lead agency" approach, which envisioned the enhancement of the MOST's coordination and monitoring role (Schüller and Shim 2010). In order to stress the prominent position

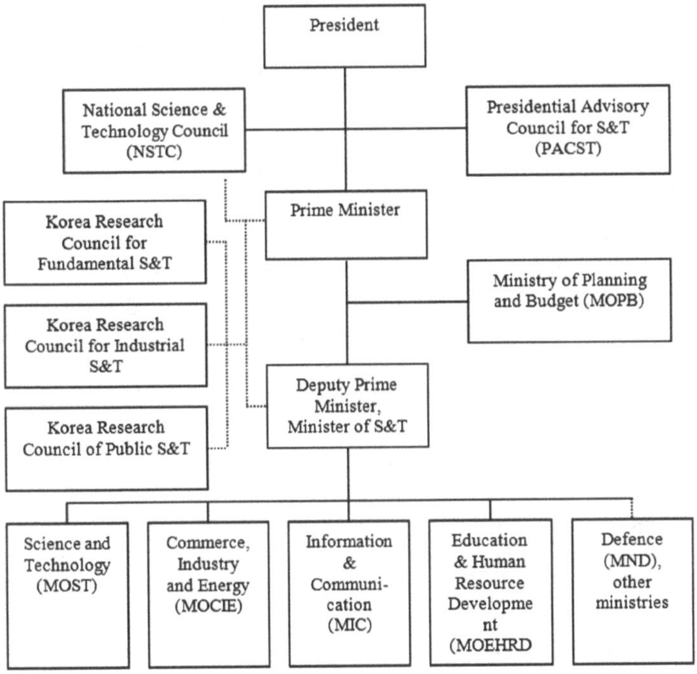

Fig. 6.1 Innovation Governance in 2004 (Source: Adapted from MOST 2006; Hong 2005)

of the MOST, its minister was promoted to the rank of deputy prime minister and became a vice chairman of the NSTC.

The MOST effectively took over a central role within the NSTC. For the purposes of more effective horizontal coordination, the Office of Science and Technology Innovation (OSTI) was established as a secretariat for the NSTC, but it was located in the MOST. OSTI was a large organization comprised of 106 officials, 40% of whom had been recruited from the MOST, another 40% from other ministries, and 20% from the private sector. Three bureaus were combined in the OSTI: the Deputy Minister in charge of R&D coordination, the S&T Policy Bureau, and the Technology Evaluation Bureau. Furthermore, three committees (the Operation Sub-Committee, the National Technological Innovation Special Committee and the Next Growth Engine Implementation Special Committee) were established within the NSTC and chaired by the minister and vice minister of MOST (Hong 2005, p. 78; Tsipouri and Patsatzis 2006, p. 11).

The influence of the MOST was also increased with regard to government research institutes (GRIs). These GRIs had been founded mainly in the 1970s and 1980s in order to compensate for the lack of research funding and human resources within the private sector. In 1970, they accounted for 84% of total R&D expenditure and for 44% of the researchers in South Korea (Kim 2001). With the expansion of large companies' R&D capacity, the role of the GRIs changed, however, and their share of R&D expenditure had declined to a mere 13.2% in 2005

(MOST 2006). In order to account for their changing role, their structure and administrative subordination were adjusted several times (OECD 2009). When the GRIs were reformed for the first time in 1999, those of them subordinated to specific ministries were transferred to research councils, with the aim of bringing about administrative innovation, improving human resource management and productivity in research projects, and ensuring more basic and long-term research and the transfer and commercialization of research results (Suh 2000; Chung 2001). In 2003, 52 GRIs existed under the auspices of the country's four research councils and the MOST.[8] The research councils were assigned the task of proposing research areas for the GRIs, evaluating their performance, submitting their budgets, and nominating their directors (Yim 2005, p. 14). The 2004 reform meant that the GRIs' supervisory organizations, the then three research councils, were removed from the Office of the Prime Minister and subordinated directly to the MOST (OECD 2009, p. 127).

As a consequence, the MOST assumed the role of a "control tower", as some of our interviewees indicated. It was the leading institution for coordinating innovation policy, directing the GRIs' research, and evaluating the effectiveness of innovation policies. Other ministries, which were also relevant for innovation-related activities, had to accept guidance by the MOST. Moreover, the S&T budgets of these ministries were now allocated by the "control tower". This was supposed to ensure that government programmes were goal oriented and not redundant.

The 2008 Reform

Lee Myung-bak's election victory in December 2007 represented a political landslide: after 10 years of liberal rule under Kim Dae-jung and Roh Moo-hyun, the conservative opposition won with the largest lead in the history of democratic elections in South Korea (Peters and Shim 2007). Many observers within South Korea and abroad related the displacement of the former government by the Grand National Party (GNP) to the many economic problems the country had faced after 2005. The fact that Lee Myung-bak was the former CEO of Hyundai Engineering and Construction and the former mayor of Seoul seems to have given credibility to his election promise to revive the South Korean economy. He had a reputation for dynamism and initiative and announced plans to rebuild South Korea on the basis of his "747" plan, which included the goals of reaching an annual economic growth rate of 7%, an increase in the per capita income to US$40,000, and South Korea's advancement to become the world's seventh-largest economy by 2017. Innovation

[8]The councils established initially were the National Research Council for Economics, Humanities and Social Sciences; the Korea Research Council of Fundamental Science and Technology; the Korea Research Council for Industrial Science and Technology; and the Korea Research Council for Public Science and Technology. Their number was cut from four to three in the following years and to two in 2008.

policy was given a prominent place in his economic revitalization programme (Schüller and Shim 2010).

Following his pledge that he would strive for a small and efficient government and pursue the improved integration of innovation and educational policies, Lee Myung-bak completely changed the institutional landscape by merging and abolishing government agencies in charge of policy making and coordination, by relocating R&D implementing units, and by changing the roles of decision-making and advisory bodies. The Lee administration disbanded OSTI – the former secretariat of the NSTC – and cut the number of GRIs from 52 to 26. One of the most striking changes made was the dissolution of the MOST, which had been assigned such a central role in innovation policy by Lee's predecessor Roh Moo-hyun. Instead of continuing the "lead agency" approach, the new administration favoured the merging of ministries in order to reduce the number of competing agencies.

The MOST's human resources–related administrative functions were merged with those of the MOEHRD to form the Ministry of Education, Science and Technology (MEST). This ministry is now responsible for both education and basic research. Akin to similar ministries in Germany (Federal Ministry of Education and Research), Sweden (Ministry of Education, Research and Culture) and other countries, the MEST is supposed to enhance the interface of education and S&T. The MOST's industry-related innovation-policy functions and those functions previously performed by the MOCIE and the MIC were shifted to a newly established Ministry of Knowledge Economy (MKE). Both the MOCIE and the MIC were dissolved. In contrast to the MEST, the MKE is responsible for supporting applied research. The MKE had a 35% share of the government's 2009 R&D budget at its disposal, whereas the MEST had a share of 21% (KIAT 2009). As a consequence, there are now two super-ministries which have been given supposedly clear-cut tasks and authorities (see Fig. 6.2).

The spheres of competence are also evident in the realignment of the 26 remaining GRIs. These were subdivided between the MEST and MKE, with each ministry controlling 13. Those GRIs subordinated to the MKE are now headed by the Research Council for Industrial S&T, while those GRIs subordinated to the MEST are coordinated by the Research Council for Fundamental S&T (see Fig. 6.3). Restructuring was also conducted among the GRIs, with several of them merged in order to arrive at a lower number of institutes. Under the MKE, for example, five R&D financing organizations (ITEP, KOTEF, KMAC, KTTC, IITA[9]) were merged into two organizations: the Korea Institute for Advancement of Technology (KIAT) was established for policy research, HR development, international cooperation and mid- and long-term planning; the Korea Evaluation

[9]ITEP (Korea Institute of Industrial Technology Evaluation and Planning), KOTEF (Korea Industrial Technology Foundation), KMAC (Korea Materials & Components Industry Agency), KTTC (Korea Technology Transfer Center), IITA (Institution for Information Technology Advancement).

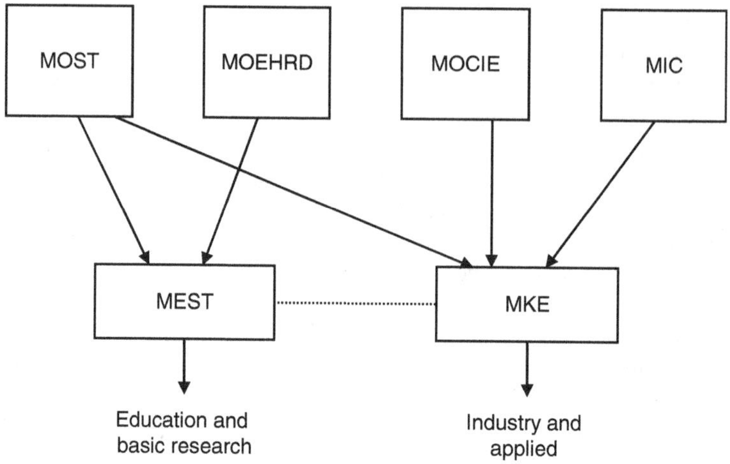

Fig. 6.2 Merging and abolishing ministries in charge of innovation policy (Source: Authors' own compilation)

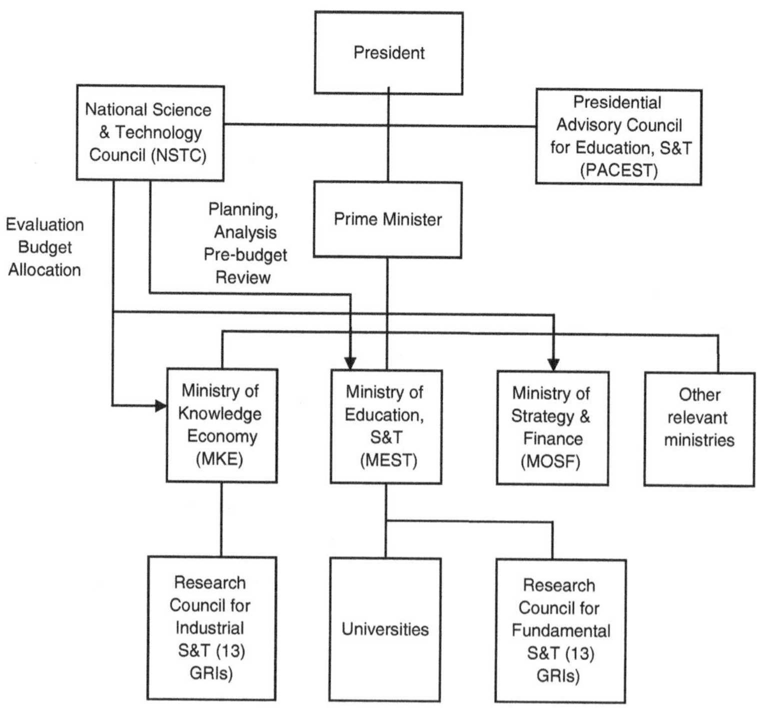

Fig. 6.3 South Korea's New S&T Administration in 2008 (Source: ERAWATCH 2009, authors' own compilation)

Institute of Industrial Technology (KEIT) for the planning, evaluation and management of national R&D programmes (KIAT 2009).

After stripping the MOST of its power with regard to horizontal policy coordination, the new administration sought to strengthen the capacity of the NSTC. In particular, five expert committees were established and staffed with experts from the private sector in order to improve the institution's knowledge base.[10] In 2008 the newly established MEST began functioning as the NSTC secretariat; the OSTI, which had previously performed this function, was disbanded. The MEST, however, did not acquire all the responsibilities of the OSTI but had to share these with the MKE. The MEST now supports the NSTC in policy planning, R&D budget planning and survey analysis. In turn, the MKE is responsible for budget coordination and policy evaluation. Yet while both the MEST and the MKE ostensibly have some authority in the R&D budgeting process, the actual budget coordination authority lies with the Ministry of Strategy & Finance (MOSF) (ERAWATCH 2010). Although the MEST functions as the NSTC's secretariat, it has not assumed the role of the former MOST; rather, it is only one of several ministries that are relevant to innovation policy. The MOSF, on the other hand, is less concerned with the effectiveness of innovation policy as with the efficiency of government spending.

In 2008, the Lee administration also introduced a new, comprehensive policy framework: the Low Carbon, Green Growth strategy was declared the national development paradigm for the next 60 years. With massive investments in "green" technologies, innovation, and industry, South Korea was set to shift its existing growth model to a more sustainable concept linking economic growth to ecological preservation (Shim 2010). However, the initiation of a sustainable growth and innovation framework was not new to Korea. In fact, predecessor Roh Moo-hyun had unveiled the long-term Vision 2030 development strategy only 2 years earlier in August 2006 (MOFE/MOHW/MOPB 2006). Compared to the Roh administration's strategy, the new strategy has a stronger institutional basis. All relevant ministries were involved in the process of planning and formulating Korea's Green Growth strategy. Moreover, members from the ministries and representatives from the private sector, academia and civil society were named to the new Presidential Committee on Green Growth (PCGG) in order to assist with the strategy's implementation of the (Shim 2009, 2010; UNEP 2010). Since the ministries basically maintained their autonomy, a novel position, the Chief Green Officer, was instituted to coordinate communication between the various agencies and the PCGG. This director-general-level position has been recommended for all public institutions (PCGG 2009).

In sum, the 2008 reform featured de-specialization on the one hand and the devolution of authority on the other. Indeed, although the merging of ministries was one of the most salient features of the reform, the vertical dimension was also given significant attention. The authors' expert interviews in Korea revealed that

[10] These committees are working on key industrial, large-scale, state-led, cutting-edge and infrastructure technologies.

performance-related pay schemes were introduced in 2008 for government officials, in particular for researchers in GRIs. Moreover, performance indicators were given more weight in the evaluation of individual researchers, while the GRIs were given more flexibility to perform their tasks. The indicators introduced to broadly guide their research activities were selected with the intention of orienting GRIs more towards specific areas (basic research) and "customer" groups (SMEs). As we have already indicated, the relative attractiveness of these NPM elements to the current administration hinged in part on the latter's "small government" ideology. This ideology was also reflected in the new government's determination to reduce administrative expenses by freezing social benefits and downsizing the merged ministries' and GRIs' workforces.

Discussion

Several OECD member countries – including Korea – supported a report on the MONIT project that proposes "third-generation" reforms for public innovation governance. This OECD (2005) report claims that increasing knowledge about innovation policy has changed the general approach to this subject. Initially, innovation was treated as a linear process from basic research via applied R&D towards the introduction of new products and processes that feature unrestricted knowledge spill-overs between the different stages (knowledge as a "public good"). Research on NIS then exposed the more complex and interactive nature of innovation, which involves more policy areas than suggested by the linear model. Governments' views on innovation policy therefore broadened to create more all-encompassing institutional support for innovation. The proposed "third-generation" innovation governance approach (hereafter the OECD approach) deals with the proper structure of the state necessary to provide a more coherent policy framework for innovation and adaptation.

Generally, the OECD approach coincides with the recently popular WOG approaches outlined in section "Public Governance." The main difference lies in its comparatively strong emphasis on policy learning, which parallels the room given to learning in innovation research. Thus, the OECD approach attempts to introduce flexibility in order to overcome and prevent path dependencies. It advocates "flexible, decentralized management practices, open learning and flexibility. A high degree of self-organisation under a broader strategic objective would support such governance" (OECD 2005). The basic features of the advice contained in the OECD report are nonetheless similar to the WOG elements. Although the OECD approach tries to wed NPM with WOG elements, the prescriptions are rather abstract and contain various possibilities.[11]

[11]Interestingly, the report contains advice which is solely based on the reform experiences of the participating countries. Most of the experiences, however, show the difficulties of and the diversity

Korea's 2004 reform is cited favourably in the OECD report as an example of the implementation of "third-generation" elements to improve horizontal coherence. Accordingly, those interviewees involved in that round of reforms considered the NPM-introduced changes in 2008 to be a step backwards. However, even these proponents of the 2004 reform acknowledged that the "lead agency" approach had some major deficiencies. Although the 2004 reform was not effective long enough to allow for a thorough evaluation, it was quite obvious that the MOST's role as the control tower was not recognized by the other ministries. This may have changed with time, but the unfavourable point of departure already created a situation where collaboration among ministries was hard to initiate. In the face of opposition from the other ministries, the MOST was not able to assert its authority over S&T budget spending. Once the MOST distributed the S&T funds to the ministries, it had little influence or supervision over the latter's spending behaviour.[12] Korea's experience with "lead agencies" indicates that this approach is not very useful in effecting horizontal collaboration. It is a step towards hierarchy, but a step that falls short of curtailing the rights and responsibilities of the other ministries. The MOST was dependent on the benevolence of those ministries that had been demoted in relation to it. Cooperation between ministries may be more effective if the parties meet each other on equal footing.

Another problem of the 2004 reform was that it was strongly geared towards policy duplication. In contrast, there was little effort to directly tackle the problems of Korea's NIS described above. The 2004 reform provided support for GRIs to enhance their knowledge base and increase the number of skilled researchers (Hong 2005). Yet the basic problem of the GRIs was that they were fundamentally maladapted to the existing configuration of the NIS. It is doubtful whether the allocation of more government resources would have significantly improved the situation. Altogether, the 2004 reform gives the impression that the "developmental state" model of former times still loomed large, as the primary concern appeared to be rendering the top-down approach more effective.

The Lee Myung-bak administration, in turn, has offensively targeted the weak points of Korea's NIS. While it has created a common vision – Green Growth – to guide policy making, the main focus of its 2008 reforms was on the vertical dimension of the state. The MEST and the MKE were formed to improve basic research and innovation in SMEs. Likewise, the two research councils and their associated GRIs were subsumed under these ministries and given performance targets to guide their research in the desired direction. Instead of putting time and

of strategies for implementing such a structure. It is quite difficult to see how concrete policy advice can be derived from these experiences. More importantly, some elements appear to contradict each other; this is due to the fact that the authors of the report try to introduce WOG elements without abandoning NPM measures. As we indicated in Sect. 2, it is quite difficult to combine both approaches. In the present case, it is unclear, for example, how a "high degree of self-organisation" on the part of the agencies should be connected with the specification of "concrete measures to be taken by each ministry or agency" (OECD 2005).

[12]This argument is based on the authors' interviews in Korea.

effort into creating common values and understandings, the current administration has established a hierarchical relationship through the merging of ministries and GRIs. In this way, the problem of coordination has been internalized. This procedure may not have the wanted effects on the reduction of administrative expenses as is usually envisioned by the supporters of a "small government". Indeed, there are many organizational challenges associated with mergers; in Korea these are aggravated by the fact that most state employees are members of powerful unions. Therefore, at least in the short term, the abolition of ministries and other agencies will very likely not produce significant savings for the state budget. Yet the despecialization at the ministerial level appears to be closer to solving the issue of horizontal coherence between different areas than the "lead agency" approach of the former administration. This, of course, is only true for those areas that are actually integrated. The problem of horizontal coordination remains unresolved outside the boundaries of these institutions.

Critics of the 2008 approach, including some of our interviewees, have expressed their concern that the merger of the MOST and the MOEHRD could jeopardize support for science and technology. However, since the new administration has committed itself to innovation, this fear may be exaggerated. Much more problematic is the division of tasks and responsibilities between the MEST and the MKE. Although their missions appear clear-cut and separate, inefficient overlapping is still a possibility. University spin-offs offer one straightforward example. In the high-tech industries (e.g., biotechnology), there is a tendency for spin-offs to be significantly involved in basic research (Nelson 2008). The questions that arise are whether this kind of research has to be assessed as "fundamental" or "industrial", and which ministry is in charge. Such cases could create conflicts of responsibility. From an NIS perspective, an even more serious point is that innovation is a non-linear process with several feedback loops. The division of responsibilities between the MEST and the MKE, however, creates a wedge between basic and applied research. The central idea of the OECD (2005)'s concept of "third-generation" governance is that a more holistic approach to the innovation process would likely be more appropriate for providing support over the whole innovation cycle. Certainly, the problem could be alleviated through stronger collaboration between both ministries. However, as all our interview partners told us, collaborative efforts between these super-ministries can hardly be expected.

In the long run, the Green Growth framework may contribute to the achievement of a more coherent approach to innovation policy. Besides introducing a common vision, the framework includes some novel institutional changes, such as the Chief Green Officer, who acts as a communication channel for the institutions involved in innovation policy making. But while the "lead agency" approach has proven problematic due to the lack of recognition granted by other ministries, the present administration's institutional approach to horizontal coherence appears to be too closely connected to its Green Growth policy framework. Horizontal cooperation requires continuous efforts at creating mutual values and common visions. Therefore, the reference of the present coordination efforts to the current administration's policy preferences may turn out to be problematic in the South Korean political context. Indeed, questions regarding the longevity of Green Growth can be raised, as South Korean politics has already seen many long-term strategies dawn and then disappear: in 2000 President Kim Dae-jung disclosed his Vision 2025 (MOST 2000), which represented a comprehensive move towards achieving the transition from a catch-up to an advanced knowledge-based economy (Schüller and Shim 2010). His successor Roh Moo-hyun developed the Vision 2030 – Korea: A Hopeful Nation in Harmony (MOFE/MOHW/MOPB 2006), which was intended to be a holistic strategy that was supposed to synchronize economic development with the demographic changes Korea will face in years to come. Only 2 years later, the long-term concept of Green Growth was presented by the current president Lee Myung-bak. Given the ephemerality of South Korean development strategies, it will be interesting to see if Green Growth will remain the central motto once the next government comes into office in 2012.

Conclusion

Within only 4 years, Korea undertook two rounds of public innovation governance reforms. Although these reforms were quite distinct, they were nonetheless targeted at the same set of problems with regard to Korea's NIS. As is usual for many other reforms as well (March and Olson 1983), the underlying rhetoric in both cases was based on efficiency. However, the two administrations responsible for the 2004 and 2008 reforms had different perceptions of the problems at hand. Whereas the Roh Moo-hyun administration felt that the lack of efficiency was primarily due to a lack of concerted efforts at government intervention and consequently to the inefficient duplication of government programmes and initiatives, the subsequent Lee Myung-bak administration saw the inefficiency as being associated with a lack of "customer" responsiveness. As a consequence, the Roh administration emphasized the horizontal dimension of the state and the Lee administration the vertical dimension.

In this contribution, we have set out to evaluate the 2008 reform, which came as a surprise to many observers of Korea's NIS. Due to the Lee administration's focus on the vertical dimension, elements of the NPM approach were rather more salient. However, NPM schemes pose the type of problems regarding horizontal coherence that had been explicitly addressed in the 2004 reform. Arguably, there exists a trade-off between the specialization and flexibility advocated by NPM and the political capacity and policy coherence emphasized by the WOG approach. The adoption of elements of one approach hence compromises the effectiveness or opposes the use of elements of the other. As a consequence, the 2008 reform fundamentally deviated from the earlier reform in both rationale and substance. Some observers have viewed the new round of reforms as a step backwards, since the recent worldwide trend in public governance is to adopt WOG approaches. Indeed, the OECD has coined the term "third-generation" innovation governance to introduce WOG into public innovation governance.

The observations cited, however, are an injustice to the 2008 reform. As a matter of fact, it could be argued that the 2008 approach is broadly in line with the OECD's (2005) general discussion of "third-generation" innovation governance. A careful reading of that report shows that it gives priority to flexibility at the agency level, even though it is ostensibly a roadmap for horizontal coherence. Since the report considers several methods to advance effective innovation policy making, the possibility that it would have given the 2008 reform the same type of positive feedback it gives the 2004 reform cannot be dismissed. The Lee Myung-bak administration has strengthened the NSTC as the central science and technology policy council and has introduced other institutional changes to enhance horizontal coherence, in particular the Chief Green Officer. Indeed, if the 2004 and 2008 reforms were considered as two alternative options, it would be difficult to assess which was better suited to the OECD concept.

This is possibly the major shortcoming of the OECD's report on innovation governance. Its main idea is to overcome rigidities (or path dependencies) which work against the formulation of a coherent policy approach to address future-related issues. Given this point of departure, the focus on governance structures inevitably suggests that major reorganization and a continuous review of organization charts is a necessary prerequisite for effective innovation policy making, even if the report itself is more nuanced in this respect. Due to its focus, the OECD report considers temporal coherence – a feature that is mentioned three times in the synthesis report but is mostly neglected throughout the analysis – to be achieved mainly by improving horizontal coherence and by maintaining flexibility and room for learning, both of which are viewed as crucial to institutional change. However, the Korean example shows quite clearly that the temporal dimension is more complex since it involves a trade-off between continuity and change. Arguably, there has been too much change and not enough continuity in Korean innovation governance to render these institutional changes relevant to the actual functioning of the system. Our analysis indicates that it is not necessarily the governance structure which is problematic. Indeed, it appears that the 2004 and 2008 reforms do equally well if evaluated from the point of view of "third-generation" innovation governance. Rather, it is the frequency of organizational change which will very likely lead to serious adverse effects. In this environment, it would not be surprising to observe that those impacted by restructuring – the staff of the agencies – either wait until the next round of reforms (and in the meantime continue with their routines) or invest resources in positioning their agency to secure a larger piece of the pie. Neither of these behaviours would be amenable to a coherent innovation policy.

References

Brandenburger AM, Nalebuff BJ (1996) Co-opetition. Currency Doubleday, New York

Christensen T, Lægreid P (2007a) Introduction – theoretical approach and research questions. In: Christensen T, Lægreid P (eds) Transcending new public management: the transformation of public sector reforms. Ashgate, Aldershot

Christensen T, Lægreid P (2007b) The whole-of-government approach to public sector reform. Public Adm Rev 67(6):1059–1066. doi:10.1111/j.1540-6210.2007.00797.x
Chung S (2001) The research, development and innovation system in Korea. In: Laredo P, Mustar P (eds) Research and innovation policies in the new global economy: an international comparative analysis. Edward Elgar, Cheltenham
Denzau AT, North DC (1994) Shared mental models: ideologies and institutions, Kyklos 47(1): 3–31
Erawatch (2010) ERAWATCH research inventory report: Republic of Korea. European Commission. http://cordis.europa.eu/erawatch. Accessed 01 Aug 2010
Freeman C (1995) The 'national system of innovation' in historical perspective. Camb J Econ 19:5–24
Hemmert M (2007) The Korean innovation system: from industrial catch-up to technological leadership? In: Mahlich J, Pascha W (eds) Innovation and technology in Korea: challenges of a newly advanced economy. Physica-Verlag, Heidelberg/New York
Holmström B, Milgrom P (1991) Multitask principal-agent analyses: incentive contracts, asset ownership, and job design. J Law Econ Organ 7:24–52, 10.1093/jleo/7.special_issue.24
Hong YS (2005) Evolution of the korean national innovation system: towards an integrated model. In: OECD (ed) Governance of innovation systems, vol 2: case studies in innovation policy. OECD, Paris
Hood C (1991) A public management for all seasons? Public Adm 69:3–19. doi:10.1111/j.1467-9299.1991.tb00779.x
Hood C (1995) The "new public management" in the 1980s: variations on a theme. Account Organ Soc 20(2/3):93–109. doi:10.1016/0361-3682(93)E0001-W DOI:dx.doi.org
Hood C, Peters G (2004) The middle aging of new public management: into the age of paradox? J Public Adm Res Theory 14(3):267–282. doi:10.1093/jopart/muh019
KIAT (Korea Institute of Advancement of Technology) (2009) Industrial R&D policy of South Korea. http://akseli.tekes.fi/opencms/opencms/OhjelmaPortaali/ohjelmat/Materiaalit/fi/Tapahtumia/Korea_Suomi08062009/Moon.pdf. Accessed 20 Dec 2009
Kim L (2001) Crisis, national innovation, and reform in South Korea. MIT Japan Program. http://web.mit.edu/mit-japan/www/product/WP0101.pdf. Accessed 19 Apr 2007
Kim L (1993) National system of industrial innovation: dynamics of capability building in Korea. In: Nelson RR (ed) National innovation systems: a comparative analysis. Oxford University Press, New York
Lee J (2010) S. Korea's R&D investment 4th highest among OECD nations. Yonhap News. http://english.yonhapnews.co.kr/n_techscience/2010/08/11/4501000000AEN20100811005800320.HTML. Accessed 11 Aug 2010
Lim Y (2000) Development of the public sector in the Korean innovation system. Int J Technol Manag 20(5–8):684–701. doi:10.1504/IJTM.2000.002888
Lundvall BA (2007) National innovation systems–analytical concept and development tool. Ind Innov 14(1):95–119. doi:10.1080/13662710601130863
March JG, Olson JP (1983) Organizing political life: what administrative reorganization tells us about government. Am Political Sci Rev 77:281–296
Milgrom P, Roberts J (1992) Economics, organization and management. Prentice-Hall, Englewood Cliffs
MOFE/MOHW/MOPB (2006) Vision 2030 – Korea: a hopeful nation in harmony. joint press release of the Ministry of Finance and Economy, Ministry of Health and Welfare, Ministry of Planning and Budget, Seoul, Republic of Korea
MOST (2000), Vision 2025 – Korea's long-term plan for science and technology development. Ministry of Science and Technology, Seoul, Republic of Korea. http://unpan1.un.org/intradoc/groups/public/documents/APCITY/UNPAN008040.pdf. Accessed March 3rd, 2010
MOST (2006) Science and technology in Korea. Ministry of Science and Technology, Seoul, Republic of Korea

Nelson RR (2008) What enables rapid economic progress: what are the needed institutions? Res Policy 37(1):1–11. doi:10.1016/j.respol.2007.10.008 DOI:dx.doi.org

OECD (1996) Reviews of national science and technology policy – Republic of Korea. OECD, Paris

OECD (2005) Governance of innovation systems, volume 1: synthesis report. OECD. http://www.mkm.ee/failid/OECD_Governance_of_innovation_systems___9205081E__2_.pdf. Accessed 20 Jan 2007

OECD (2009) OECD reviews of innovation policy: Korea. OECD, Paris

PCGG (2009) Green growth: a new path for Korea. Presidential Committee on Green Growth, Seoul, Republic of Korea

Peters BG (1998) Managing horizontal government: the politics of coordination. Public Adm 76 (2):295–311. doi:10.1111/1467-9299.00102

Peters M, Shim D (2007), Konservativer Wahlsieg: Hintergründe und mögliche Auswirkungen der Präsidentschaftswahl in Südkorea, GIGA Focus Asien 1/2008. http://www.giga-hamburg.de/dl/download.php?d=/content/publikationen/pdf/gf_asien_0801.pdf. Accessed April 16th, 2008

Pollitt C (1995) Justification by works or by faith?: evaluating the new public management. Eval The Int J Theory Res Pract 1(2):133–154. doi:10.1177/135638909500100202

Pollitt C (2003) Joined-up government: a survey. Political Stud Rev 1:34–49

Roness PG (2007) Types of state organizations: arguments, doctrines and changes beyond new public management. In: Christensen T, Lægreid P (eds) Transcending new public management: the transformation of public sector reforms. Ashgate, Aldershot

Schüller M, Shim D (2010) The innovation system and innovation policy in South Korea. In: Schüller M, Frietsch R (eds) Competing for global innovation leadership: innovation systems and innovation policies in the USA, Europe and Asia. Karlsruhe, Fraunhofer

Shim D (2009) Green growth, green economy und green new deal. Die 'Vergrünung' nationaler Politik in Südkorea (The Greening of National Policy in South Korea), GIGA Focus Asien, 10/2009

Shim D (2010) Green growth–South Korea's panacea? In: Frank R, Hoare JE, Köllner P, Pares S (eds) Korea 2010–politics, economy and society. Brill, Leiden

Scott WR (2004) Reflections on a half-century of organizational sociology. Annu Rev Sociol 30:1–21. doi:10.1146/annurev.soc.30.012703.110644

Suh J (2000) Korea's innovation system: challenges and new policy agenda. Discussion paper maastricht. Institute for New Technologies http://www.intech.unu.edu/publications/discussion-papers/2000-4.pdf. Accessed 20 May 2007

Tsipouri L, Patsatzis V (2006) Annual innovation policy trends report for SE Asia countries Japan, Korea, China, India, Taiwan, Singapore, Malaysia, Thailand, Indonesia. European Trend Chart on Innovation. European Commission. http://trendchart.cordis.lu/reports/documents/Country_Report_ASIA%20COUNTRIES_2006.pdf. Accessed 2 June 2007

UNEP (2010) Overview of the Republic of Korea's national strategy for green growth. United Nations Environmental Programme. http://www.unep.org/PDF/PressReleases/201004_UNEP_NATIONAL_STRATEGY.pdf. Accessed March 3rd, 2010

Yim DS (2005) Korea's national innovation system and the science and technology policy. Science and Technology Policy Institute (STEPI). http://www.unesco.org/new/fileadmin/MULTIMEDIA/HQ/SC/pdf/yim-korea.pdf. Accessed March 3rd, 2010

Chapter 7
Comment on Schüller, Conlé, Shim

Eunsook Yang

Introduction

The need to stay part of a globalized society makes restructuring national innovation policy an appropriate undertaking for the Lee Myung-bak administration. The paper, "Innovation Policy under Lee Myung-bak: A Critical Analysis of Governmental Actor's New Division of Labour" provides a thorough and descriptive introduction to fieldwork with a very up-to-date analysis of recent developments. The South Korean government's restructuring policy is very appealing to foreign firms and the world of academia. Detailed descriptions on how to set up priorities under the new structure and how these goals can be achieved methodologically are of major interest. However, the limitations mentioned primarily on two resources, financing (the budget) and personnel could be extended more. Finally, bearing in mind that we are living in a globalized society with active commercial exchanges, it is important to identify the trends in this administration's restructuring policy and how foreign countries and firms can benefit from them.

Historical Review of Innovation Policies

In the history of Korean economic development, the government played a very important role in innovation programs through its centralized decision making system. Three decades of extraordinary growth transformed Korea from one of the poorest agrarian societies to the 11th largest economy and exporting country in the world, mainly due to successful government economic transformation policies. Other factors that cannot be ignored in Korea's rapid development are its high rates of savings and investment and strong emphasis on education. Confucianism is also an important element as it dominates the daily lives of Koreans as a source of personal and national ethics as strong as or stronger than any of the officially recognized religions. Korea's economic growth has benefitted from the education

J. Mahlich and W. Pascha (eds.), *Korean Science and Technology in an International Perspective*, DOI 10.1007/978-3-7908-2753-8_7,

level of its human resources, which have played a key role in absorbing advanced technology from developed countries.

The Government's Innovation Policies in Korea's Economic Development[1]

Five-Year Economic Development Plans in the 1960s

As mentioned above, Korea's economy was once predominantly based on agriculture. Korea's industrialization began in the early 1960s with the introduction of the first Five-Year Economic Development Plan. The government made a conscious policy shift from domestic development and the importation of products to an outward-looking growth strategy of export promotion for light manufactured goods. The government utilized various macroeconomic mechanisms in implementing this strategy, such as maintaining high interest rates to mobilize domestic savings and enacting the Foreign Capital Promotion Act to encourage the inflow of foreign investment. By the end of the first two Five-Year Economic Development Plans, Korea had clearly left the ranks of the poorest countries in the world.

The Restructuring of Korea's Economy in the 1970s

In the 1970s, the changing international economic environment had a significant impact on Korea. Due to the Nixon Doctrine in 1971, the possibility of an eventual withdrawal of U.S. forces from the Korean Peninsula pushed the Korean government to develop its own defense industry in order to build a self-sufficient military force. Another important phenomenon arrived: the introduction of innovation in economic transformation due to the 1973–1974 oil shock. Korea had to modify its export promotion strategy in order to correct the trade imbalance.

The measures undertaken by the government were to restructure the composition of commodity exports in favor of a more sophisticated, higher value-added product to diversify trade partners. With all these changes in external conditions, and to adjust its exports accordingly, Korea turned to the Heavy-Chemical Industry Drive (HCI), giving it priority in the Third Five-Year Plan (1972–1976).[2] The HCI Drive policy produced impressive results. Between 1972 and 1978 the GNP growth averaged 10.8% annually, and the annual growth rate from 1976 to 1978 reached

---

[1]Overview of Korea's economic development. www.asianinfo.org/asianinfo/korea/economy.htm. Accessed Jan 14

[2]With the announcement of the Heavy-Chemical Industry Development Plan in 1973, the government set forth an accelerated development schedule for technologically sophisticated industries.

11.2%. The share of HCI products in total exports rose from 21.3% in 1972 to 34.7% in 1978.[3]

This progress however, came at the cost of high inflation. In addition, Korea's industrial structure was distorted by over investment in HCI and not enough investment in light industries. The government controls also distorted prices and stifled competition. At the same time, real wages were increasing faster than productivity, weakening export competitiveness.

Economic Reform in the 1980s

By 1979, the government realized the dangers that these structural imbalances posed. Accordingly, it initiated a comprehensive stabilization program designed to control excess liquidity, realign credit priorities, eliminate price distortions, and promote competition. However, external and internal circumstances such as the second oil shock and the assassination of President Park Chung-hee in 1979 jolted the Korean economy. Korea's economic performance in 1980 was the worst in more than 20 years as the economy contracted 5.2% while wholesale prices soared more than 38% and the current account deficit reached USD5.3 billion.[4]

To address the excess liquidity problem, the government forced firms to merge. The firms that produced engines, heavy electrical equipment, electronic exchangers, and copper smelting were ordered to specialize according to particular product lines, or to merge with others. On the other hand, government restrictions on foreign direct investment (FDI) were relaxed substantially in recognition of the FDI's role in promoting competition and adopting advanced foreign technologies.

Globalization and Korea's Economic Transformation in the 1990s

With the wave of growing regionalism in the 1990s, the need that the government felt to change its economic strategy from exporting goods using cheap labor while protecting domestic markets from foreign competition came to an end. The government had to apply new reform policy in order to cope with an international commerce atmosphere and the previous system of irrational procedures.

The government's first step in the early 1990s was the reform of the financial sector; establishing the Real Name financial transaction system. The Real Name system was the government's effort to do away with business transactions made under false names that fostered institutionalized corruption and illegal financial dealings.

For its globalization policy, the government participated actively in the Uruguay Round of the General Agreement on Tariffs and Trade and in Asia-Pacific

[3]Overview of Korea's economic development

[4]Ibid

Economic Cooperation talks. The government's effort culminated in Korea's entry into the OECD in 1996.[5]

The Economic Crisis of 1997

Since 1960, Korea's economic performance has largely recorded extraordinary growth rates, but that has also brought structural imbalances. First of all, the economic plans concentrated economic and political power in large industrial conglomerates and retarded the development of small and medium-sized enterprises essential for entrepreneurship and technological innovation, resulting in an inadequate private financial services sector and costly labor strife. Three years before the 1997 Asian Financial Crisis, Dr. CHO Soon, former vice prime minister and minister of economy, warned that Korea must undertake a "great transition" in order to sustain growth and democratization.[6]

Throughout the 1990s, the structure of the Korean economy had become vulnerable and exposed to unexpected international economic changes. This problem came from two sources: the first, an overly short-term oriented external debt structure that drained foreign exchange reserves. Beginning in 1994, rapid increases in private sector borrowing, primarily direct borrowing by corporations and bank borrowing to finance corporate investment, accounted for most of the external debt increase. By the end of 1996, the share of short-term debt stemming from the total external debt peaked at 58%, while the foreign exchange reserve remained low Kim (2006).

The second factor that Dr. Cho mentioned that contributed to Korea's economic vulnerability was a highly leveraged corporate financial structure – largely driven by over-investments by Korean conglomerates, or *chaebol*. As a result, the corporate sector had become increasingly vulnerable to unfavorable external shocks.

The three previous decades of a government-led growth had created a very close tie between the government and the *chaebol*. The business firms profited by expanding in size instead of producing healthy profits. To finance the expansion of businesses, firms chose the option of debt-financed growth rather than equity-financed growth. The high debt-equity ratio that resulted from such a strategy exceeded 400% by the end of 1997, and the average ratio for the 30 largest *chaebol* reached 518%.

---

[5]The Korean economy made a remarkable comeback in the first half of the 1990s. With increased investment and export, the economic growth rate increased from 3% in 1992 to 8.6% and 8.9% in 1994 and 1995, respectively. The GNP per capita surpassed the USD10,000 mark in 1995, and in 1996, the unemployment rate was an unprecedented 2%. With such high economic growth, inflation remained relatively stable around 4% throughout the 1990s.

[6]Cho (1994) The dynamics of Korean economic development. ISBN paper 0-88132-162-1, p. 216. He recommends policies to reduce industrial concentration, establish a competitive pricing system (particularly in the financial sector), promote small and medium-sized enterprises, improve labor-management relations, and encourage investment in human capital.

The Southeast Asian Financial problem in 1996 severely damaged the profits of Korean corporations in 1997, producing a series of corporate bankruptcies. The deterioration of the corporate sector brought about the weakening of the financial sector (Baliño and Ubide 1999).

At the end of his term, the administration of President Kim Young-sam was unable to provide much-needed reforms. The Finance and Economy Ministry's inadequate handling of the developing foreign exchange crisis finally forced Korea to ask for relief from the IMF. In other words, indebted corporations and financial institutions damaged Korea's credibility abroad, leading to foreign capital departure. The vicious cycle of the foreign exchange shortage and deterioration of Korea's credibility developed into a full-fledged foreign exchange crisis at the end of 1997.

Out of the Financial Crisis

During the first 6 months after the Korean government and the IMF agreed on the provisions of the IMF standby loans in December 1997, the administration of President Kim Dae-jung implemented numerous reforms to stabilize the situation. By implementing overall economic structural reforms in finance, corporate governance, labor, liberalization, and the public sector, Korea once again demonstrated its reform strategy was efficient and rapid. The country emerged from the financial crisis much quicker than most people had expected.

The Role of Confucianism in Korea's Economic Development

Korea's economic performance is considered particularly impressive because it has been achieved despite the burdens of Japanese colonial rule, the devastation of the Korean War, political turmoil, and heavy military expenditures necessitated by the national partition.

In fact, Korea's path to economic development was a matter of national survival.[7] Certain aspects of the economic policies and development that led to the rapid transformation of the Korean economy could be exemplified through its culture and history, which are separate from its export policies.[8] Religious and ideological

[7]With all the difficulties mentioned above, limited natural resources, and an insufficient domestic market with virtually no tradition of economic development experience, Korea set out to insure its sovereignty by establishing a stable and independent economy.

[8]Exchange rate, import liberalization, factor price distortions, and changing comparative advantage; production technology and total factor productivity; macroeconomic perspectives; conglomeration, business concentration and management practices; income distribution and labor issues, and urban and regional policies. Note: Without an explanation of the relevant cultural influences,

beliefs; principles; and rules that regulate the behavior of individuals, families and communities contributed greatly to Korea's rapid economic development.

Korea, over a long historical period, has developed a cultural sphere bound by Confucianism. Although Buddhism has had major impacts on ancient Korea along with Taoist traditions and certain aspects of shamanism, Confucianism has been the most influential in shaping personal behavior patterns and the structure of the family and community.

The philosophy of Confucianism has enabled Korea to take on a different type of capitalism and a different path to modernity than the West. Furthermore, Confucianism has been the source of such traits as activism, hard work and prudence, to name a few, that have proven conducive to economic development. These Confucian virtues have been monitored by the state for social engineering, modernization and for economic development (José 1998).

Not only has Confucianism served as a philosophical doctrine to guide the daily life of the people, it has also imposed the self-sacrificing work ethic and enthusiasm for education that has played a key role in Korea's economic growth.

The central pillar of Confucianism is the family. Indeed, family cohesion and continuity are considered the foundation for sustaining social bonds and the state. These values can be observed in Korean hierarchical social relations, such as those between ruler and subject, parent and child, and husband and wife. The family is the fundamental unit of society, incorporating the economic functions of production and consumption as well as the social functions of education and socialization, guided by moral and ethical principles.[9]

In its teachings, Confucianism has traditionally deified ancestors, institutionalizing ancestor worship, and delegated the duties of ritual master to the head of the male lineage - that is, to the father and husband. In this respect, Confucianism may be viewed as a familial religion, and it seems that no other culture has placed such emphasis on the family as the Confucian culture of Korea has (Coale et al. 1980)

In particular, "filial piety" is considered the core religious duty expected from Confucianists. These values came to be a powerful economic motivator during the process of institutionalization.

The economic effect of filial piety in relation to the economic growth in contemporary Korean society is a very unique feature. The religious tendency of filial piety which attempts to "remember" and "represent" one's ancestors acted as an important spiritual philosophy in Korea's economic growth-centered social development. Filial piety did not cease to be an ethical standard; rather it was the

---

the above is just a random list of economic terms. It would be better to pick the two main factors and describe the specific cultural influences.

[9]Park and Cho (1995).

Note: Sometimes the family name comes first (14), sometimes not (12 and 13). Please consult the Springer guidelines for authors, p. 6

fundamental basis for macro-social dynamism that was closely linked to the development of capitalism in Korea.

Comment on the Innovation Policy Under President Lee

To cope with the world economic crisis at the beginning of 2009, the administration stressed cooperation, efficiency and job creation. The first step to overcome the economic crisis involved the close collaboration of 22 ministries and government offices and the private sector. The Digital New Deal and Energy New Deal policies, collectively called the "Green New Deal," were pushed as the top priority.[10]

The Digital New Deal, designed to reform industry with IT and software green technologies, and the Energy New Deal, aimed at promoting low-carbon green growth suggest two axes to overcome the economic crisis (Kim and Jung 2009). They provide an opportunity to examine a new form of integration between the socio-cultural aspects of S&T innovation and STI policy formation (Kim 2009).

For the successful outcome of the project, the *Korea* Communications Commission (KCC) and the communications industry worked to increase the size of investments in networks, while in 2009 the Ministry of Knowledge and Economy invested KRW1.3 trillion – about KRW143 billion of that for the R&D of IT fusion systems, which is one of the nation's new growth engines, and KRW121 billion for robots. The primary objective was to positively attract investment from related private sectors. The expected amount of private investment regarding IT new growth engines, including the IT fusion system and robots reached KRW10.6 trillion.

In addition, the ministry invested over KRW100 billion, including KRW52.8 billion for the Digital New Deal related to the development of fusion software to spur innovation in smaller IT companies, LEDs and RFID.

On the other hand, the Ministry of Education, Science and Technology has focused on overcoming the most recent financial crisis and strengthening related competitiveness. For this, the ministry tries to create new jobs in schools and research fields to relieve the problem of young job seekers.

All these strategies are aimed at preparing for the future, while maximizing advantages, developing new investment opportunities and overcoming the ongoing crisis. These strategies will provide the country with more variety in the upcoming

[10]Government institutions such as the Ministry of Knowledge and Economy; the Korea Communications Commission; the Ministry of Education, Science and Technology; the Ministry of Culture, Sports and Tourism; and the Small and Medium Business Administration revealed that they plan to generate domestic demand by activating investments in the digital industry putting IT, software, broadcasting and communications and content together to cultivate them as strategic export industries. The digital industry, which was the key player in pulling the country out of the IMF crisis in the late 1990s, has become a relief pitcher called upon to rescue Korea once again.

economic system and the era of development to come after going through the tunnel of depression. Both plans were made with the assumption that they can generate many high-quality jobs by supplying renewable energy, commercializing software service models, and supporting the expansion of outsourcing such as design consulting, that is also closely connected with the Digital New Deal.

As an ongoing project in 2010, the Knowledge Ministry has a plan to prepare for the future through preemptive R&D investments in order to develop green energy technologies and carve out new growth engine technologies. Large companies will push ahead with the transformation to a low-carbon economy by improving energy efficiency and localizing new renewable energy technologies by checking the possibility of improving thermal energy efficiency, targeting 10,000 small and medium-sized firms. For this purpose, the Green IT industry and green standards will be developed and resource productivity will be raised (Chun 2009).

New Opportunities and Challenges for the Future

South Korea will assume leadership of the rich nations for the first time during the 2010 G20 in Seoul. South Korea became the first member of the OECD to emerge from the global recession with a 0.4% growth rate in the third quarter of 2009. In 2010, the OECD expects South Korea's GDP to expand by 4.4%, the highest growth rate of any of its members.

Now, President Lee Myung-bak wants to turn the end of this economic crisis into an opportunity. His goal is to transform South Korea from a successful economic power into a respected, developed medium power to mediate between rich and poor nations on global issues such as climate change and financial regulation.

South Korea's successful management of the economic crisis has become the model for many Asian countries. When the South Korean won depreciated by 30% in the first 3 months of the crisis, the stock market dropped by half, and foreign investors left in droves. The Korean government handled it with efficient analysis. Many of its current financial leaders are veterans of the Asian crisis that crippled the country's economy in 1998, and these technocrats know how to manage a new economic crisis. Lee's team immediately moved to save threatened banks and companies by setting up $200 billion in various funds to guarantee payment of their debts and for other forms of emergency aid.[11] A government focused on protecting jobs kept consumer sentiment relatively high, and the Bank of Korea cut interest rates by 3.25% points to 2%.

[11]They struck currency-swap deals with major economies like the U.S. to secure dwindling reserves of foreign currency and front-loaded public spending so that 65% of the country's $250 billion budget was spent during the first half of 2009, ensuring that the money got into the economy rapidly – but without adding new debts.

President Lee, now out of this crisis, is positioning South Korea within Asia as a dynamic alternative. Watching Korea's performance, the current U.S. Ambassador to South Korea, Kathleen Stephens mentioned that South Korea is "the best example in the post–World War II era of a country that has overcome enormous obstacles to achieve this kind of success." From now on, South Korea, geographically surrounded by bigger powers – China, Russia, and Japan – needs to take a global role for itself to ensure its prosperity and security.

References

Baliño TJT, Ubide A (1999) The Korean financial crisis of 1997-a strategy of financial sector reform. IMF Working Paper. www.imf.org/external/pubs/cat/longres.cfm. Accessed 6 Jan 2010

Cho S (1994) The dynamics of Korean economic development. ISBN paper 0-88132-162-1. p 216

Chun G-E (2009) Lee Myung-bak's operational policy direction for 2009. Korea Times, Friday, 16 Jan 2009

Coale AJ, Cho L-J, Goldman N (1980) Estimation of recent trends in fertility and mortality in the Republic of Korea. National Research Council, Committee on Population and Demography, Report No. 1, National Academy of Sciences. See also, Report of Kim U, Park Y-S, Confucianism and family value. Dept. of Psychology, Chung-Ang University, Seoul

José LL (1998) Culture, the state, and economic development in Korea and Mexico. Instituto Matías Romero de Estudios Diplomáticos. www.orpheus.ucsd.edu/las/studies/pdfs/leon.pdf. Accessed 12 Jan 2010

Kim K (2006) "The 1997–98 Korean financial crisis: causes, policy response, and lessons" at the high-level seminar on crisis prevention in emerging markets organized by The International Monetary Fund and the Government of Singapore, 10–11 July 2006. www imf.org/external/np/seminars/eng/2006/cpem/pdf/kihwan. Accessed 7 Jan 2010

Kim G-T (2009) An analysis of green technology innovation: a case study of green new deal policy initiative in South Korea 2009. In: Atlanta conference on science and innovation policy. Conferences.library.gatech.edu. Accessed 15 Jan 2010

Kim SJ, Jung H (2009) The Korean strategy for green technology development and role of IT. Presentation at the Information Technology and Innovation Foundation, Washington, DC

Park IH, Cho L-J (1995) Confucianism and the Korean family. J Comp Fam Stud 26:2–10

Chapter 8
Three Potential Role Models for the Korean Innovation System: USA, Japan and Germany

Markus Pudelko and Jörg Büechl

This article analyzes the Korean innovation system and presents suggestions for its further improvement on the basis of three other country models, that of the U.S., Japan and Germany. We suggest that innovation systems are embedded in their respective managerial, economic, socio-political, and ultimately cultural context and that those contextual factors exert a significant influence on a national innovation system. Furthermore, we propose that the influence of those contextual factors result in innovation systems that are either more transformational or incremental in nature. We conclude that Korea has (like Japan) due to its societal context a particular strength in incremental innovations but will increasingly need to build up its ability to generate also transformational innovations (a particular strength of the US model). Due to distinct differences in the societal context, it might be difficult, however, for Korea to follow the U.S. model. We suggest, therefore, that Korea might find some valuable lessons in the German innovation system, particularly regarding small and medium enterprises.

Current State of the Korean Economy

The growth rates of the South Korean economy over the last decades represent one of the most impressive economic success stories of the twentieth century. As late as 1970, living standards in Korea[1] were roughly comparable to those in Ghana ($260 GNP per capita compared to $250). By 1995 the situation has changed dramatically. Korea had a GNP per head of $9,700, while Ghana's was still at only $390, reflecting vastly different degrees of innovativeness of both economies and resulting differences in growth rates. While Ghana had between 1968 and 1995 annual growth rates of under 1.4%, Korea achieved in the same time span an average annual growth

[1]When we mention here and in the following Korea, we refer only to the Republic of Korea, better known as South Korea.

J. Mahlich and W. Pascha (eds.), *Korean Science and Technology in an International Perspective*, DOI 10.1007/978-3-7908-2753-8_8,

of about 9% (Hill 1999) in a process commonly known as the "Miracle on the Han River." Over the last 40 years Korea experienced only 3 years of negative growth: 1980, 1998 and 2009. Today, Korea is a developed country, the fourth largest economy in Asia and the 13th largest in the world. Its GDP of $1.3 trillion (2008) is comparable to that of an average EU country[2] and its per capita GNP of $27,700 (2008) is similar to Italy's and higher than Poland's or Turkey's. Furthermore, Korea's growth rates are still significantly higher than those of Europe[3] or the U.S.[4] The growth rate for 2010 is predicted to be between 2.5% and 4%.[5] Nevertheless, the current economic crisis has impacted Korea as well. The Korea Development Bank recently published, for example, a survey which indicated that small and medium enterprises (SMEs) in the manufacturing industry will decrease their investments for assets and equipment by 40%, large enterprises by 12% and the automotive sector by 25%.[6]

The Input-Based Economy of Korea and its Consequences for Innovativeness

Korea's impressive economic surge over the last decades depended to a large extent on two factors: the input of capital and the input of labor. The strong focus on capital input is demonstrated by Korea's rate of business investments which is the highest of the entire OECD area (Jones 2007). As for labor input, Korea is the country with the longest working hours per year in the entire OECD (over 2,300 h), significantly ahead of the U.S. and Japan (less than 1,800 h) and Germany (approximately 1,400 h).[7]

We would argue that while this input-based strategy served Korea very well during the decades of catching-up with more advanced economies, it has lost much

---

[2]International Monetary Fund (IMF)

[3]"Wachstumsrate des realen BIP pro Kopf – [tsdec 100]"(Luxemburg: Statistische Amt der Europäischen Gemeinschaft (eurostat), 2010), http://epp.eurostat.ec.europa.eu/tgm/table.do?tab=table&plugin=0&language=de&pcode=tsdec100 (file accessed March 16, 2010)

[4]"Gross Domestic Product: Fourth Quarter 2009 (Second Estimate)" (Washington, D.C.: Bureau of Economic Analysis: National Economic Accounts, 2010), http://www.bea.gov/newsreleases/national/gdp/gdpnewsrelease.htm (file accessed March 16, 2010)

[5]"Korea (Republik Korea, Südkorea): Wirtschaft" (Berlin: Auswärtiges Amt, 2009), http://www.auswaertiges-amt.de/diplo/de/Laenderinformationen/KoreaRepublik/Wirtschaft.html (file accessed November 17, 2009)

[6]"Wirtschaftstrends kompakt Korea (Rep.) Jahresmitte 2009"(Berlin: Germany Trade and Invest, 2009), http://www.gtai.de/DE/Content/__SharedDocs/Links-Einzeldokumente-Datenbanken/fachdokument.html?fIdent=MKT200906178016 (file accessed November 3, 2009)

[7]"Workforce: The World's Hardest Working Countries" (New York City: Forbes, 2008), http://www.forbes.com/2008/05/21/labor-market-workforce-lead-citizen-cx_po_0521countries.html (file accessed November 25, 2009)

of its merits since Korea has reached the status of a developed country. With Korea now already possessing an advanced and modern stock of capital goods, the growth potential stemming from further capital investment is limited. This is in particular so as Korea will soon hit the international technological frontier, leaving little room for further growth simply by bringing its technology up to already elsewhere existing standards. A survey, conducted in cooperation with 6,000 Korean companies, has shown that the majority of the sample has reached the international technological frontier to at least 80% (ibid). Consequently, the opportunities for easy technological "catching-up" by imitation or reverse engineering are continuously diminishing. As a result, the share of capital input has already declined from 37% of GDP during the 1990s to less than 30% in the past few years.

As for Korea's second source of economic development, the input of labor, demographics are restraining its further growth potential as Korea is becoming an increasingly aging society, resulting in a decreasing share of its working population. In fact, Korea has the lowest birth rate of the OECD area with 1.1 children per woman (1.4 in Europe and 2 in the U.S.). As a consequence, its labor force will reach its peak in 2015. The percentage of the population over the age of 65 will double from 7% to 14% in 2018 and increase further to 20% in 2026 (ibid).

Physical capital and hard work can lead a country only to a certain point. As an economy approaches the technological frontier and labor, in particular cheap labor, is reaching its limit as a source for continued economic growth as well, an economy has to redirect its strategy and focus more on innovation, effectiveness and efficiency. Korea is a case in point as it needs to move beyond the sheer accumulation of capital and work. It needs to leave this input driven economic growth strategy and its still prevailing developing country mentality and reflexes behind and redirect its energies toward increased productivity and efficiency. A survey conducted by the OECD indicated that the output per hour worked in Korea was only 46% of that in the United States, 50% of that in Germany and 65% of that in Japan, making Korea (only) as efficient as Hungary, Portugal or Poland.[8] From this, we conclude that for Korea, with its traditional focus on (quantitative) capital and labor *input*, a transition toward a (qualitative) *output*-based economy will pose a decisive challenge on which much of Korea's economic future will depend. This implies significant changes in the country's industrial policy and innovation system.

Over the decades, Korea entrusted the role of its economic growth engine to its mighty conglomerates, the chaebols, such as Samsung and LG, following to a significant degree the model of the Japanese keiretsu system with company networks such as Mitsubishi and Mitsui. Much of the initially scarce financial resources were channeled by the government to the chaebols and the government continued to put their interests before those of any other stakeholder. The chaebols have stood at the forefront of the Miracle on the Han River, representing the pride of the Korean nation. Only with the Asian crisis (1997–1998) did the all-mighty

[8]"OECD estimates of labour productivity levels" (Paris: OECD.StatExtracts, 2008), http://stats.oecd.org/Index.aspx?DatasetCode=LEVEL (file accessed on March 17, 2010)

position of the chaebols start to slowly wane: Hanbo and Kia were even allowed to go bankrupt in 1997, but chaebols are generally still the most preferred places of employment.

While the chaebols were forced to restructure themselves in the aftermath of the Asian crisis, they continued to persevere, even if to a somewhat lesser extent, their leading role in the Korean economy. With the chaebols having adapted to a large extent to a more open and less protected economic environment, it becomes now increasingly important that also the rest of the Korean economy undergoes a process of restructuring and reform. This is particularly the case with the relatively unproductive service sector (e.g., financial services, retail and real estate).

In fact, the extent to which the productivity of Korea's heavily regulated service sector is lagging the productivity of the industrial sector is, in comparison to other OECD countries, striking. Even though the service sector employs by now three fourths of Korea's workforce (1990 it was only 50%), it is responsible for only 50% of the economic output (Melser 2007). Restrictions on competition in several sectors, rigid labor laws and tight retail zoning rules are some of the underlying reasons for the weaknesses of Korea's service sector. As a consequence, Korea needs to focus on productivity gains in particular of its retail, banking, insurance, transportation and educational institutions (ibid). With the gradually decreasing importance of the manufacturing sector, a process every developed country is experiencing, this problem becomes even more acute (Kang 2007).

The Korean Innovation System

Studies of Korea's technological development have shown that the country's catching-up during the past 60 years can be divided into three phases: The first phase, which lasted from the 1950s to the 1970s, was characterized by the introduction and imitation of foreign technology, a period during which the country shifted from a mainly agricultural to a mainly industrial economy. As innovativeness meant in Korea at that time mainly imitation and adaptation of foreign technology, the R&D intensity was as low as 1% of GDP (Hemmert 2007).

During the second phase of Korea's technological development, in the 1980s, Korean firms began to pursue a rigorous R&D strategy, based on the improvement of existing products and the development of new products, mostly in middle tech technologies, through incremental innovation. R&D expenditure doubled during this period to almost 2% of GDP. As with the previous imitation and adaptation phase, also this phase was strongly shaped by the country's major industrial groups, the chaebols. During this period Korea continued to rely heavily on foreign technological know-how, which it sought to gradually improve, still reflecting the country's "catching-up" strategy of innovation.

In the third and current phase of technological development, which began in the 1990s, some Korean companies have achieved global technological leadership in a few R&D intensive industries, such as microelectronics and telecommunications, creating a considerable knowledge base in engineering (ibid). During this phase, R&D expenditures have reached almost 3% of GDP which is one of the highest percentages in the world. In the upcoming years, this percentage is even expected to increase to 5%. The nation's impressive R&D efforts are further supported by the country's strong and culturally deeply embedded focus on education and learning. However, despite pockets of impressive technology leadership, Korea still struggles in many sectors and industries to generate innovations that lead to high value added competitive advantages.

This problem becomes even more acute with Korea's rising wages due to increasing wealth. With resulting rising labor costs, low to middle tech industries, characterized by incremental technological innovations, will be more and more transferred to low cost economies that are equally gaining technological know-how, such as China. With China being able to out-compete Korea in terms of labor costs and Japan still possessing more advanced technological know-how than Korea, the latter finds itself in a very uncomfortable position, running the risk of being sandwiched between both of its larger neighbors. As returning to lower labor costs is not a viable solution, the only way forward for Korea is to improve its innovativeness.

Consequently, despite the admirable progress Korean multinationals were able to achieve, Korea's innovation strategy has not been without pitfalls and continues to need upgrading. Even though it enabled Korea's relatively small number of chaebols to gain international competitiveness in several sectors and industries, it has also led to a highly imbalanced capital allocation for R&D, given the concentration of spending on few industries by few large conglomerates. A full 60% of the corporate R&D spending is done by large companies in information and communication technology and the automotive sector (Jones 2007). This concentration led to a distinctive and arguably detrimental dualism that is characterizing the Korean economy with few, highly innovative and competitive conglomerates on one side and a large amount of small and medium sized enterprises which are frequently mere suppliers to the bigger companies. The latter are often left with such meager profit margins that they have no sufficient resources to invest themselves in R&D in any significant way, thus lacking innovation potential and international competitiveness.

A second imbalance of corporate R&D spending lies in the fact that a full 75% takes place in the manufacturing sector, with only 10% going to the service sector, a number which is significantly below the OECD average of 25%. A third imbalance lies in the distribution of corporate versus university R&D spending. Although 75% of the researchers with a Ph.D. degree work for universities, they are responsible for only 10% of the total national R&D spending which is only half of the OECD average. This indicates a strong focus in Korea on applied R&D – done by a few large companies in a few sectors – while basic research – mostly executed by public institutions – is comparatively neglected. In addition, the link between basic and applied R&D is only relatively poorly established, given the underdeveloped cooperation between corporate, government and university R&D activities (ibid).

The narrow and selective focus of R&D spending and the resulting strengths in only a few sectors have their origin in the nation's traditional strategy to promote the development of a few industries the government considered of strategic importance. Similar to government support of the chemical and heavy industry in the 1970s, in 2003 the government passed the Growth Engine Plan, supporting ten key industries for the next 5–10 years: bio-medical products, next-generation semiconductors, next-generation computer displays, next-generation batteries, future automobiles, digital TV and broadcasting, intelligent robots, next-generation mobile communications, intelligent home networks and finally digital content and software solutions. Only recently, did the government decide to invest heavily in green technologies.

While this industrial policy in which government is essentially attempting to preempt market outcomes might have worked in past decades, under the complexities of a globalized market it has proven increasingly difficult to outsmart market forces, leading to distortions and sub-optimal capital allocation. As also the more studied case of the Japanese METI has indicated, in the time of rapid global change, the risk of picking the wrong industries to invest in becomes ever stronger. Consequently and similar to the situation in Japan, a frequent criticism of the Korean industrial policy is that government should refrain from choosing industries it thinks should become the nation's growth engines, but invest instead in the development of human capital and generic technologies, leaving investment and innovation decisions to the corporate sector, thus also avoiding the crowding out of private investment.

Furthermore, government should promote more the integration of innovation activities of applied and fundamental research; of corporations, universities and other public sector institutions; of institutions and individual researchers; and of innovators inside and outside Korea. In particular cross-border cooperation and manpower exchanges in the area of research and innovation among universities, public research institutes and the corporate world are possible ways to promote an exchange of ideas and initiatives with foreign parties. Another approach is the liberalization of foreign direct investment (FDI) policies as FDI in Korea still is relatively small in comparison to other OECD countries (ibid). This would allow for more foreign investments (not only in form of capital but also know-how), providing additional stimuli to improve innovations.

So far this government guided approach has helped certain sectors to grow at the expense of others, thereby limiting entrepreneurship and innovativeness in non-funded sectors and businesses. In particular Korean SMEs, which are hardly present on the world markets due to their lack of global competitiveness, have been the ones who suffered most from this approach. As SMEs account for 85% of employment and 50% of manufacturing output (Kang 2007), increasing the innovativeness and profitability of SMEs should be a key priority. An important step in this direction would be to facilitate access to capital SMEs need to finance necessary investments in R&D. The prospect of additional cash flow based on securities markets, venture capital and equity based capital could help start-ups and already established SMEs to invest in new business ideas and technologies, thus improving their innovativeness. However, so far the Korean financial system is still heavily based on bank lending. In this system entrepreneurs often have to provide collateral, usually in the form of real estate. Leaving them with the burden of a great amount of personal risk is counter-productive when it comes to promoting innovativeness.

Currently and fortunately, the Korean financial system is addressing this bottleneck factor and is in the process of facilitating the provision of capital to SMEs. In addition, the government could also offer direct subsidies to SMEs in order to encourage them to spend more on R&D. Ultimately, not only market entry barriers (insufficient access to capital) impede SMEs' business, but also market exit barriers play a significant role. Strict bankruptcy laws and the incentive for banks to roll over debts lead to high exit costs (Melser 2007).

Having briefly reviewed the Korean innovation system, important strengths have been identified, in particular concerning incremental innovations by the major chaebols in selected industries and technologies. However, as indicated, important sectors and businesses, primarily the service sector and SMEs, still lack sufficient innovativeness and, thus, global competitiveness. Given these shortcomings, the question of potential sources of inspiration for the Korean innovation system becomes of increasing relevance.

Potential Role models: The U.S., Japan and Germany

We have identified three potential role models for the Korean innovation system: the U.S., Japan and Germany. These countries are the three largest developed economies in the world; the three main representatives of the triad of regions which dominates the global economy: North America, East Asia and Europe; and the main representatives of the three leading variants of market economies: the free-market economy of mainly Anglo-Saxon countries (U.S.), the government guided market economy of mainly Asian countries (Japan) and the social market economy of mainly European countries (Germany).

The starting point of our analysis is the assumption that innovativeness does not flourish in a socio-cultural void, but, to the contrary, is in large parts determined by socio-cultural influences. Consequently, potential lessons to be learned from other country models can only be developed under close consideration of such contextual factors. In the following we will demonstrate the extent to which the innovation models of Korea, the U.S., Japan and Germany are embedded in their respective socio-cultural context. More specifically, we will show how the particular socio-cultural context in Korea and Japan drive the Korean and Japanese innovation systems toward being largely based on *incremental innovations*. These are innovations that are taking place in small, gradual steps, leading only in their sum to larger effects. They are occurring in large parts within the existing paradigm of technology and ways of doing business. By contrast, the entire socio-cultural context in the U.S. is more geared toward an innovation system characterized by break-through or *transformational innovations*. These are innovations that are transforming the previously existing paradigm, leading to radically new technological solutions and business approaches. We will also indicate that the German innovation model stands in many ways between the Korean and Japanese system on one side and the US system on the other side.

Innovativeness of Korean, American, Japanese and German Companies in Their Respective Socio-Cultural Context

While national innovation systems are the ultimate focal points of this analysis, we consider them to be subsystems of their respective management system. The management system is, in turn, embedded in the respective economic system which is again a subsystem of the socio-political system which is ultimately a subsystem of the most comprehensive system, the cultural system. In order to describe these four layers of contextual sub- and super-systems and their impact on innovativeness, we will look for each of them at three different criteria. Consequently, in the following 12 contextual criteria and their impact on the innovation systems of Korea, the U.S., Japan and Germany will be briefly outlined. This will assist us in drawing lessons for the Korean innovation system under specific consideration of the socio-cultural context. The model that underlines our argumentation is depicted by Fig. 8.1 (adapted from Pudelko 2006a).

It goes without saying that due to space constrains an illustration of such a wide array of topics for four very different countries can only be delivered in a highly stylized and simplified way, largely neglecting nuances and gradations. Nevertheless, we hold that the central thrust of our argumentation, different socio-cultural contexts leading to either more incremental or transformational innovations, will help us to better understand the strengths and weaknesses of innovation systems and how they originate. Only with this deeper contextualized understanding can applicable inspirations be developed.

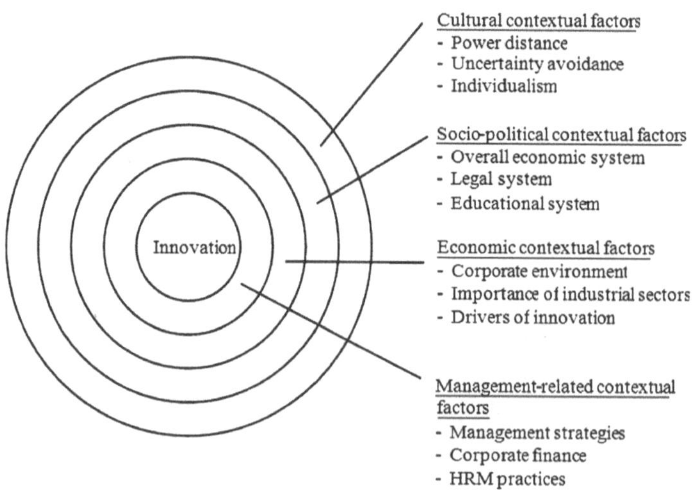

Fig. 8.1 Competitiveness system: innovation embedded in its socio-cultural context

Cultural Context

The most comprehensive contextual system determining the innovation system of a country is its *culture*. Culture and its influence on innovativeness will be briefly explained by those three of Hofstede (2001) dimensions we consider the most significant ones in influencing innovation: *power distance, uncertainty avoidance* and *individualism.*

Power distance can be defined as the "extent to which the less powerful members of institutions and organizations within a country expect and accept that power is distributed unequally" (Hofstede and Hofstede 2005). Countries with a low level of power distance are prepared to question authority systems such as traditionally held beliefs, institutions or individual power figures, something high-level power distance cultures prefer to avoid. This indicates that innovations in low-level power distance countries, such as Germany and the U.S., tend to be more transformational, as there is far greater willingness to challenge existing paradigms. By contrast, high-power distance countries, such as Korea and Japan, are more hesitant to go against authority systems and prefer, therefore, more incremental innovations which add to existing knowledge and improve existing approaches more than they challenge them.

Uncertainty avoidance is "the extent to which the members of a culture feel threatened by ambiguous or unknown situations" (ibid). People from low uncertainty avoidance cultures, such as the U.S., are consequently much more willing to take risks and envisage uncertain outcomes, a requirement to engage in transformational innovations (see, for example, Galbraith 1982). On the other hand, high uncertainty countries, such as Japan, Korea and to a lesser degree Germany, prefer to avoid risk and ambiguity and favor more secure, foreseeable and controllable approaches, something incremental innovations are much more able to guarantee.

Individualism pertains to "societies in which the ties between individuals are loose: everyone is expected to look after himself and his immediate family. Collectivism as its opposite pertains to societies in which people from birth onwards are integrated into strong, cohesive in-groups, which throughout people's lifetime continue to protect them in exchange for unquestioning loyalty" (Hofstede and Hofstede 2005). Cultures with high degrees of individualism, such as the U.S. and to a lesser degree Germany, stress the potential and value of ideas emanating from the single individual. This favors transformational innovations which are often realized by few creative individuals and often by outsiders who exhibit strong beliefs in themselves and their abilities. By contrast, collectivistic cultures, such as Korea and Japan, exhibit more trust in group efforts and outcomes. As a result, they show strengths in incremental innovations which are often the results of collective efforts and tend to be more focused on optimizing existing processes than on inventing new ones and, hence, are more status quo oriented (Galbraith 1982).

Socio-Political Context

The *socio-political context*, as the second sub-system, will be divided into the *overall economic system*, the *legal system* and the *educational system* – all of which play a significant role regarding the form of innovation a country promotes.

The **economic system**, as it pertains to free market economies such as the U.S., is based on each economic actor's individual freedom to decide what is best for him- or herself and a laissez-faire-ideology calling for minimal government interference in the market (Dore 2000; Pieterse 2003). The social market economy model, for which Germany stands exemplary, also promotes market forces and largely rejects government interference in industrial investment decisions. However, it does see a role for government in restraining market forces and the degree of freedom of companies in order to protect weaker market participants. Japan as well as Korea have largely been portrayed as having government-induced market economies. The government decided which sector it wanted to support and concentrate financial and intellectual resources in, favoring heavily big companies networks, i.e. the *keiretsu* and chaebols, respectively. With this system more and more in decline, Korea will need to find a new balance between market forces and government interference (Kang 2007). Given the different degrees of government involvement in corporate and entrepreneurial decision making, bold investments and transformational innovations are least constrained in the U.S., followed by Germany but are substantially more constrained in Japan and Korea.

Regarding the legal and the educational systems it is worthwhile noting that much of the Japanese systems were modeled after those of Germany and that Korea's systems were later fashioned by colonial Japan. The legal system in the US is geared toward promoting entrepreneurialism and risk taking. This becomes particularly evident with the bankruptcy laws which protect mainly the individual entrepreneur from the fall-outs of bankruptcy.[9] By contrast, in Germany, and even more so in Japan and Korea, bankruptcy laws traditionally show much more favor to the interests of capital lenders. While these countries have recently modified their laws in order to stimulate more entrepreneurialism, this difference still remains, primarily, intact. This is also reflected in the societal attitude toward bankruptcy. While Americans who go bankrupt easily get a "second chance," Japanese and Korean (and to a lesser extent German) entrepreneurs who fail are socially stigmatized. Consequently, high-risk transformational innovations are comparatively easy to implement in the U.S. legal context. By contrast, in Germany, and in particular in Japan and Korea, the legal system is more geared toward incremental innovations, as they are less uncertain and more controllable.

[9]"U.S. Bankruptcy Laws Encouraging Risk-Taking and Entrepreneurship" (Washington, D.C.: America.gov, 2006), http://www.america.gov/st/econ-english/2008/July/20080814223511XJyrre P0.102215.html, (file accessed November 25, 2009)

The educational system in the U.S. is characterized by considerable differences in quality among secondary schools and universities. On one hand, it offers at the very top of the pyramid a world-class education to the brightest students from all over the world and produces world-class research at elite universities. The Economist (2005) praises the U.S. educational system for its independence from the state, its competitive spirit and its ability to make academic scholarship relevant to society. According to Salmi (2009) it supports competitiveness, unrestrained scientific inquiry, critical thinking, innovation and creativity, and can respond instantly to the demands of a quickly changing global market. On the other hand, there is a relative neglect of mass-education (Thurow 1996).

This contrasts to the German, but in particular to the Japanese and Korean educational systems which are much more homogeneous in their quality (while in Japan and Korea major differences exist in the prestige of educational institutions, based very much on the difficulty with which to enter them, we argue that comparatively less differences exist in their de facto quality). Broadly speaking, these educational systems are not providing world-class elite education and only in pockets world-class research; however, they are successful in delivering a solid education to most of their respective populations, whereas the average educational level of U.S. students might not be considered as very high. Furthermore, according to cultural values, the German and, in particular, the U.S. educational systems focus on creativity and the development of independent thinking, whereas the Japanese and Korean educational systems stress more conformity with authorities, memorization and repetition of established truths and imitation of good practices (Whitehill 1991; Emmott 1992). In conclusion, the U.S. system is geared toward producing an elite of highly educated, independent-minded and creative individuals who are capable of transformational innovations; the Japanese and Korean systems, by contrast, have their strength in generating a large amount of well-trained people who together as engineers or workers are able to achieve collectively incremental innovations; and Germany stands, again, somewhat in between these two poles.

Economic Context

The next level of our analysis is the *economic context*, which can be divided by the *corporate environment, the importance of industrial sectors* and the *drivers of innovation.*

The corporate environment in both the U.S. and Germany is largely characterized by stand-alone, independent companies which have highly flexible relations with suppliers, customers and business partners (Case 1992). SMEs are sources of major innovation for both countries, next to large MNCs. For the U.S., this is exemplified by the start-ups in the highly dynamic IT sector of Silicon Valley or the Boston area. These companies are generally very much research driven, as their regions' close connections to Stanford University or MIT and Harvard University indicate. As a result, they generate highly transformational innovations.

The mobility in this sector is so strong that companies are constantly being launched, merged, acquired or declared bankrupt, depending on the short-term success of transformational innovations. In Germany, SMEs, the *Mittelstand*, are certainly less astounding than the Silicon Valley IT start-ups but are nevertheless considered to be the backbone of the German economy. Many of those companies have been labeled "hidden champions" (Simon 2009) as they are, despite their small size, often world market leaders (i.e. "champions") but operate in investment-goods sector niches, such as machinery, where companies and products are often little known to the public (i.e., "hidden"). These companies are frequently family owned and passed on through the generations. They invest highly in R&D and are, as a result, very innovative. However, given the mature industries in which they often operate; their long-term, at times even generational, time perspective; the little influence short-term-oriented venture capital plays on them; and their overall conservative business strategies, they focus much more on prudent, incremental innovations than the transformational.

The corporate environment in Japan and Korea is, by contrast, very much characterized by a dual structure with, on one side, influential, prestigious, innovative and comparatively profitable large corporate networks (the keiretsu or chaebol); on the other side are SMEs which are mostly mere suppliers in a rigid supply chain of major companies and which lack sufficient profit margins to invest in innovation-generating R&D. The keiretsu and chaebol companies relied, until recently, almost exclusively on organic growth, based on incremental innovations, whereas transformational changes such as mergers and acquisitions or even the recruitment of specialists from outside the company network were more the exception than the rule. The constant refinement of products, the optimization of production processes and stable relationships with suppliers largely continue to be the basis of this model and reinforce the incremental innovation approach (Inohara 1990).

Also with regard to the relative importance of industrial sectors the by-now-familiar pattern re-emerges, with the U.S. on one side and Japan and Korea on the other side, with Germany somewhat in the middle. One might argue that to some extent the strengths of the U.S. are the weaknesses of Japan and Korea and vice versa. The U.S. economy has its comparative strength in high-tech and high-value added manufacturing industries, such as IT and biotechnology, and in service industries, such as financial services, legal services, consulting, communication and advertising, all of which involve a high degree of transformational originality in technology, organizational structuring and corporate strategy. This originality is provided by a few outstanding and creative high-performers, reflecting the strengths of the aforementioned U.S. educational system (Emmott 1992).

By contrast, the German, Japanese and Korean economies have their competitive strengths in middle-tech manufacturing industries (e.g., mechanical engineering, chemicals and automotives for Germany and consumer electronics and automotives for Japan and Korea), displaying a high average standard of product quality and production technologies that are largely based on incremental innovations (ibid). These strengths mirror the already described high educational standard of the German, Japanese and Korean workforce. German products, even though often 'only' from middle-tech industries, are frequently positioned in high-end niches (either produced by large MNCs such as car manufactures or the "hidden champions" of SMEs) and can therefore generate high profit margins. By contrast, Japanese and Korean companies often follow a competitive strategy that is based on competing head on with their domestic rivals, often all operating in the same industries, with very similar products.

The drivers of innovation are in the U.S., since the rise of the IT sector, not only big MNCs anymore, but increasingly also small start-up companies. Since they tend to be more risk oriented, basing their business model on transformational innovations, many of them fail; but some succeed and become highly profitable. Some of them are so successful that they become in a relatively short period of time giant MNCs themselves (Google, for example). Consequently, start-ups are increasingly responsible for the creation of new wealth, jobs, and technological but also management innovations. In Germany, large companies as well as SMEs are known for high quality, innovative but also relatively expensive products, mostly in mature, middle-tech industries that are less characterized by transformational and more by incremental innovations. German SMEs, in particular the "hidden champions," frequently employ a strategy that combines product specialization, focusing on a particular market niche, and geographic diversification, being strongly commitment to global expansion (OECD 2001). Despite their small size they often achieve global market leadership in their respective segments. Traditionally, they invest significant portions of their earnings in R&D, assuring their innovativeness and market leadership status. Despite operating in middle-tech industries and focusing more on conservative, incremental innovations, they are, given their global market leadership, able to generate high profit margins.

In Japan and Korea, mostly the *keiretsu* or chaebols with their well-educated work force drive innovations (Herbig and Jacobs 1998; Melser 2007). These large MNCs invest heavily in R&D and are able to dominate through an innovative combination of excellent product quality, efficient production processes and cheap prices in certain middle-tech sectors, frequently in the consumer goods industries. It is because of their size and their interwovenness in larger company networks that their innovation processes tend to be more conservative. Both, the Japanese and the Korean economies have traditionally been divided between MNCs, which drive innovations, and the significantly less innovative SMEs. While this pattern is gradually changing in Japan, with more and more SMEs becoming highly innovative and producing not only for one key customer but for customers around the world (Schaede 2008) it still holds largely true for Korea.

Management Context

Moving on to the *management context* of innovation, we are able to differentiate between *management strategies*, *corporate finance* and *organizational structures*.

Regarding management strategies, both U.S. and German companies employ differentiation strategies, frequently offering products to a segmented customer base but reaping higher profit margins because of perceived uniqueness. U.S. firms achieve this mainly through substantial, if not transformational technological, organizational, managerial and marketing innovations, while German companies tend to focus on quality leadership, requiring constant but gradual innovations. Regarding Japan, Porter et al. (2000) noticed that Japanese firms often invest in innovations resulting in the best possible product for the largest amount of potential customers, a strategy which includes adopting features of successful products offered by direct competitors. This strategy leads to reliable, high-quality products, which suffer, however, from lack of differentiation, as customers view them as rather exchangeable (in particular in the electronics industry). With little product differentiation, the key differentiator becomes price. As a result, Japanese manufacturers put much emphasis on optimizing operational efficiency which is largely based on incremental innovations. While this helps to reduce costs, given the largely exchangeable character of the products, companies have to pass the cost reduction advantage on to customers. By doing so, Japanese companies might be able to increase their market share, however, they find themselves in a downward spiral of reduced costs, reduced prices and reduced profit margins.

Korean companies are in many instances not quite able to match their Japanese competitors in terms of product quality and innovation. However, they can usually produce at even lower costs than the already cost-conscious Japanese. Placing priority on cutting costs and improving the quality of largely undifferentiated products can be a dangerous strategy, though, given that competitors, like China, produce at even lower costs and are increasingly able to meet quality standards as well.

With regard to corporate finance, it is comparatively easy for entrepreneurs and start-ups in the U.S. to obtain the necessary funding for investments in innovations, also potentially transformational ones which are significantly more risky. The financial industry is offering, next to conventional bank loans, a well-established range of venture capital, capital partnerships and IPOs (Initial Public Offerings) (Stratos and Panos 2002). Also larger companies do not have to rely on bank loans, as they can refinance themselves on the securities market. The same instruments are, in principle, also available now in Germany and, to a lesser degree in Japan and Korea. However, they are still less accessible to companies, in particular for riskier investments. Consequently, Germany, and to a much more significant degree Japan and Korea, have to rely much more on bank lending (Kang 2007). While German banks are also willing to extend credit after a meticulous investigation of the business plan, they are still more risk adverse compared to specialized financial institutions or the financial markets in the U.S. In Japan and Korea, it is even more difficult to obtain the required funding for riskier investments, despite efforts to develop the supply of private risk capital (Imai and Kawagoe 2000). Entrepreneurs often have to provide collateral, for example in the form of private real estate, which is significantly restraining their financial possibilities and making such investments much riskier for them. As a consequence, the rate to which in Japan and Korea innovation-oriented start-ups are launched is comparatively low. As start-ups are, however, increasingly important sources for innovation, in particular the transformational variety, both countries are missing important opportunities to reinvigorate their economies. Even larger companies have to rely on bank loans due to underdeveloped securities markets.

The HRM practices in the U.S. are highly flexible and strongly performance oriented. Due to these characteristics the labor market, particularly in highly innovative sectors, is very mobile. This helps bring critical human resources, such as product developers, into the companies and jobs where they are the most needed and will be the most productive. Due to this constant recombination of key personnel, creativity is enhanced substantially which is particularly helpful for transformational innovations. By contrast, German companies are rather long-term oriented, emphasize technical expertise and do not favor the hire-and-fire mentality (Child et al. 2001). Consequently, in Germany the job market, also for those employees who play a major role in innovations, is much more stable. With seniority still playing an important role next to performance, fluctuation is further constrained. As a result, innovators know intimately the company they work for, the products the company has developed in the past, the market they are producing for, the customers they are delivering to and the rival companies they are competing with. This detailed knowledge does not necessarily lead to groundbreaking new developments, but to a constant flow of incremental innovations which, in their sum, ultimately lead in many cases to technological leadership.

In Japan and Korea the labor markets and the HRM policies are noticeably differentiated, due to the dual structure of the Japanese and Korean economies. In large companies employees still enjoy many privileges, including good training, relatively high salaries, a high degree of job security and a foreseeable career path. The economic malaise in Japan in the 1990s and the Asian crisis of 1997 that affected Korea strongly led to certain changes, in particular an increased performance orientation (Lee 2003; Pudelko 2006b), but the basic principles of Japanese and Korean HRM are for large companies still mostly intact. By contrast, in smaller companies, due to their lower profit margins, training is limited, salaries are significantly lower and job security is not provided which also makes career paths much more difficult to plan. As a consequence, employees at major corporations, including those who are in charge of product innovations, feel very loyal to their companies, are well qualified and work hard to guarantee a constant outflow of improved products, while considering at the same time the optimization of the production process. By contrast, smaller Japanese and Korean companies mostly do not have the financial and human resources to do proper research and the product development they are engaged in often takes place in very close coordination with their key customers, who are at times even their only customers. Korean SMEs have, like their Japanese counterparts, the difficulty of not being considered as appealing employers in terms of prestige and real benefits. Attracting employees with high potential who could drive innovations is, therefore, a major challenge for them. Consequently, in Japan and Korea innovations have so far been very much a prerogative of larger companies. In Japan, however, this long held pattern is in the process of change, as more and more smaller companies employ highly talented people, become very innovative and, consequently, "hidden champions" in their own right, quite similar to their counterparts in Germany. This has, however, not occurred in Korea yet to any comparable degree, where smaller companies are mostly still limited to being part of the supply chain of larger companies.

Having provided a brief and admittedly simplified overview of how 12 different contextual criteria, pertaining to the cultural, socio-political, economic and management context, influence the innovation systems of Korea and our three reference countries, the U.S., Japan and Germany, the following three key conclusions can be drawn:

1. In all four countries, the various contextual systems and sub-systems are closely interlinked and mutually reinforcing.
2. Taken together, these contextual factors have, in all four countries, a significant impact on the respective innovation system.
3. The combined influences of those contextual factors operate in a way that is geared toward transformational innovations in the U.S.; at the other side of the spectrum both Japan and Korea are geared toward incremental innovations; and Germany is located somewhat in between these two opposite poles, with an innovation system that is more incremental- than transformational-oriented, and this even though it is culturally and institutionally closer to the U.S. than to Japan or Korea.

Lessons and Inspiration for the Korean Innovation System

Having compared Korea with three reference countries, the U.S., Germany and Japan, it became evident that Korean innovation practices resemble most those of Japan, in terms of both, strengths and weaknesses. This should not come as a surprise, given the relative similarities in the contextual factors of both innovation systems. Consequently, while it would be comparatively easy for Korea to follow even more the Japanese model, the scope for new stimuli from Japan is, due to the already existing similarities, relatively limited. In addition, given that also in Japan entrepreneurship and the capability to generate transformational innovations are (still) rather underdeveloped, it also represents in this regard a relatively poor role model for Korea. Specifically regarding these criteria, Korean companies could learn the most from the U.S. innovation model by a wide margin, at least in principle. However, while in the U.S., the entire network of interlinked cultural, socio-political, economic and managerial context factors is strongly geared toward those transformational innovations, it should not be forgotten that the contrary is the case in Korea.

We would, therefore, propose that arguably the most interesting source of inspiration for the Korean innovation system might in fact be the least considered option among the three proposed models, namely the German system. This is particularly the case with regards to the "hidden champions" in the SME sector, and it is in the SME sector that Korean innovativeness needs to catch up most. We are by no means suggesting that the German innovation model is the "best" of the three models reviewed. However, it is in our view the best suited for Korea, given the lesser scope for inspiration from Japan, due to already existing similarities, and the largely unrealistic option to adopt the U.S. model, due to the completely opposite socio-cultural contexts in which both innovation systems are embedded. Given that the German model is, as is its Korean counterpart, more geared toward incremental innovations, Korea could build on its already existing strengths in this area (instead of building from scratch a completely different set of strengths which would go against the grain of its entire socio-cultural context).

While it is obviously not possible to reform *national culture*, Korean society should be aware of its specific cultural strengths (for example, orientation toward hard work, learning, improvement, discipline, respect) but also its competitive disadvantages arising from cultural traits (for example, risk adversity, status quo orientation, reluctance to challenge authorities, little tolerance for deviations from standards and norms). Acknowledging such culturally induced competitive disadvantages can be a first step to counteracting them, for example in the definition of educational norms, the content of curricula, etc.

In terms of the *socio-political context*, Korea's government should refrain from giving preferential treatment to a few economic sectors or a few chaebols. In the wake of the 1997–1998 Asian crisis, the ties between government and chaebols already became looser. By not promoting certain industries, the government can promote a level playing field, so that market forces can prevail in investment decisions, etc. Given the relative weakness of the Korean SME sector, the government could, however, provide structural assistance to those companies, while simultaneously avoiding overly influencing market outcomes (for example, changing regulations could help SMEs find better access to financial markets; modifying bankruptcy laws could reduce the risks entrepreneurs face; supporting training efforts could improve SME's skill set). Also the educational system could encourage more creativity and independent thinking and more financial resources should be allocated to higher education (Jones 2007).

In terms of the *economic context*, the dichotomy of large innovative companies and less innovation-oriented suppliers should make way for a more balanced corporate sector in which smaller companies also drive innovation. This could be in the form of U.S.-style start-ups which often generate transformational innovations or, possibly more likely, in the form of German-style "hidden champions" which produce, due to constant incremental innovations, high value-added products in specific niches. Korean companies should play their strengths in middle-tech manufacturing industries, in particular electronics, but improve at the same time the so far still underdeveloped service industries where probably the most catching-up can be achieved.

Regarding the *management context*, Korean companies will have to escape the trap of getting sandwiched between undifferentiated high-quality products from

Japan and undifferentiated low-cost products from China. Differentiation strategies, which target a segmented customer base and are able to reap higher profit margins due to the perceived uniqueness of products, could provide one possible solution. U.S. companies follow this strategy successfully in high-tech manufacturing and service industries and German companies in middle-tech manufacturing industries. Again, the German example might be probably closer to the Korean reality. While the corporate finance and HRM practices in Korea are all very much geared toward supporting major corporations, they should facilitate more the creation of innovations in the SME sector as well. An increasing degree of flexibility in both areas would assist in the generation of more innovations in this sector.

Our comparative analysis demonstrated that Korea can benefit from seeking inspirations from other countries' innovation systems. However, role models need to be picked carefully, since only those innovation practices can be successfully transferred which are in line with Korea's overall cultural and societal environment.

To conclude, we regard it as a key challenge for the Korean innovation system to defend its position and prevail against the two approaches squeezing Korean companies' market share: the production of undifferentiated high-quality products, represented by Japan, and the manufacturing of undifferentiated low-costs products, followed by China. As Korea will hardly be able to reduce substantially its cost base (i.e., wages), a particular threat will be a significant improvement of Chinese product quality, as Chinese wages will continue to be considerably below the Korean ones. A decisive upgrading of the Korean innovation system appears to be the only solution to this problem. This requires further continued improvements of product quality and production processes (a lesson best learned from Japan). Yet, in order to pursue more differentiation strategies which should lead to higher profit margins, inspiration from other countries should be more helpful. Such inspiration could come with regards to transformational innovations in particular from the U.S., but given the largely opposing socio-cultural contexts of both Korea and the U.S., this will probably only be a realistic option in selected niches. Further and more relevant inspirations might be sought from German companies which tend to pursue, like Korea and Japan, more incremental innovations, but in combination with differentiation strategies, unlike both countries. Given that Korea will have to improve especially the innovativeness of its SMEs, particular important lessons might be drawn from the German so-called "hidden champions."

The ultimate lesson to be learned from our investigation is that improving a nation's innovation system is not merely a question of some institutional or technical changes. Instead, this is a challenge going far beyond the narrow domain of product or process innovation or even the entire business sphere, involving ultimately the whole of society with all its complex and interrelated facets. This implies that any inspiration from abroad cannot be studied in isolation and absorbed in total but needs to be seen in its wider context and modified to fit the adopting nation's own cultural and societal environment.

References

Case J (1992) From the ground up: the resurgence of American entrepreneurship. Simon & Schuster, New York

Child J, Faulkner D, Pitkethly R (2001) The management of international acquisitions. Oxford University Press, New York

Dore RP (2000) Stock market capitalism: welfare capitalism Japan and Germany versus the Anglo-Saxons. Oxford University Press, Oxford

Economist (The) (2005) Secrets of success. The Economist 376(8443):6

Emmott B (1992) Japan's global reach. The influences, strategies and weaknesses of Japan's multinational companies. Random House, London

Galbraith JR (1982) Designing the innovating organization. Organ Dyn 10(3):4–25

Hemmert M (2007) The Korean innovation system: from industrial catch-up to technological leadership? In: Mahlich J, Pascha W (eds) Innovation and technology in Korea: challenges of a newly advanced economy. Physica, Heidelberg/New York, pp 11–32

Herbig P, Jacobs L (1998) Culture as an explanatory variable for the Japanese innovation process 1998. Cross Cultl Manag 5(3):5–30

Hill CWL (1999) International business: competing in the global marketplace. McGraw-Hill, Boston

Hofstede G (2001) Culture's consequences: comparing values, behaviors, institutions and organizations across nations, 2nd edn. Sage, Thousand Oaks

Hofstede G, Hofstede GJ (2005) Cultures and organizations: software of the mind. McGraw-Hill, New York

Imai Y, Kawagoe M (2000) Business start-ups in Japan: problems and policies. Oxf Rev Econ Policy 16:114–123

Inohara H (1990) Human resource development in Japanese companies. Asian Productivity Organization, Tokyo

Jones R (2007) R&D key to sustaining rapid growth in Korea. In: The Korea Herald (ed) Insight into Korea. Herald Media, Seoul, pp 174–188

Kang DC (2007) Korea's growth strategy: work smarter, not harder. In: The Korea Herald (ed) Insight into Korea. Herald Media, Seoul, pp 146–154

Lee Y-I (2003) From revolution to evolution: Korea's new market development. Ashgate, Hampshire

Melser D (2007) Korean economy needs new playbook for growth. In: The Korea Herald (ed) Insight into Korea. Herald Media, Seoul

OECD (2001) Enhancing SME competitiveness. In: The Bologna ministerial conference: enterprise, industry and services. OECD, Paris

Pieterse JN (2003) Die ausnahmestellung der supermacht: globalisierung auf Amerikanisch ("The exceptional position of the super power: globalization the American way"). In: Beck U, Sznaider N, Winter R (eds) Globales Amerika? Die kulturellen folgen der globalisierung (Global America? The cultural consequences of globalization). Transcript, Bielefeld, pp 93–132

Porter ME, Sakakibara M, Takeuchi H (2000) Can Japan compete? MacMillan, Houndmills/Basingstoke

Pudelko M (2006a) A comparison of HRM systems in the US, Japan and Germany in their socio-economic context. Hum Resour Manag J 16:123–153

Pudelko M (2006b) Universalism, particularism and singularism in cross-national management. Int Stud Manag Organ 36(4):9–37

Salmi J (2009) The challenge of establishing world-class universities. World Bank, Washington, DC

Schaede U (2008) Choose and focus: Japanese business strategies for the 21st century. Cornell University Press, Ithaca/London

Simon H (2009) Hidden champions of the twenty-first century: success of unknown world market leaders. Springer, Berlin

Stratos P, Panos M (2002) Bridging the start-up equity financing gap: three policy models. Eur Bus Rev 14(2):104–110

Thurow LC (1996) The future of capitalism: how today's economic forces shape tomorrow's world. Warner Books, New York

Whitehill AM (1991) Japanese management: tradition and transition. Routledge, London/New York

Chapter 9
National Innovation Systems: An Institutional Perspective

Markus K. Hödl and Jonas F. Puck

Introduction

Innovation capabilities are argued to be crucial for the growth and sustainability of organizations, industries, and nations. Consequently, a large number of existing studies analyzed the determinants of innovativeness on different levels. On the individual level, studies such as Amabile and Gryskiewicz (1989), Scott and Bruce (1994), and Abbey and Dickson (1983) investigated the effects of workplace climate on innovative behavior. Scott and Bruce (1994) found, for instance, that individual factors such as leadership, support for innovation, managerial role expectations, career stage, and systematic problem-solving style explained the innovative behavior of R&D lab employees. Moreover, Abbey and Dickson (1983) evaluated the effects of different dimensions of work climate such as autonomy, flexibility, level of reward, etc. on the innovation process in the U.S. semiconductor industry.

Other studies (e.g., Baldridge and Burnham 1975; Pierce and Delbecq 1977) analyzed antecedents of innovativeness on the organizational level. Baldridge and Burnham (1975) found, for example, empirical evidence that the size and complexity of the organization had a significant positive impact on innovative behavior and Pierce and Delbecq (1977) derived theoretical propositions for organizational innovativeness. Moreover, in their study of hospitals, Kimberly and Evanisko (1981) analyzed individual, organizational, and contextual factors and concluded that the hospital administrator's education, the organization size, and the competitive pressure in the environment were important determinants of the adoption of administrative and technological innovations.

Finally, studies such as Williams and McGuire (2010) aim towards an analysis of innovativeness on the country level by investigating the effect of culture on national innovation and prosperity. Besides those studies that focus on a single dimension, some few studies such as Steenkamp et al. (1999) and Drazin and Schoonhoven (1996) applied an integrative approach and analyzed innovativeness as a system using a multi-level analysis. Steenkamp et al. (1999) examined, for instance, the effects of personal, consumer-related, and national culture on

J. Mahlich and W. Pascha (eds.), *Korean Science and Technology in an International Perspective*, DOI 10.1007/978-3-7908-2753-8_9,

consumer innovativeness in several European countries and Drazin and Schoonhoven (1996) combined organizational context and industry-level dynamics in their study to obtain a more refined understanding of innovation phenomena.

Due to its remarkable technological development within the last four decades (Hemmert 2007), Korea has been frequently used as a research object in the existing literature, with important edited books, monographs (e.g., Kim 1997; Amsden 1992), and research articles (Hobday et al. 2004; Yoo et al. 2005; Choung and Taejon 1998) focusing on innovation in this context. In their contribution to this edited book, Pudelko and Büechl (2011) move beyond existing research and use a socio-cultural perspective to explain the development, strengths, and weaknesses of the Korean innovation system. Furthermore, by comparing the Korean environment with the environment of Germany, Japan, and the United States, they derive suggestions to further improve the innovative environment in Korea. Specifically, they analyze the socio-cultural context based on the model of Pudelko (2006) and examine influences on the cultural, socio-political, economic, and management-related context to evaluate whether the environment tends to favor incremental or transformational innovations. We believe that Pudelko and Büechl (2011) present a valuable approach that expands our knowledge in the field of national innovation systems. Without intending to downplay their contribution, we suggest that a more theoretically founded analysis of the "environment" may lead to further implications to enhance Korea's innovativeness. Overall, in our comments, we focus on alternatives and extensions of their model of the socio-cultural context and suggest dimensions derived from institutional theory and the innovation literature as alternative approaches for analyzing innovation systems.

Institutional Theory

Institutional theory has achieved considerable prominence in recent years and has been applied to different international management fields (for a review, see Kostova et al. 2008) and innovation literature (e.g., Dosi 1999; Freeman 1995; Nelson and Rosenberg 1993). In general, institutional theory highlights the role of institutions for individual and organizational behavior and decision-making. The theory's rationale is that all individuals and organizations are embedded in an institutional environment, which ultimately guides their behavior and decision-making. Moreover, as opposed to purely economic theories such as transaction cost economics, institutional theory takes into account the social construction of organizational behavior and recognizes the limits imposed by social constraints. Specifically, individuals and organizations are assumed to face pressures to conform to the institutional context and may thus behave in a way that is not consistent with efficiency criteria as implied by purely economic theories. Instead of achieving efficiency, individuals and organizations rather strive for legitimacy within their social constraints with legitimacy being "a generalized perception or assumption that the actions of an entity are desirable, proper, or appropriate within some socially constructed system of norms, values, beliefs, and definitions" (Suchman 1995, p. 574).

Therefore, individual and organizational behaviors are affected by societal expectations, which are termed isomorphic pressures (Puck et al. 2009) according to institutional theory. These isomorphic pressures are assumed to shape national innovation systems through their influence on the behavior and decision-making of individuals and organizations, which are the drivers of innovations. Thus, we suggest in line with past research (e.g., Dosi 1999; Freeman 1995; Nelson and Nelson 2002) that institutions may exert a substantial influence on national innovation systems since institutions define the "rules of the game" and affect the environment in which innovations take place. Nelson and Nelson (2002, p. 265) even state that "sophisticated scholars of technological advance always have understood the important role of institutional structures in supporting and molding efforts to advance technology." Moreover, Berry et al. (2010, p. 5) highlight the role of innovation systems as being an important institutional part of the environment, referring to innovation systems as "configurations of institutions that foster the development of technology and innovation" (Nelson and Rosenberg 1993). While institutional environments have often been viewed as a determinant of "institutional drag" in the past, scholars have increasingly begun to recognize the importance of institutional innovations for technological and economic advance (Freeman 1995, p. 8).

In order to contribute to our understanding of how institutions interact with national innovation systems, we propose dimensions derived from the institutional and innovation literature. Institutional theorists such as Scott (1994) have suggested that the institutional environment can be divided into regulative, normative, and cognitive dimensions. While the regulative dimension refers to the laws, formal rules, and enforcement mechanisms prevalent in a society, the normative dimension concerns legitimate expectations and societal practices. Finally, the cultural-cognitive dimension refers to the shared, internalized cognitive beliefs and values in a society (Peng 2003). In a similar vein, North (1990) conceptualizes the institutional environment as the interaction of formal (i.e., regulative) and informal (i.e., normative, cognitive) constraints (Cantwell et al. 2010). More recently, Berry et al. (2010) combined different strands in the institutional literature and used a more "pragmatic" approach, concluding that institutional environments mainly vary in terms of "national business systems," "national governance systems," and "national innovation systems." Based on this distinction, the authors derive nine dimensions of distance along which institutional environments or countries differ. In this article, we combine the suggestions of institutional theorists such as Berry et al. (2010) and the existing innovation literature to derive extensions and alternatives of Pudelko and Büechl's (2011) model. In the following, we thus present selected dimensions as examples, which have been suggested in both the institutional and innovation literature and which we think may entail important implications for national innovation systems. In this article, we draw on Berry et al. (2010)'s approach which distinguishes between national business, national governance, and national innovation systems. Moreover, to complement Pudelko and Büechl

(2011)'s rather qualitative study, we also present measures, which have been frequently used to quantify and operationalize these rather abstract dimensions.

National Business Systems

Financial Environment

With respect to national business systems, institutional scholars have contended that a country's financial system shapes the institutional environment as it defines the boundaries for individuals and organizations to access financial resources (e.g., Whitley 1992; La Porta et al. 1998). Therefore, individuals and organizations need to conform to these "financial constraints" in order to produce knowledge and drive innovations. Miller and Parkhe (2002) highlight, for instance, that countries differ in their financial orientation. They argue that while some countries rely on banks to source capital, other countries primarily use private capital markets for funding. While the private capital market predominantly provides access to financial resources for comparatively large firms, it is often inaccessible for entrepreneurial start-ups owing to the stringent and costly regulations on stock exchanges. From an innovation perspective, entrepreneurs are thus often dependent on the funding of external sources such as private equity or bank loans.

In the innovation literature, scholars have particularly paid attention to the development of venture capital markets, postulating that the development of venture capital markets is an important driver for national innovation systems as it permits small entrepreneurial firms to finance risky investments without providing collateral (e.g., Allen and Song 2002; Pudelko and Büechl 2011). To characterize and compare venture capital markets across several countries, Allen and Song (2002) measured the amount of venture capital investments by percentage of GDP. In a similar vein, the OECD Report "Science and Technology Statistical Compendium 2004" (OECD 2004) provides figures for venture capital investment (in% GDP) between 1998 and 2001. While Korea reaches a value of 0.20%, Germany's values amounts to 0.13%. Due to its highly developed venture capital market, the United States displays, not surprisingly, the highest value with 0.50% of all countries included in the study.

Despite being considered an important driver for national innovation systems, the development of venture capitals market might only be one of several factors that determine the favorability of a financial institutional environment since funding does not necessarily need to come in the form of private capital. In his study, Chesbrough (1999) compared, for example, the venture capital markets in the United States and Japan, discovering that while new U.S. businesses are predominantly start-ups financed through external capital, their Japanese peers often represent subsidiaries funded by parent firm's internal capital markets. While Chesbrough (1999) acknowledged that these distinct companies do not necessarily pursue the same objectives in terms of pioneering innovations, it is important to note that funding through an internal capital market may similarly address national innovation requirements.

Cultural Environment

In addition, culture has been viewed as a key determinant of the national business environment and has been extensively used to capture differences between countries (e.g., Inglehart 2004; Hofstede 1980; Whitley 1992). From an innovation perspective, culture may – through its influence on the cognitive and normative mindset of individuals and organizations – considerably influence national innovation systems as outlined by Pudelko and Büechl's model (2011). Most studies have drawn on Hofstede's (2001, p. 9), defining culture as "the collective programming of the mind that distinguishes the members of one group or category of people from another." Yet, Hofstede's work has also been subject to criticism. For instance, authors have stated that his four dimensions of culture are not exhaustive because the study was not initially geared toward identifying dimensions along which countries differ. Moreover, critics have complained that his data were confined to one company and that his questions exclusively focused on work values. Furthermore, scholars have criticized that his research framework was biased towards Western standards. Finally, his cultural values may be outdated since the study was conducted between 1967 and 1973 and since then, major cultural shifts have taken place worldwide (for a review, see Drogendijk and Slangen 2006; Erez and Early 1993; Javidan et al. 2006).

Despite these concerns, his study continues to be the largest empirical study connecting cultural orientation with observable differences between countries within a single framework. Moreover, Hofstede's dimensions of distance have been empirically validated in studies before (e.g., Van Oudenhoven 2001; for a review, see Sondergaard 1994). Furthermore, Drogendijk and Slangen (2006) found that Hofstede's measures of cultural distance outperformed the explanatory power of managerial perceptions in explaining establishment mode choices by parent firms. Overall, we contend that these findings indicate that Hofstede's (2001) approach is a reliable and valid measure for analyzing the effect of culture on national innovation systems. Pudelko and Büechl (2011)'s reasonably used Hofstede (2001)'s framework to investigate the effects of cultural differences on national innovation systems. However, the authors only included three dimensions – power distance, uncertainty avoidance, and individualism – to capture cultural differences between Korea, Japan, the United States, and Germany. We recommend incorporating the fourth dimension "masculinity" – defined as the extent to which masculine or feminine values prevail in a society – in order to grasp the full effect of culture on national innovation systems and to further improve the value of their approach.

Besides Hofstede (1980)'s study, alternative constructs may provide additional insights for further analyses. For instance, Schwarz (1994)'s value survey attempts to overcome some of Hofstede (1980)'s limitations. Compared to Hofstede (1980), Schwartz employed a larger set of cultural values along which countries are assumed to differ and theoretically derived his values from the existing literature. In the end, he identified seven interpretable cultural dimensions, which he named "conservatism," "intellectual autonomy," "affective autonomy," "hierarchy," "egalitarian commitment," "mastery," and "harmony" (Schwarz 1994; for a review, see Drogendijk and Slangen 2006). Moreover, since Schwartz conducted his survey between 1988 and 1992 (Ng et al. 2007), his data is more recent than Hofstede's. Due to Korea's remarkable development in the last four decades (Hemmert 2007), it seems reasonable to assume that cultural values have rapidly evolved and that Hofstede's values may not accurately reflect the current situation. Therefore, especially for the case of Korea but also for other emerging economies, we recommend including additional, more recent cultural value surveys such as Schwarz (1994) in order to corroborate implications for national innovation systems.

National Governance Systems

Regulatory Environment

Institutional theorists have contended that the regulatory environment or systems of governance substantially affect the institutional environment as strong forces of governance mitigate uncertainty for individuals and organizations (e.g., Henisz 2000; Henis and Williamson 1999). Drawing on prior innovation research, we believe that within the regulatory environment, the protection of intellectual property rights is a particularly important determinant of the national innovation infrastructure since the safeguarding of intangible knowledge defines the way innovations can be exploited. Furman and Heyes (2004) postulate, for instance, that the strength of intellectual property rights protection determines the quality of the national innovation infrastructure, ultimately affecting innovation productivity. Conversely, an insufficient protection of intellectual property rights weakens incentives for an individual or organization to invest in innovations. Especially in emerging economies, the enforcement of intellectual property rights often differs considerably from a country's legal regulations. Economists such as Williamson and Masten (1995, p. 130) argue that it is an illusion to believe that "the legal system enforces promises in a knowledgeable, sophisticated and low-cost way (. . .) this convenient assumption is commonly contradicted by the facts." Therefore, it is evident that the actual enforcement of laws does not always correspond to "official" regulations and guidelines in the legal system.

To provide an example, the Global Competitiveness Report 2010–2011 (Schwab 2010) ranks 139 countries in terms of their potential to protect intellectual property rights. While Germany is ranked 9th and Japan 21st, the United States and Korea are at the 24th and 44th positions, respectively. The level of intellectual property rights protection is based on executives' responses. Following the results of the report, it seems that Germany provides a highly advantageous regulatory environment for innovations to be exploited. Pudelko and Büechl (2011) differentiate in their article between incremental innovations (i.e. gradual technological advancements) and transformational innovations (i.e. break-through technological advancements). We believe that a reliable enforcement of intellectual property rights may particularly favor transformational innovations since the amortization period of these break-through advancements is often substantial. The ranking in the Global Competitiveness Report 2010–2011 (Schwab 2010) has to be viewed, however, with caution, given that the measure used in the Report is a perceptual one and may thus not reflect actual intellectual property right practices.

Political Environment

It has been well acknowledged in institutional theory that the political environment makes up an important part of the formal institutional environment that considerably determines the favorability of an investment regime (e.g., Henisz 2000; Henis and Williamson 1999; Whitley 1992). Prior research contends that political instability increases the unpredictability and transaction costs on part of individuals and organizations (e.g., Anderson and Gatignon 1986; Lu and Hebert 2005; Henisz 2000; Roy and Oliver 2009). From an innovation perspective, political volatility enhances the level of uncertainty for organizations since it may complicate the exploitation of innovations in the desired way and therefore endangers the economic viability of these investments. Similar to a weak regulatory framework, political instability may deter individuals' and organizations' incentives to invest in innovations with a long-term time horizon since it is difficult to predict how these innovations can be exploited in the future. Conversely, a high predictability of the political environment may particularly facilitate transformational innovations because these break-through advancements often tie up substantial resources and involve a long amortization period.

To compare the favorability of political environments and to assess political stability, institutional scholars have frequently used the POLCON V index (Henisz 2000). The POLCON V index is a quantitative measure of political constraints that computes how feasible a change in government policy is likely to occur. Specifically, POLCON identifies the number of independent government branches with (de jure) rights of veto over policy change. Moreover, the index accounts for the extent of alignment across government branches and the extent of preference heterogeneity within each branch. The final index ranges from 0 to 1, where 1 represents maximum political discretion and 0 maximum political constraints. For example, Korea displays a value of 0.74 in 2007, while Japan reaches a value of 0.75 in the same year. The values for Germany and the United States both stood at 0.85 (Henisz 2000). The index suggests that the political environment in all four countries is comparatively favorable, with Germany and the United States being slightly ranked ahead of Korea and Japan. Overall, based on this data, it appears that the political stability prevalent in these countries substantially contributes to the quality of their innovation infrastructure.

National Innovation Systems

Besides the financial, cultural, regulatory, and political environment the innovation literature has suggested many additional dimensions, which influence national innovation systems. Scholars have, for example, used the economic development (often measured in GDP per capita) to characterize national innovation systems (e.g., Furman and Heyes 2004). However, economic development has – similar to the number of patents – been predominantly viewed as an outcome variable given that it displays "the extent to which ideas are embodied in valuable goods and services" (Furman and Heyes 2004, p. 15) even though not all values produced necessarily stem from innovations. In addition, Furman and Heyes (2004) included the total amount of R&D expenditures and the amount of public spending on secondary and tertiary education in their model to assess the national innovation infrastructure. Moreover, Freeman (1995) provided an historical perspective on national innovation systems, viewing education systems, industrial relations, and research institutions, among others, as crucial drivers for national innovation systems. Specifically, besides different R&D measures he used, for example, the percentage of people in third-level education, the percentage of engineering students in the population, growth rates in electronics, and telephone lines as quantitative factors to compare national innovations systems. Chesbrough (1999) argued that the technical labor market, buyer-supplier relationships as well as the development of venture capital markets are among important institutional factors that explain technological change. Furthermore, in their study on institutional drivers of innovation in the information technology industry, King et al. (1994) concluded that governments are the most powerful institutions in shaping national innovation systems as government entities heavily intervene in innovation processes. They argue that while the U.S. government has, for example, deliberately intervened in the nuclear power industry, the Japanese government has particularly focused on the electronics industry. Finally, given that the importance of clusters and industries considerably differs among countries, previous studies have also taken the industry- or cluster-specific dimensions of national innovation systems into account. For example, Furman and Heyes (2004) incorporated the percentage of R&D funded by the private sector and Pudelko and Büechl (2011)'s investigated the importance of industrial sectors and the corporate environment in the respective countries.

Conclusion and Implications

Combing institutional theory and research on innovation, we aimed to derive extensions for and alternatives to Pudelko and Büechl (2011)'s model. Specifically, we proposed selected dimensions, which have been suggested in the literature and which we deemed important for analyzing national innovation systems.

Furthermore, to complement Pudelko and Büechl (2011)'s qualitative approach, we also presented measures, which have been frequently used to quantify and operationalize these rather abstract dimensions. Altogether, we found that there is a wide array of factors, which have been suggested by prior research to be important determinants of national innovation systems. However, to advance theory and derive implications, scholars appear to disagree whether one should focus only on a few dimensions or opt for a comprehensive approach when analyzing the relationship between institutional factors and innovation systems. Due to the richness of factors, Chesbrough (1999, p. 455) decided in his study to "strive for parsimony" by focusing only on selected dimensions to derive clear and interpretable implications for both theory and practice. While this approach addresses the criticism that many dimensions suggested in the literature to characterize institutional environments are not mutually exclusive, it certainly runs the risk of neglecting some important explanatory factors. Pudelko and Büechl (2011)'s approach, on the other hand, employed in its model a comprehensive approach and investigated the respective countries' environmental context at large. While the authors take a large set of factors into account, we believe that many of their dimensions are not mutually exclusive, given that culture, for instance, influences the economic or management-related context that individuals and organizations are embedded in and vice versa. Apart from this difficult trade-off, we recommend future research to combine quantitative and qualitative factors to assess and compare national innovation systems. Freeman (1995) argued, for example that both quantitative and qualitative indicators need to be controlled when characterizing national innovation systems. He further reported that while the former Soviet Union heavily invested in R&D (predominantly in the military and space industry) in separate research centers, the country failed, however, to ensure links between these research centers and corporate organizations on the one hand and manufacturers and users on the other hand – all factors which were deemed important for the development of national innovation systems. Therefore, the sole reliance on R&D as a quantitative indicator to assess national innovation capacity would have hidden the fact that the former Soviet Union put – compared to other countries – insufficient emphasis on the diffusion of innovations into the civil economy (Freeman 1995).

Moreover, due to the comprehensive set of different factors, we recommend scholars to base their selection of criteria on theoretical models. In particular, we showed that institutional theory provides a valuable framework for analyzing the interaction of the institutional environment with national innovation systems. Therefore, we believe that a more theoretically founded analysis of the "environment" may lead to further implications for the improvement of national innovation systems. Finally, another way forward in institutional theory and the innovation literature is to look beyond how institutional factors and changes thereof influence national innovation systems. Institutional scholars recently argue that in view of the growing importance of MNCs, we need to recognize that these global organizations may actively trigger institutional change and thus ultimately influence the development of national innovation systems (e.g., Cantwell et al. 2010). Therefore, we argue in line with these scholars that an analysis of how MNCs (as key drivers of innovation) and the institutional environment co-evolve would provide a more comprehensive understanding of national innovation systems. From a practical perspective, this approach would also point out to policy makers that national innovation infrastructures can only be developed when taking the agents – the MNCs – into account. Overall, we believe that an analysis based on established frameworks such as institutional theory may further expand the implications of Pudelko and Büechl (2011)'s model and provide a valuable starting point for improving the development of national innovation systems.

References

Abbey A, Dickson JW (1983) R&D work climate and innovation in semiconductors. Acad Manage J 26(2):362–368

Allen F, Song W-L (2002) Venture capital and corporate governance. Wharton Working Paper Series, Wharton Financial Institutions Center

Amabile TM, Gryskiewicz ND (1989) The creative environment scales: work environment inventory. Creat Res J 2(4):231–253

Amsden AH (1992) Asia's next giant: South Korea and late industrialization. Oxford University Press, New York

Anderson E, Gatignon H (1986) Modes of foreign entry: a transaction cost analysis and propositions. J Int Bus Stud 17(3):1–26

Baldridge JV, Burnham RA (1975) Organizational innovation: individual, organizational, and environmental impacts. Adm Sci Q 20(2):165–176

Berry H, Guillen MF, Zhou N (2010) An institutional approach to cross-national distance. J Int Bus Stud 41(9):1460–1480

Cantwell J, Dunning JH, Lundan S (2010) An evolutionary approach to understanding international business activity: the co-evolution of MNEs and the institutional environment. J Int Bus Stud 41(4):567–586

Chesbrough HW (1999) The organizational impact of technological change: a comparative theory of national institutional factors. Ind Corp Change 8:447–485

Choung J-Y, Taejon E (1998) Patterns of innovation in Korea and Taiwan. Eng Manag 45(4):357–365

Dosi G (1999) Some notes on national systems of innovation and production, and their implication for economic analysis. In: Archibugi D, Howells J, Michie J (eds) Innovation policy in a global economy. Cambridge University Press, Cambridge, pp 35–48

Drazin R, Schoonhoven CB (1996) Community, population, and organisation effects on innovation: a multilevel perspective. Acad Manage J 39(5):1065–1083

Drogendijk R, Slangen A (2006) Hofstede, Schwartz, or managerial perceptions? The effects of different cultural distance measures on establishment mode choices by multinational enterprises. Int Bus Rev 15(4):361–380

Erez M, Early CP (1993) Culture, self-identity and work. Oxford University Press, New York

Freeman C (1995) The 'national system of innovation' in historical perspective. Cambridge J Econ 19(1):5–24

Furman JL, Heyes R (2004) Catching up or standing still? National innovative productivity among 'follower nations'1978–1999. Intellectual property research institute of Australia working paper, 8(4)

Hemmert M (2007) The Korean innovation system: from industrial catch-up to technological leadership? In: Mahlich J, Pascha W (eds) Innovation and technology in Korea: challenges of a newly advanced economy. Physica Verlag, Heidelberg/New York

Henis WJ, Williamson OE (1999) Comparative economic organization – within and between countries. Bus Polit 1(3):261–277

Henisz WJ (2000) The institutional environment for economic growth. Econ Polit 12(1):1–31

Hobday M, Rush H, Bessant J (2004) Approaching the innovation frontier in Korea: the transition phase to leadership. Res Policy 33(10):1433–1457

Hofstede G (1980) Culture's consequences: international differences in work-related values. Sage, Newbury Park

Hofstede G (2001) Culture's consequences: comparing values, behaviors, institutions, and organizations across nations, 2nd edn. Sage, Thousand Oaks

Inglehart R (2004) Human beliefs and values. Siglo XXI, Madrid

Javidan M, House RJ, Dorfman PW, Hanges PJ, Sully de Luque M (2006) Conceptualizing and measuring cultures and their consequences: a comparative review of GLOBE's and Hofstede's approaches. J Int Bus Stud 37(6):897–914

Kim L (1997) Imitation to innovation: the dynamics of Korea's technological learning. Harvard Business School Press, Boston

Kimberly JR, Evanisko MJ (1981) Organizational innovation: the influence of individual, organizational, and contextual factors on hospital adoption of technological and administrative innovations. Acad Manage J 24(4):689–713

King JL, Kenneth VG, Kraemer L, McFarlan FW, Raman KS, Yap CS (1994) Institutional factors in information technology innovation. Inf Syst Res 5(2):139–169

Kostova T, Roth K, Dacin T (2008) Institutional theory in the study of MNCs: a critique and new directions. Acad Manage Rev 33(4):994–1006

La Porta R, Lopez-de-Silanes F, Shleifer A, Vishny RW (1998) Law and finance. J Polit Econ 106(6):1113–1155

Lu JW, Hebert L (2005) Equity control and the survival of international joint ventures: a contingency approach. J Bus Res 58(6):736–745

Miller SR, Parkhe A (2002) Is there a liability of foreignness in global banking? An empirical test of banks' x-efficiency. Strateg Manag J 23(1):55–75

Nelson RR, Nelson K (2002) Technology, institutions, and innovation systems. Res Policy 31:265–272

Nelson RR, Rosenberg N (1993) Technical innovation and national systems. In: Nelson RR (ed) National innovation systems. Oxford University Press, New York, pp 3–21

Ng S, Lee JA, Soutar G (2007) Are Hofstede's and Schwartz's value frameworks congruent? Int Mark Rev 24(2):164–180

North DC (1990) Institutions, institutional change and economic performance. Cambridge University Press, Cambridge

OECD (2004) Science and technology statistical compendium. Meeting of the OECD committee for scientific and technological policy at ministerial level. www.oecd.org/dataoecd/17/34/23652608.pdf. Accessed 22 Sept 2010

Peng MW (2003) Institutional transitions and strategic choices. Acad Manage Rev 28(2):275–296

Pierce JL, Delbecq A (1977) Organisation structure, individual attitudes and innovation. Acad Manage Rev 2(1):27–37

Puck JF, Holtbrügge D, Mohr AT (2009) Beyond entry mode choice: explaining the conversion of joint ventures into wholly owned subsidiaries in the People's Republic of China. J Int Bus Stud 40(1):388–404

Pudelko M (2006) A comparison of HRM systems in the US, Japan and Germany in their socio-economic context. Hum Resour Manag J 16:123–153

Pudelko M, Büechl J (2011) Three potential role models for the Korean innovation system: USA, Japan and Germany. In S. Mahlich and W. Pasoha (eds.), Koreah Science and Technology in an international prespeotive. Springer-Verlag Berlin Heidelberg 2011

Roy J-P, Oliver C (2009) International joint venture partner selection: the role of the host-country legal environment. J Int Bus Stud 40(5):779–801

Schwab K (2010) Global competitiveness report 2010–2011. World economic forum. http://www.weforum.org/en/initiatives/gcp/Global%20Competitiveness%20Report/index.htm. Accessed 22 Sept 2010

Schwarz SH (1994) Beyond individualism/collectivism: new cultural dimensions of values. In: Kim U, Triandis HC, Kagitcibasi C, Yoon G (eds) Individualism and collectivism: theory, method, and applications. Sage, Thousand Oaks, pp 85–119

Scott WR (1994) Institutions and organizations. Sage, Thousand Oaks

Scott SG, Bruce RA (1994) Determinants of innovative behavior: a path model of individual innovation in the workplace. Acad Manage J 37(3):580–607

Sondergaard M (1994) Hofstede's consequences: a study of reviews, citations, and replications. Organ Stud 15(3):447–456

Steenkamp J-BEM, Hofstede F, Wendel M (1999) A cross-national investigation into the individual and national cultural antecedents of consumer innovativeness. J Mark 63(2):55–69

Suchman MC (1995) Managing legitimacy: strategic and institutional approaches. Acad Manage Rev 20(3):571–610

Van Oudenhoven JP (2001) Do organizations reflect national cultures? A 10-nation study. J Intercult Relat 25(1):89–107

Whitley R (1992) Business systems in East Asia: firms, markets, and societies. Sage, London

Williams LK, McGuire S (2010) Economic creativity and innovation implementation: the entrepreneurial drivers of growth? Evidence from 63 countries. Small Bus Econ 34(4):391–412

Williamson OE, Masten SE (1995) Transaction cost economics volume I: theory and concepts. Edward Elgar, Hants

Yoo Y, Lyytinen K, Yang H (2005) The role of standards in innovation and diffusion of broadband mobile services: the case of South Korea. J Strat Inf Syst 14(3):323–353

Chapter 10
The Role of the State in Qualitative Growth – A Consideration of Regional Innovation Clusters in Gangwon Province (South Korea)

Bernhard Seliger

Introduction

The search for qualitative growth has been a preoccupation of Korea and Korea-related economic research not only in the last years, but even long before the current crisis and indeed the Asian crisis of 1997 and 1998. The spectacular growth of Korea, like its fellow "tiger economies", triggered a debate about the reasons for this growth. Growth accounting established that quantitative growth played a large role in East Asia's economic miracles. However, the extent of this role has been widely debated. Was it "inspiration or perspiration", as Krugman put it, what fueled the growth of the East Asian NIC?[1] While qualitative growth is understood here in the more narrow economic sense, the wider definition of quality components of growth as systems that combine economic, social and ecological factors in order to achieve more sustainable and inclusive growth, as it is used in the debate on inclusive development, also played an important role in Korea's growth history. In Korea, this has been seen as both a social problem, particularly before democratization, and a spatial problem. Biased regional policies have long been blamed for alleged income disparities under past autocratic governments.[2] The concentration of people and resources in the greater Seoul Metropolitan Area, which accounts today for nearly half of the South Korean population, has led to a number of pressing policy issues, among them congestion, environmental problems and real estate speculation. These issues in turn have driven the policy debate regarding the study of qualitative growth in Korea.

A core question is if policy makers can do more than acknowledge these problems and search for solutions; can they find remedies to change the situation. Looking at the "wish list" from policy makers, growth should ideally be inclusive,

[1] Krugman (1994), p. 70.

[2] However, Kim (2003) showed that these effects are only imagined; they are not mirrored in regional income disparities.

J. Mahlich and W. Pascha (eds.), *Korean Science and Technology in an International Perspective*, DOI 10.1007/978-3-7908-2753-8_10,

regionally balanced, sustainable, "clean and green", and, naturally, should bring a high return from the public and private investment. However, some of these desired features are potentially conflicting, e.g., the wish for growth that is inclusive and green. (Market inclusion can sometimes only be achieved through simple technologies which rely on cheap labour, but these are not green technologies.) The wish for regionally balanced growth and high returns is also contradictory (Letting growth centers lead ahead of lagging peripheral regions would bring higher returns). In other cases, the desire for such policies simply looks unrealistic, e.g., when comparing policy makers' expectations to the actual numbers in regard to the job effects of "green growth". In particular, the question how to make concepts of qualitative growth operational for economic policy purposes is tricky. And this is a question, which has to be answered on the national policy making level as much as on the regional or local level, all the more so, since otherwise policies will easily result in beggar-thy-neighbour situations, where the attraction of investment is a zero-sum game.[3]

Innovation policies are among the most attractive economic activity in terms of the wish list cited above. The concentration of innovative activity will benefit companies, regions and people. Innovative activity promises high returns and skilled, sustainable jobs. The concentration of innovative activity in clusters as prominent as Silicon Valley has led to numerous attempts to recreate similar development in other regions. Therefore, clusters have become a well-researched object of economic policy making. However, the problem remains how to translate the recognition that clusters sharply augment qualitative growth into concrete policy actions; how to promote or even design clusters. And here, an important policy paradox arises: while clusters, as they were originally defined, naturally occur in places which fulfill preconditions for fast growth in terms of resources – a large and available pool of trained specialists or research institutes able and willing to cooperate with industry – regional economic policy makers follow a quite different agenda. For them, clusters serve a balancing function and should ideally be placed in locations disadvantaged in terms of human resource development, technology, etc. Can government incentives be large enough to lure companies to structurally weak areas and make them economically viable? Moreover, do regional governments that subsidize cluster development realize sufficient returns? Will parachuted clusters fulfill their function and survive?

This paper discusses these problems in relation to regional development strategies in South Korea, and particularly, in Gangwon province. The second section explores the economists' case for (and against) cluster policies. Section "Quantitative vs.

[3]A different though related problem is that of the alleged "race to the bottom". Deregulation around the world, some critics maintained, would result in a competition to offer the lowest standards. However, though there are highly visible and widely discussed single indicators for the quality of a location (e.g. certain tax rates), investors decide not on the basis of one such indicator, but on a complex "bundle of institutions", which means that the simple race to the bottom, as forecast in partial models, in reality has not occurred.

Qualitative Growth – the Korean Experience and Its Regional Dimension" looks at the Korean experience with regional growth. The investment environment of Gangwon province will be discussed in section "Gangwon Province Development Planning," followed by a conclusion (section "Conclusion)".

Theoretical Considerations

The Problematic Task of "Picking the Winners"

State industrial policy has to cope in general with the problem that "picking the winners", i.e., supporting those industries which emerge as industrial champions (in the European terminology) to lead economic growth processes, is an attractive, but very complex and often impossible task for policy makers.[4] In theory, several reasons have been forwarded for the need to support such champions, among them, in particular, the "infant industry argument": young, but strategically important, industries should be protected until they grow to a size where economies of scale allow them to compete with established competitors – and, in the framework of new economic geography, the idea of "rent shifting" in incomplete competition.[5] For innovation policy, this means that the state should identify strategic sectors and should nurture companies in these sectors, e.g., through subsidization of R&D, exclusion from competition policy rules, etc. In practice, this policy has to cope with problems that result in failure more often than not.

The first problem has already been mentioned in a slightly different context by Hayek as early as 1937 in his famous article on economics and knowledge. Competition produces relevant market data, which are ex ante unknown and which cannot be centralized. As Hayek says, competition works as a discovery procedure: those goods which meet the preferences of consumers, those innovations which are marketable, are discovered, others are rejected. Given this, subsidizing firms becomes the exact opposite of competition: A central bureaucratic institution must decide if ex ante unknown innovations or innovative sectors are to be subsidized or not. So from an informational point of view, the strategy of "picking the winners" is highly questionable.

[4]For a discussion from the point of view of evolutionary economics see Boschma (2004). For a theoretical model, see Den Butter and Jo (2009).

[5]A complete discussion of industrial policy development and problems cannot be carried out here. But for the problems of "picking the winners" in the European context, see Maincent and Navarro (2006), for an economic policy perspective see Rodrik (2004). For the ongoing policy debate, see the special issue of the Journal for Industry, Competition and Trade, Special Issue on the Future of Industrial Policy, Vol. 7, 2007.

But there is a second and related problem: which projects are to be subsidized by bureaucrats? The general conclusion is that the bureaucrats' *maximand* is not equal to the private entrepreneurs' *maximand*. The latter tries to equal marginal gains and costs (risks from investment) to maximize utility. For the bureaucrat, the potential gains of an investment of subsidies are relatively uninteresting. Successful investment of tax money can perhaps (but not surely) help them in their careers. On the other hand the risks of investments are interesting in a special way: the risk of a (technical) failure are as uninteresting as the gains, since the bureaucrats' future success is not dependent on the profits from the investment. But if there is failure, the public will examine if the bureaucrats spent the money wisely. Consequently, bureaucrats tend to be conservative in their spending. This means that small, risky, newly established firms, even if they are highly innovative, are not the bureaucrats' first choice for R&D subsidies. This is the reason, why industrial policy in many cases sets aside additional resources for small and medium enterprises for R&D; most of the money flows to the large firms. If their investments fail, the bureaucrats are still on sure footing: The large firm's reputation remains as solid as it was before the subsidies were awarded. For small firms, if they fail they are only risky, without reputation. So there is a bias to subsidize large, well-known and potentially less innovative firms. A second factor working in large firms' favor is the influence of lobbying. Only large, established companies have the resources to steer industrial policies toward their interests while they are still in the planning phase.

There are two more implications arising from bureaucrats' conservativeness: one is related to their own lack of information on "innovations". Only when advancements are already successfully running in other countries, especially in the US and Japan and other leading economies, are they seen as innovative. The other implication is that bureaucracy does not like to admit failure, which leads to cost-plus contracts. In other words, sunk costs are not considered as sunk costs but as decisive costs. Consequently, industries can be awarded ever new subsidies from bureaucrats who are reluctant to declare a project that has already received a considerable amount of state funds a failure, even if the whole project does not seem to be working.

With industrial policy facing these problems, it is even more problematic to change the allocation of innovative activity geographically. Here, additional distortions regarding economic decision-making exist. First, the "cluster policy paradox" mentioned above means that state-designed clusters might be allocated to places with particularly weak economic foundations. Second, the designation of clusters and related incentives (for example, infrastructure development, tax breaks, etc.) are not free, but costly, and the costs are an additional burden on enterprises elsewhere, in the form of asymmetric regulation unfavourable to other regions or in terms of higher taxes.

So, from a nation-wide perspective three challenges arise:

- The innovation cluster strategy has to define the right industries (government failure can arise from failure to pick the winners).
- The innovation cluster strategy has to select the right location (government failure can arise from failure to pick the most advantageous location).

– The total costs of the innovation cluster has to be taken into account (government failure can arise from too costly regional planning initiatives, including the additional burden put on taxpayers).

In short, a market allocation might be superior to government-led allocation and there is no a priori case for the promotion of innovative activity. This does not discard all such activities, but there must be a good justification for cluster policies, and there might be certain activities which are more likely to lead to the described potential for government failure.

Analysing Clusters and Cluster Policies

Since the seminal analysis of Porter (1990) the role of clusters has been prominent in the spatial analysis of the growth processes. Industry clusters are sectoral, geographical, horizontal or vertical agglomerations of businesses and related institutions, like research institutes and universities. As agglomerations, clusters profit from localization and urbanization economies, i.e., increasing returns to scale in terms of localization (e.g., joint use of a pool of experienced workers and knowledge spill-overs), forward and backward linkages (related to the effects of agglomeration on the price index and home market), and urbanization economies (related to economies of scale realizable in urban areas).[6] In clusters, the network effects of increasing economies have been named the cluster effect. It should be noted that industry concentration alone does not make a cluster and that there might also be the case of over-concentration, i.e., an inverse U-shaped curve between the degree of industry concentration and innovative performance.[7]

The contribution of clusters to growth, not only in a core region but also the surrounding periphery, has been highlighted by many studies. Among the important studies are those from the "Porter school of thought" in Harvard, where intensive efforts at cluster mapping and the cluster meta-study resulted in the Cluster Initiative Green Book (Sölvell et al. 2003), which analyses 250 clusters in OECD member countries; the Global Cluster Initiative Survey of 2005, which analyses 450 clusters worldwide; and the study on Cluster Initiatives in Developing and Transition Countries (Ketels et al. 2006), which looks at 1,400 clusters in developing and transition countries. Policy aspects became more and more important, as in the OECD cluster policy study of 2007 (OECD 2007). The European Union in 2008 established a so-called European Cluster Policy Group with the aim to make recommendations on how to better design cluster policies in the Community, to assess international trends in cluster development, to identify future challenges for cluster policies in response to globalisation, to explore tools for the removal of

6For an introduction see Fujita et al. (1999).

7Hornycha and Schwartz (2009) show this for East Germany.

existing barriers to transnational cluster cooperation and to analyze complementarities between the main Community level policies and the financial instruments that support clusters.

From the beginning of Porter's analysis, the value of clusters as an instrument of economic policy has also been discussed. While the original analysis of clusters focused on growth regions which emerged spontaneously and grew historically, like Silicon valley or the car manufacturing cluster in the USA and the chemical and metal clusters in the Rhine and Main valleys in Germany, respectively, clusters today are an important instrument of growth and innovation policy. This is not only the case in OECD member countries: successful examples of clusters in developing countries, like Bangalore in India, have led to cluster initiatives for developing and transformation countries. However, cluster policies face the same criticism, industrial policies faced from the beginning: the active promotion of clusters by public policy initiatives like the change of incentive structures for certain types of investment, the public provision of supporting infrastructure or the creation of specific rules (e.g., relaxation of environmental regulations) faces the problem of limited knowledge within the bureaucracy. This is even more the case, since often regional policy considerations favour locations for new clusters according to political rather than business logic. So the question remains if successful clusters can really be created "from above" by political decisions.

From Cluster Analysis to the Analysis of Regional Innovation Systems

From a political economy point of view, clusters are attractive as tools of economic policy, since they seem to allow for concrete political activities to support regional economic development, they promise "future industries" for regions lagging behind and burdened with old industries and they result in politically highly visible outcomes (a cluster, an industrial park, etc.). New Economic Geography, which analysed regional competition policies in terms of imperfect competition processes, seemed to enlarge the policy options available, compared to a world of perfect competition (Potter 2009). However, as the economics of innovation points out, innovation is a complex, path-dependent process characterized by the interdependence and interaction of a variety of heterogeneous agents able to learn and react creatively with subjective and procedural rationality (Antonelli 2009). Therefore, the isolated view of clusters easily leads to policy failures, if complexity, feed-back processes, and path dependency are not taken into account.[8] To avoid a simplistic

[8]In this respect, it is also important that clusters themselves, like companies, undergo life-cycles a process of emergence, growth, decline and renewal, which is determined, among others, by the technological heterogeneity of firms and the relative absorptive capacity of firms for new knowledge. (Menzel and Fornahl 2010). Therefore, cluster policies must also be tailored to fit the relevant phase in the life-cycle of the cluster.

view of clusters, clusters have been analysed in relation to and as part of the relevant regional innovation system. Regional innovation systems describe the interaction of actors in the innovation process, the role of institutions and policy.[9] Among the actors in the regional innovation system, there is a division of labour between the government, which provides innovation-conducive institutions and finance for research; foreign investors, which provide knowledge and technology transfer, management skills as well as links to foreign markets; regional and national scientific institutions, providing the scientific-technological basis; and human resource development and the regional industry, providing entrepreneurship. Given such a set-up for successful cluster strategies, certain policy tasks on various levels can be identified.[10] As a facilitator of growth and innovation processes, there is an important role for the government (national and regional), however, the limitations to their role have to been kept in mind.

On the national level, globalisation and increasing competition have led to efforts around the world to provide certain services conducive to growth, e.g. investor one-stop-services, which today are a standard in most countries. Technological change and a more mobile, ubiquitous infrastructure are also pointing in a direction, where regions, which formerly might have met higher obstacles, might be more successful at attracting investment. Where human resources are a key point in the success of clusters, policies should probably focus on providing an attractive environment for highly-skilled workers (regarding the living environment, housing, education, culture) instead of changing the industry structure. In these fields the scope for government policies is particularly wide. So, there is a case for cluster policies and there is a case for regional cluster strategies. However, these cannot be derived globally (one size fits all), but have always to be seen as part of the concrete regional innovation system and challenges.[11] The next section will discuss the specific Korean growth experience and its spatial implications.

Quantitative vs. Qualitative Growth – the Korean Experience and Its Regional Dimension

The Korean economy is known to be among the global showcases for successful export-oriented catch-up, exhibiting one of the highest growth rates worldwide between 1960 and 1990. This is not the place to discuss extensively the Korean and

---

[9]See the overview of Cooke (2001) and the critical review of Doloreux and Parto (2005).

[10]Ramstad (2009) points out that these policies, while today mainly focusing on innovation in science and technology, should also include organisational innovations. Similarly, Rosiello (2008) stresses that besides technological factors and independent from them, knowledge, interconnected with questions of industrial development, economic viability, public safety and social acceptability, contributes to the emergence of innovative products and processes.

[11]For example, for some emerging markets like China incentives for the return of migrated scientists played an important role; see Prevezer (2008).

East Asian growth experience, but some remarks are warranted. Today, there is a wide consensus that the Korean growth "miracle" can be explained by a mixture of free enterprise and a hard-working and -saving population plus a state policy conducive to growth, mainly through export orientation, but also through a policy of selective and gradual liberalization. While certainly not neo-liberal in the sense the word was employed in the 1990s, Korean policy was far from the failed interventionism and import-substitution employed elsewhere, e.g. in many Latin American states and in Korea itself before the 1962 military coup. Openness for technology imports from Japan and acceptance of the "dictates" of the world markets complemented this policy. It was one of the major cases made by the World Bank in its 1993 study "The East Asian Miracle". Among the findings of the study important for this paper were the importance of institutional and macro-economic soundness. Stability, among other things, guarantees a higher likelihood that investments pay off and is a precondition in particular for risky investment processes. Due to a relatively equal income distribution and productivity growth in agriculture, inclusive growth was reached, while at the same time increasing the middle classes – another factor increasing stability. Having a relatively efficient, less corrupt public administration was not essential for reducing business costs, and the system proved sufficiently strong enough to cope with obvious failures, such as the aborted Heavy and Chemical Industry Drive of the late 1970s. The state role in providing infrastructure and long-term finance through a state-led banking system was crucial as investment incentives. While this investment was not always efficient and sometimes outright political, market orientation, in particular world-market orientation, corrected quasi-automatically domestic policy failures.

Box 10.1: Five Important Factors in East Asian Growth Processes
Factors contributing to the East Asian Growth Miracle

1. Getting the basics right – macroeconomic soundness.
2. Reducing social tensions through relative equal income distribution.
3. Productivity growth in agriculture as a precondition for inclusive growth.
4. Relative efficient public administrations (relatively low degrees of corruption, and flexibility with regards to policy goals).
5. Market-friendly interventions, in particular export orientation.

Source: adapted from World Bank (1993).

This account of the East Asian growth experience was originally very much contested. Critics in the 1970s saw market-orientation as a submission to global capitalism and many did not believe in sustainable growth processes in East Asia. However, the 1980s and 1990s provided overwhelming evidence for the success of the East Asian strategies and by the time, the World Bank study was published, belief in the East Asian growth experience became an orthodoxy; admired, copied and no longer widely questioned. Like every orthodoxy, it still had its critics. Paul

Krugman (1994) in his "Myth of Asia's Miracle", saw these growth processes mainly as a result of increased factor input, in particular labour, but also capital, through high savings rates. The succinct formula of "perspiration instead of inspiration" denotes a growth process, which is, like in Singapore, almost entirely based on factor inputs, and thereby the same kind of growth experienced in the early phases of centrally-planned economies. However, Krugman's formulation that Singapore exhibited growth "Stalin would have proud about" goes too far. Indeed, while the increased input of factors played an important role for growth, it did have distinctly different characteristics than Soviet growth. First, mobilization of increased resources itself was an important fact, since it was not forced recruitment of resources like in centrally-planned economies, but rather the successful growth of new industries, which brought labourers from the less productive agricultural sector or previously dormant households into the factories. Second, investment, which was possible from voluntary savings, was basically free (though there was some intervention in credit allocation). Most notable, industries were highly adaptive to world markets and productivity was increased through the free import of machines and tools. The capital stock was built up according to market allocation and this is the most important difference from growth processes in centrally-planned economies. High reinvestment rates also provided Korea with a very young capital stock. Krugman's critique, which was indeed important, also predicted some problems likely to rise to the point of an Asian crisis. However, Krugman did not foresee the rapid technological catching-up process that had begun even before the East Asian crisis and even accelerated afterwards. During this time, Korea became not only an OECD member state, but also a leading technology exporter.

While Korea successfully sustained growth over a long period of time, it was a highly unequal process. Clusters of economic activity and also of research activity emerged, for example in the shipbuilding industry and automotive industry in the southeast. These clusters mostly had an industrial core built by one of the country's large conglomerates. Small contractors grew up and settled down sometimes wholly dependent on one or a few of these chaebols. More important, metropolitan Seoul (Seoul, Incheon and the surrounding Gyeonggi province) grew to become the all-dominating center of the country. While it comprised around 20% of the Korean population in 1970, it approached 50% in the 2000s. This growth came at a cost: the OCED (2006) estimated that congestion in the capital region shaved 3–4% off of Korean GDP. Regional policies tried to remedy inequalities in development, but the attempts were often politically-focused, due to the historic and intense divide between the peoples of the southeastern and southwestern provinces. Traditionally, there has long been fierce competition for power among factions from these regions, the southeastern region, from where authoritarian ruler Park Chung-Hee hailed, seemed to have been preferred by central government policies. Only the election of Kim Dae-Jung in 1998 changed this. But the real regional split was not between east and west, but rather between the capital region and the rest of Korea. This split was very difficult to address. Though a number of initiatives have taken place in the last decade and a half, the trend for centralization in the capital region continues. Decentralization policies included the relocation of public companies, among them those generating sizeable taxes and employing a large number of white-collar and highly qualified employees, like the state electricity company KEPCO to Gwangju. Regional free enterprise zones or free economic zones had been in operation since the 1960s, but they gained new importance and attractiveness as business hubs in the last decade. The planned relocation of the capital from Seoul to South Chungcheong province was the most radical of the decentralization plans, but if failed due to political and judicial opposition. Now, even the downsized relocation of certain administrative offices to that new Sejong City site is in jeopardy, not least because of a poor and clearly politically-dominated planning process. The major criticism is that the selected area is much too close to Seoul (160 km) and was chosen for political reasons, since the respective governments hoped to win important swing votes from Chungcheong residents. Also, regulatory approaches, for example a non-commercial green belt around Seoul, have been enacted to try and push business activities away from the capital.

Among the most relevant issues of the centralization-decentralization dynamics is the problem of university access. In the Korean system, a national admission test is the primary determinant for university enrollment. Secondary and cram schools in Seoul were considered the only way to access the top-ranked universities, which also happen to be located in Seoul (in particular, the big three, Seoul National, Korea and Yonsei). This is one reason the capital region holds such a strong attraction despite the high living and real estate costs. Companies adjust to this fact by staying in the capital region, even though production might be cheaper elsewhere. To counter this strong educational centralization, the government has tried for two decades to foster new well-respected universities in the outlying provinces by massive central government investment. However, while the NURI (New University for Regional Development) initiative funded considerable modern facilities, it has not yet shifted the perception that graduation from major Seoul-based universities is a precondition for career success.

With few exceptions, major Seoul universities have opened campuses out of Seoul. One such case is the Yonsei medical faculty in Wonju, site of the medtech cluster. Professors at this and other provincial campuses, which are often in the vicinity of Seoul, usually commute to teach their classes, leaving the chances of building up academic life there rather poor.

While decentralization policies have not been very successful, centralizing tendencies came not only from population trends and business decisions, but equally from new government initiatives. One such initiative was the attempt to establish Seoul as the business, commercial, R&D and financial hub of East Asia (Seliger 2004). Though this policy had its own challenges and is no longer pursued vigorously, it again concentrated resources and attention on the capital. The last administrations also saw the proliferation of new decentralization committees and agencies, among them the high-profile presidential commission on balanced growth. Under the Lee Myung-bak administration, since 2008, the strategy of "new regionalism" includes the creation of larger regional units able to compete with the strong capital region. While this has not lead yet to the merging of provinces, this option remains open. Five new "mega-regions" able to compete

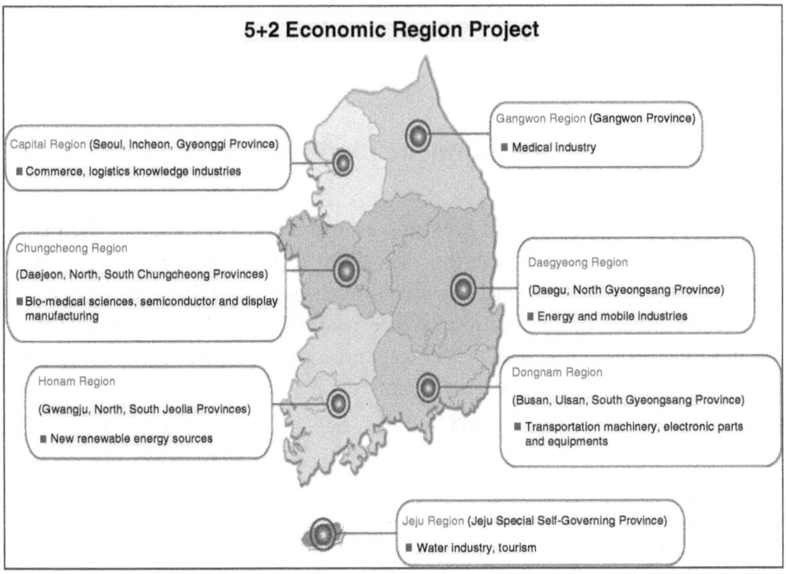

Fig. 10.1 The 5+2 economic region project

with the capital region with a population of between five and ten million inhabitants each have been defined. The plan also includes the two relatively sparsely populated provinces of Jeju and Gangwon (see Fig. 10.1).

At the same time, different industrial bases were identified as core industries or clusters. The new policy aims to foster regional industries, human resources, science and technology; build and maintain social overhead capital (SOC); and boost tourism and cultural activities. Currently, the government plans to spend KRW126 trillion (USD104.2 bn.) over 5 years for this project, taking it 2013. Funding would come mainly from the central government (KRW71.2 trillion), regional governments (KRW24.3 trillion) and private investment (KRW30.9 trillion). While key projects will be developed autonomously by regions, some large infrastructure and land reclamation projects will be started by the central government, (Korea Times 2009, p. 10). It is questionable if the aims of the project can be achieved through this spending. The basic idea, to overcome fragmentation and create possibilities to exploit economies of scale, is understandable. However, interprovincial cooperation is not only necessary, but has to be mandatory and will lead to greater intraregional inequality in these mega-regions. It cannot be expected, that the clusters will all settle on provincial borders or the commercial growth be equally divided among the provinces of one mega-region, so it is questionable how much provincial support these designated regions will have from the provinces gaining the smaller share. Moreover, the identification of lead industries themselves is very questionable. While some choices might be obviously dictated by geography or economic history, others are not and the question is, if the state really can dictate the emergence of a new cluster. Gangwon province's experience in this field is very enlightening and not very encouraging, so far. It will be discussed in the next section.

Gangwon Province Development Planning

Gangwon was the largest Korean province before it was divided, like the peninsula itself, as a result of the Korean war. It is now the second largest province in the South (North Gyeongsang being the largest) and comprises 17% of South Korea's territory, but has only 2.8% of its population. While its regional GDP outstrips that of the smaller but vital Seoul-embracing Gyeonggi province, Gangwon lags clearly behind in indicators of technological achievements, as Table 10.1 below shows. Also, its industry is characterized by a smaller number of larger and medium sized enterprises and a higher number of mini-enterprises than the capital region.

In terms of regional policy, Gangwon province faces a number of challenges: the Northern part of the province, adjacent to the inner-Korean border, has very limited development prospective due to inner-Korean relations. North and South Korea are divided by a so-called Demilitarized Zone (DMZ), reaching principally along the border 2 km each into the North and South, though actually it is often smaller, due to gradual changes made by the respective armies, in particular on the Northern side. In South Korea, a zone of 5–15 km below the DMZ is classified as a Civilian Control Zone (CCZ), in which all economic activity is subordinated to military

Table 10.1 Basic data for Gangwon province in comparison

	Gangwon-do	Seoul	Gyeonggi-do	Korea
Space (km^2)	16,612.89	605.25	10,132.24	99,720.39
... as percentage of total	17%	0.6%	10%	100
Population.(07)	1,503,806	10,192,710	11,106,211	49,268,928
... as percentage of total	2.8%	22.8%	20%	100
GRDP per capita (USD)	18,487	22,307	17,823	20,259
GRDP growth rate	4.4	5.2	6.0	5.0
Patents (No.)	1,314	43,654	39,188	131,400
Industrial intagible properties (No.)	1,211	44,245	41,598	128,589
... as percentage of total (patents)	1%	34.3%	30.8%	100
R&D spending (in Mio. USD)	145,837,881	4,013,019,261	9,026,423,756	21,946,792,937
... as percentage of total	0.6%	18%	47%	100
Engineers	4,766	68,075	87,246	199,990
No. of people working in R&D field	10,523	100,801	107,862	365,794
... as percentage of total (engineers)	1.8%	26.5%	34%	100
Firm size 1–9 employees (in %)	52	39	42	42
Firm size 10–499 emp. (%)	42	50	47	48
Firm size >499 (in %)	6	14	8	10

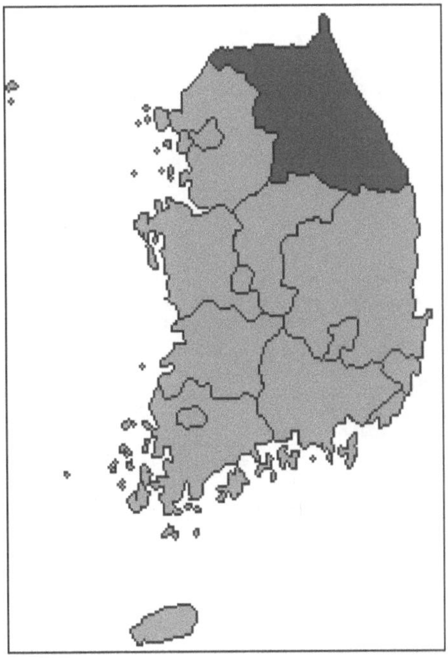

Fig. 10.2 The peripheral location of Gangwon province (Source: Gangwon province)

planning and where local administrations have no say in planning. This zone has been partially shortened and opened up in recent years and plans for the development of certain activities there have been made, but still it means that the border counties of Gangwon province – Cheolwon, Hwacheon, Yanggu, Inje, and Goseong – face severe restrictions on land use and planning.

One of the results of the peripheral location is a substantial loss of population over time, in particular in the northern part of the province. While on the eastern edge of the country, compared to the more southern regions, Gangwon province is relatively close to the capital and thereby a potentially attractive location for production as it combines closeness to markets with less congestion than Seoul (Fig. 10.2). In locational marketing this factor is praised as a major advantage over other provinces. In fact Wonju, a major city in southeast Gangwon province is only about a 90-min drive from Seoul and some workers even commute to work to and from there.

Major infrastructure projects have improved access between Gangwon province and the capital region and strengthened intra-provincial traffic, particularly along the east coast (see Fig. 10.3). The East Sea (Sea of Japan) region is a major tourist destination for mostly domestic mass tourism and suffers from the overuse of infrastructure in peak times like summer or autumn. The central government's "5+2" development plan mainly envisions Gangwon as a province for leisure activities, though the Wonju medical complex also has a place in the central government strategy.

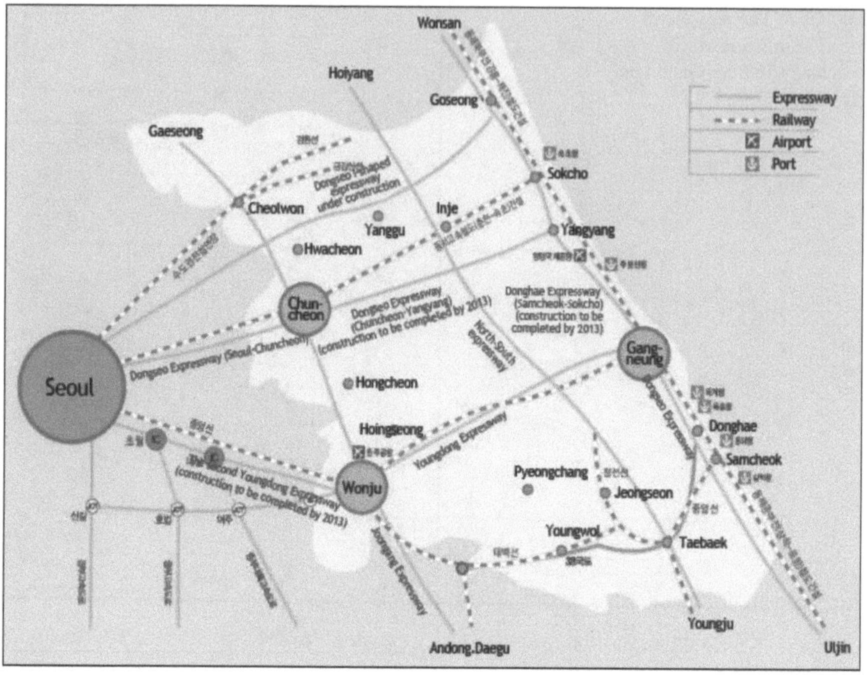

Fig. 10.3 Infrastructure projects of Gangwon province (Source: Gangwon province)

Gangwon province itself is trying to develop various industrial clusters, centered in its three major cities, namely in Chuncheon a biotechnology cluster, in Gangneung a new materials cluster and in Wonju the aforementioned medical science cluster. Recently, Chuncheon has begun additional preparations for an "animation cluster", focusing on the thriving animation industry in Korea, since the reservoir system surrounding the city is crucial for national water production and hazardous substances, like those needed for biotechnology research, are prohibited,. The formulation of the three clusters in Gangwon province is obviously triggered by regional policy considerations, achieving an equal distribution of innovation activities across the province (see Fig. 10.4). But can such a plan succeed?

To answer this question, first of all the development of the three clusters has to be seen as part of a larger development strategy; one to develop seven "growth axes" and six "life zones", according to regional comparative advantages or historic patterns of specialization.

The seven growth axes (see Fig. 10.5) are, according to the regional planning strategy, supposed to

- Secure and agglomerate economies that concentrate resources into strategic strongholds to effectively distribute them into specialized local strengths.
- Create new opportunity for growth based on local specificity by interaction between the cities and local areas.

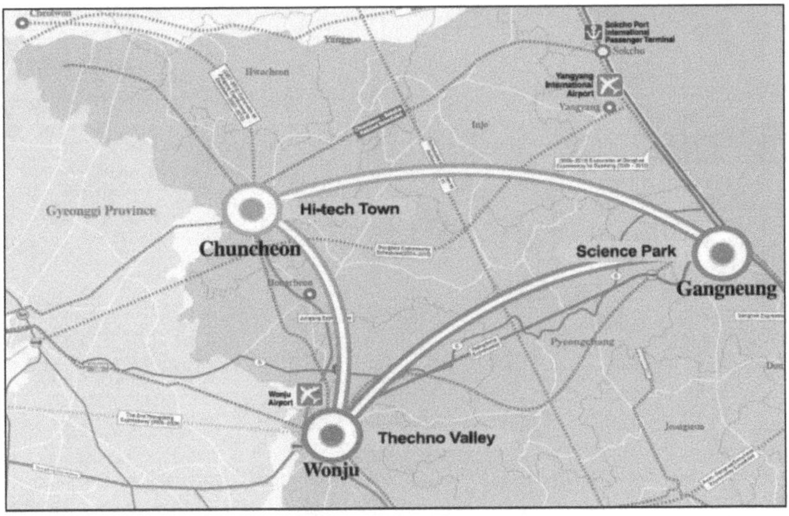

Fig. 10.4 Gangwon province growth triangle (Source: Gangwon province)

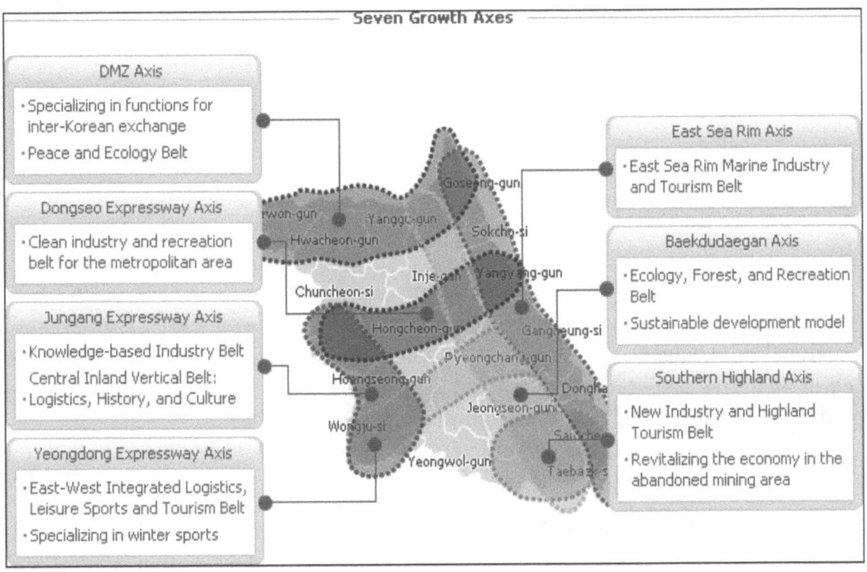

Fig. 10.5 Gangwon province growth axes (Source: Gangwon province)

– Promote functional strongholds that play the role of an artery across Gangwon based on the traffic network and local peculiarities; and
– Serve as a strategic axis that responds to the development of the metropolitan area, the administrative complex city, and the new cities.

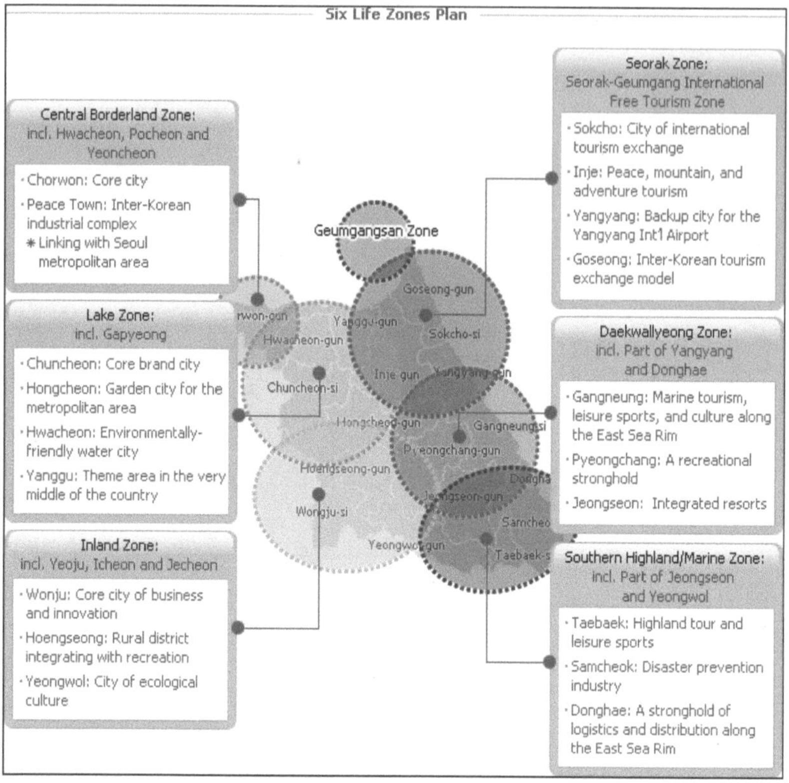

Fig. 10.6 Gangwon province life zones plan (Source: Gangwon province)

In this strategy, obviously the focus on local advantages is very important. Among these are the physical features (development of maritime or winter sports), locational advantages (nearness to the capital region and the new administrative city in western Gangwon province), and historical features (use of the inner-Korean border for new forms of eco- and cultural-historical tourism).

Additionally, planning has formulated six so-called "life zones" (see Fig. 10.6), which are intended to

- Encourage independent localization based on the progress of provincial development, growth axes, local peculiarities, land utilization, and residential, natural, and living environments.
- Build a network of zones designed to cooperate with each other in order to avoid the individualization of infrastructure.
- Implement an independent industrial infrastructure around a core city to realize an "Advanced Economy and Quality of Life"; and
- Promote cohesive power to develop the growth axes.

Both strategies are closely interrelated then, and the role of the aforementioned three clusters is as well formulated in the growth axes as in the life zone concept, as

Box 10.2: Benchmarking a Biotech-Medtech Cluster in Germany

The Berlin/Brandenburg biotech-medtech cluster.

Gangwon province, with the help of outside experts and Gangwon National University professors, benchmarked the Berlin/Brandenburg region for its decade-and-a-half agglomeration of around 170 bio-tech companies and 170 med-tech companies, located in more than 100.000 sqm of laboratories in Buch, Charlottenburg, Adlershof, Hennigsdorf, Luckenwalde and Potsdam. Together these companies employed 2,000 new employees in high technology and knowledge-intensive jobs. The companies attracted 600 mil. euro through venture capital or the capital market. A dozen stable, i.e. long-term research networks acquired around 150 mil. euro in funding from science funding institutions and international institutions like the European Union.

Source: own contribution.

agglomerations in the first, as industrial infrastructure in a core city in the second. To develop the three clusters, Gangwon province benchmarked various existing clusters in the US and Europe (see Box 10.2).

Gangwon province tried to emulate successful cluster policy strategies by providing industrial zones, overhead services (management, marketing, technical support, quality certification) and investment incentives like tax breaks. For the Wonju Medical Cluster, KRW90 billion (USD72 mil.) were invested, starting with the construction of laboratories and offices. Currently 47 firms have been attracted, plus a new campus for the renowned Yonsei University medical faculty. However, results are mixed. Among the most important problems facing the cluster is a lack of critical mass, in particular few producing units compared to small research companies, therefore weak links to export markets, limited interaction and experience with large and foreign companies, limited attractiveness for highly-qualified human resources and at the same time only limited possibilities to educate in the province (brain drain). Linkages and networks among companies, universities and research institutes are weak, and their initiation by the government through various incentives and platforms faces hurdles due to a business culture not prone to free information sharing. Venture capital is lacking, as well as experience attracting outside funding for the regional universities. The provincial government has few possibilities (through materials or legislation) to improve the investment situation beyond the existing incentives. The peripheral location of Gangwon province offers certain advantages, like more affordable industrial space and housing, less congestion, a healthy environment, but administrative and market concentration in the capital area means that transportation is a crucial issue and congestion costs cannot be completely escaped. "parachuted" infrastructure does not always pay off, as the case of Yangyang airport shows: built for USD382 million, it accumulated losses of USD10 million for 3 consecutive years and served an average of 26 passengers daily with a staff of 146 before closing down.

Compared to successful examples from abroad and from within Korea, like the Ulsan-Busan automotive cluster, Suwon LCD cluster or Taejeon IT cluster, there is a lack of leading companies, either conglomerates or technology leaders growing at least to a medium size and incorporating not only prototype development, but also larger production units. Also, while elements for a cluster are theoretically in place – an appropriate infrastructure, an agglomeration of firms and at least several research and university institutions – there is a lack of commitment to joint research and the formation of research networks, not the least due to the clear orientation of employees of companies and research and university institutes alike towards Seoul, where most of them live. In other words, there is no real stable regional innovation system. This also means that talent cannot be induced to move to the region, where development perspectives for highly-talented people are lacking, compared to the capital region. Such conditions can lead to an adverse selection process regarding the quality of companies and human resources and to the gradual degradation of the cluster.

Given these challenges, the strategy to focus on "green", "service-oriented" and "tourism" development rather than industrial, high-technology based development in Gangwon province's larger development plan and the central government's "5+2" approach seems commendable and realistic. Green technologies, like those being developed in the Chuncheon biotech cluster, and the newly emerging focus on renewable energy (wind, water) and maritime industries (e.g. deep sea research in Goseong) are in demand so funding opportunities exist. In addition, they sometimes require certain geographic features (sea, wind), and they are less dependent on historic development patterns and therefore not so bound to the old industrial centers. However, their contribution to total growth and regional development will remain small for a considerable period of time. They cannot overcome the regional problems outlined above without concerted effort by all the actors to strengthen regional innovation systems.

The central government has to create the preconditions for a change in academic and research concentration in the metropolitan area. While infrastructure development has been strong in regional universities, talent still is not willing to leave the metropolitan area. Changes in university admission and the improvement of public secondary schooling can alleviate the problem. The central government also can change the calculation of costs by rigorously internalizing external effects in the capital area. If costs for location in the urban area would include, for example, congestion costs (like costs of traffic jams) and pollution costs (including medical treatment for widespread allergies), the relative attractiveness of peripheral regions would increase.

Regional government can support this through local investment in education.[12] The fostering of regional talent by schools, universities, governments and the

[12]The study of six Norwegian innovation clusters by Isaksen (2009) finds a strong link between regional innovation activity and local higher education institutions' specializing in areas that cater

private sector should include development strategies for regional entrepreneurship. This well-defined crucial step in the growth processes has been largely underdeveloped.[13] Collaboration strategies have to be developed by the participants themselves to transform a group of companies into networks and clusters.[14] Networking itself is crucial, in particular for smaller, less experienced companies who have a relatively low starting point in terms of relevant social capital, limited business, and managerial experience and can be enhanced by institutional efforts made by the cluster management.[15] Local strengths and weaknesses have to be analysed and subsequently the strengths fostered and the weaknesses removed. For this, knowledge is also important and not always available. The fostering of a regional innovation system lacks focal points. Also, the nature of a cluster as a functional network more than a location-based network can help to overcome certain difficulties. On the national level, the opening of a "virtual cluster" system, an online-forum for cooperation, already seeks to overcome geographical barriers. In Gangwon province itself, cooperation between Chuncheon biotech and Wonju medtech cluster could achieve synergies and help overcome problems stemming from the lack of critical mass.

Conclusion

The experience of Gangwon province shows that the attraction of companies to peripheral regions is possible, given enough subsidies. Wonju medical cluster to date has attracted close to 50 companies. Realistic expectations about the size and success of cluster strategies are warranted, however. Agglomerations are not automatically clusters, and government activity alone cannot create networks. There is a danger that the survival of these "clusters" could remain permanently dependent on government subsidies. There should be a clear perspective that after the initial financing of a cluster its growth in the life-cycle should allow for self-financing. Regional innovation systems cannot be completely designed and implemented from above, if preconditions are not favourable. Instead of trying to

to the needs of key regional industries. However, as Ponds et al. (2010) point out the knowledge spillovers from academic research on regional innovation activities are crucially dependent on successful university-industry collaboration networks.

[13]Certainly, this is a very complex task, requiring appropriate education about entrepreneurship, beginning in primary school, appropriate technological education, and incentive systems among other needs.

[14]The issue of regional collective learning and possible biases towards mistrust and rivalry instead of collaboration has been recently analyzed in the case of a South German surgical instrument cluster. Staber (2009) points out that from an evolutionary point of view with the limited cognitive abilities of agents results of collective learning might be either functional or dysfunctional for the cluster.

[15]See the study of Kaufmann and Schwartz (2008).

change incentives for spontaneous allocation of economic activity within a space, the provision of preconditions for successful innovation activities seems to be the better strategy. In this sense, interventions to achieve growth on the regional level have to be market-conforming or even market-enhancing.

Also, until recently, research has mainly focused on the potential or existing positive effects of clusters. Cluster policies are never a free lunch; resources are drawn from other economic activities, usually in the form of taxes on successful businesses. A comprehensive cluster strategy should therefore include the costs of subsidizing or supporting the cluster in the calculation.[16]

Ultimately, clusters depend on successful private investment. Certainly, the state has an important role in providing necessary public goods, and a potential role as a facilitator of growth processes, in particular in peripheral regions where there might be a lack of appropriate expertise, but without private investment, without private risk capital, clusters will not be successful. Cluster policies are for various reasons attractive to administrations: they allow for certain high-profile political events (like opening ceremonies), they display progressiveness in technological terms, and they are a "me too" product on the political market. However, more humility with regard to market decisions on locations would likely enhance the success of existing clusters and diminish the creation of costly white elephants which drain resources from the whole population and provide nothing in return.

References

Antonelli C (2009) The economics of innovation: from the classical legacies to the economics of complexity. Econ Innov New Technol 18(7):611–646, October 2009

Boschma R (2004) Some reflections on regional innovation policy, paper prepared for the expert group meeting on "constructing regional advantage", Brussels, 7 Dec 2004. Internet file: http://econ.geo.uu.nl/boschma/brusselmeetingpolicy.pdf. Accessed 10 Oct 2010.

Cooke P (2001) Regional innovation systems, clusters, and the knowledge economy. Ind Corp Change 10(4):945–974

Den B, Frank AG, Seung-Gyo J (2009) Pros and Cons of "Backing Winners" in innovation policy, MPRA Paper no. 17658. Internet file: http://mpra.ub.uni-muenchen.de/17658/1/MPRA_paper_17658.pdf. Accessed 10 Oct 2010

Doloreux D-, Parto S (2005) Regional innovation systems, current discourse and unresolved issues. Technol Soc 27:133–153

Fujita M, Krugman P, Venables AJ (1999) The spatial economy: cities, regions and international trade. MIT Press, Cambridge

Hornycha C, Michael S (2009) Industry concentration and regional innovative performance: empirical evidence for Eastern Germany. Post-Communist Econ 21(4):513–530

Isaksen A (2009) Innovation dynamics of global competitive regional clusters: the case of the Norwegian Centres of Expertise. Reg Stud 43(9):1155–1166

[16]The anti-monde problem means that we do not really know what could be achieved with a different resource allocation. The impossibility to measure this effect does not mean that the effect has no importance.

Kaufmann D, Schwartz D (2008) Networking: the 'Missing Link' in public R&D support schemes. Eur Plann Stud 16(3):429–440

Ketels C-, Lindqvist G, Orjan S (2006) Cluster initiatives in developing and transition countries. Center for Strategy and Competitiveness, Stockholm

Kim W-B (2003) The evolution of regional income disparities in Korea. Korea J 43(2(summer)):55–80

Korea Times (2009) W126 Tril. Set for regional development, 17 Sep 2009, p 10

Krugman P (1994) The myth of Asia's miracle. Foreign Aff 73(6):62–78

Maincent E, Navarro L (2006) A policy for industrial champions: from picking winners to fostering excellence and the growth of firms. Industrial Policy and Economic Reforms Papers No. 2, Enterprise and Industry Directorate-General, European Commission. Internet file: http://ec.europa.eu/enterprise/enterprise_policy/competitiveness. Accessed 1 Sep 2009

Menzel M-P, Fornahl D (2010) Cluster life cycles—dimensions and rationales of cluster evolution. Ind Corp Change 19(1):205–238

OECD (2007) Competitive regional clusters: national policy approaches. OECD, Paris

OCED (2006) OCED territorial reviews: Seoul, Korea. OCED policy brief, April 2006

Ponds R, van Oort F, Frenken K (2010) Innovation, spillovers and university–industry collaboration: an extended knowledge production function approach. J Econ Geogr 10(2):231–255

Potter J (2009) Evaluating regional competitiveness policies: insights from the New Economic Geography. Reg Stud 43(9):1225–1236

Prevezer M (2008) Technology policies in generating biotechnology clusters: a comparison of China and the US. Eur Plann Stud 16(3):359–374

Ramstad E (2009) Expanding innovation system and policy _ an organisational perspective. Policy Stud 30(5):533–553

Rodrik D (2004) Industrial policy for the 21st century. Internet file: http://ksghome.harvard.edu/~drodrik/unidosep.pdf. Accessed 1 Sep 2009

Rosiello A (2008) Rethinking innovation systems in life sciences: implications for regional and innovation policy. Eur Plann Stud 16(3):329–335

Seliger B (2004) Südkorea als wirtschaftliche Drehscheibe Ostasiens? Kritische Anmerkungen zu einem aktuellen Konzept (South Korea as the hub of East Asia? Critical remarks on an actual concept). In: Patrick K (ed) Korea 2004 – Politik, Wirtschaft, Gesellschaft. Institut für Asienkunde, Hamburg, pp 67–89

Sölvell Ö, Lindqvist G, Ketels C (2003) The cluster initiative greenbook. Ivory Tower, Stockholm

Staber U (2009) Collective learning in clusters: mechanisms and biases. Entrep Reg Dev 21(5–6):553–573

Chapter 11
Knowledge Based Economy, Excellence, Clusters and Regional Development from a Systemic Perspective: Relevant Aspects for South Korea

Silvo Korez

Many nations and regions are struggling to remain or to become competitive in the context of globalization. Analysing recent documents dealing with innovation and qualitative growth in the industrialized countries, among them the EU-Strategy 2020, one can find considerable congruity in structure and goals. Most countries try to reach, on the one hand, an inclusive innovation, which is based on education factors and policies directed towards small and medium sized enterprises. On the other hand, they try to reach an inclusive growth, usually based on employment factors and being environment friendly through CO_2 and energy consumption reduction. The classical mechanism of inclusion predominantly covers individuals or individual units, but not regions as such.

In Europe this is to be reached by investing 3% of GDP in R&D and following the two principles "convergence" and "cohesion". This should result in the implementation of far reaching regional investments and the creation of incentives, including regional innovation clusters[1] in order to reduce differences in economic development levels. Nevertheless, it is more likely that the gap in GDP and R&D investment levels will increase in different regional settings.

The factors, innovation on the one side and convergence and cohesion – variables for regionally balanced growth – on the other, are in fact natural enemies. This will be explained in the following paragraph: In the knowledge-based economy, which to a large extent is a "metaphysical" one, the set of rules is different than in classical economics. It is because ideas are traded differently than classical goods. In a "metaphysical economy", there is no market for second-best ideas, whereas in a "physical economy", there is a very large market for second-best products. This is due to the fact that the price mechanism does not work with goods that are not consume-exclusive.

[1]Clusters are geographic concentrations of interconnected companies and institutions in a particular field. Clusters encompass an array of linked industries and other entities important to competition (Porter 1998).

J. Mahlich and W. Pascha (eds.), *Korean Science and Technology in an International Perspective*, DOI 10.1007/978-3-7908-2753-8_11,

This can be illustrated by comparing two products, a piece of soap and a piece of software. The soap can be consumed only once and for sale it has to be produced anew every time. Its price is determined respectively by production costs, demand, quantity and quality. In contrast to this, the software is produced only once while the reproduction costs are practically zero. So, the creator of the best software (idea) can easily drive all second best competitors out of the market by lowering the price, thus increasing his market share and setting the de-facto standards. The production costs of the second best software (idea) are usually similar to the best one. The innovation market tends to follow the principle "the winner takes it all", and it is based on either time (patent) or quality. The same applies also to the R&D market (see Scherer and Harhoff 2000)[2]. The skew-distributed returns to innovation led in Europe to the "excellence"-discussion. Specifically, in Austria the concepts of excellence clusters were developed. Accordingly, in my understanding, "excellence" in a knowledge-based economy is not caused by a big difference in the production of an idea (best and second best) but by a very big difference in market success.

The above mentioned skew-distribution is usually bypassed by high specialisation, including finding a niche or niche-strategy. Another way to avoid economic asymmetry is applying fiscal redistribution: taxing a successful individual or region and sharing the revenue with less successful ones.

The most common way to try to overcome the harsh distribution effects of the knowledge-based economy is by bundling, i.e., the idea with its realisation are combined so that the price mechanism works again. This approach has also the advantage that the second-best idea can be brought to market effectively. Another efficient tool to influence market access and also success is to set legal standards and limitations.

Regarding the case of South Korea in general, it is important to take into consideration that it can be classified as a de facto island economy. This has a limiting impact on infrastructure design (energy supply, traffic, etc.) and the flow of goods, making a high-export orientation necessary. Furthermore, the country cannot profit from the activities of surrounding regions and work on synergies. In this respect, South Korea has a special position considering the application of policy variations and analogies.

Considering past economic success, one can observe bundling in the above mentioned sense and limited market access through biased business practices, the exclusion of foreigners and protectionism. For an island economy this is a logical and likely development. In the meanwhile, transition to high specialisation has already begun. It can be expected that during this process the traditional success factors will loose weight due to shifts to lower-cost locations for production, e.g. China, on the one hand, and international trade agreements on the other hand.

[2]See Scherer and Harhoff (2000). 575t.

New trends show that for achieving optimal economic results the most efficient way is to separate ideas (design) and production.[3]

Taking the convergent and coherent economic development of regions into consideration, it is necessary to elaborate the mechanism of fiscal redistribution, which is a difficult task. In order to support development on (inter-)regional levels, cluster approaches try to explain how economic development can be influenced by comprehensive identification of economic and geographic interdependencies, including history and country specificities, types of knowledge and networking practices. Fiscal redistribution and cluster policies are a multidimensional balancing act between analysis and highly practical policy actions, bottom–up initiatives and top–down steering. Finally, established and newly emerging structures, various geographical levels and various policy roles or capacities have to be taken into consideration.[4] Vital for success are customisation and accountability in the production of the policies. It is also promising to develop polyvalent structures that build on the favours of a region and allow serendipity.

References

den Hertog P (2003) The role of cluster policies in economic growth and competitiveness. Presentation at the European Seminar on Cluster Policy Copenhagen, June 10th 2003

European Commission (2010) Europe 2020 – A strategy for smart, sustainable and inclusive growth

Porter M (1998) Clusters and the New Economics of Competition in Harvard Business Review

Scherer FM, Harhoff D (2000) Policy implications for a world with skew-distributed returns to innovation. Res Policy 29(4–5):559–566

[3]This has been shown by Petri Rouvinen in the case of Nokia 95.

[4]den Hertog (2003)

Chapter 12
From Learning Knowledge Outside to Creating Knowledge Within: Korea's Mobile Phone Industry Compared with Those of Japan, Taiwan and China

Keun Lee and Jia Jin

Introduction

Korea is one of the few countries currently poised to make a successful transition from a newly industrializing economy (NIE) to an advanced economy. The book *Innovation and Technology in Korea* (Mahlich and Pascha 2007) used the term newly advanced economy (NAE) to refer specifically to Korea's status. As the authors noted on page 2 of their book, reaching the level of a newly advanced and eventually an advanced economy requires the ability to generate innovation and technological change at the frontier of global knowledge. Technological innovation can be regarded as the exploitation of the available stock of knowledge to generate new knowledge. Then, if we pursue that distinction further, we can see that while NIEs tend to still be occupied with learning knowledge from others, an NAE should be able to create new knowledge independently within its own territory. This paper examines to what extent Korea has made such a transition, namely become able to generate its own knowledge.

In this paper, learning from others is measured in terms of international diffusion of knowledge, while creating knowledge is measured in terms of intra-national diffusion of knowledge, that is the extent that knowledge is generated within the territory of a country rather than from outside of it. Using the mobile phone industry as an example, we will examine, first, from whom Korean industry has been learning, and, then, to what extent it has internalized the knowledge-creating mechanism within its boundaries. In this respect, Korea's development will be compared with the experiences of Japan, Taiwan and China.

Methodologically, this paper relies upon Jaffe et al. (1993, 2000) and Lee and Yoon (2010). Using the patent citation data, the latter investigates the dynamic random access memory chip (DRAM) industry, a sub-sector of the semiconductor industry, to find that the latest comers tend to depend upon knowledge diffusion from their immediate forerunners. Consequently, the Chinese (the latest entrants) tend to cite the Taiwanese, the Taiwanese cite the Koreans, the Koreans cite the Japanese, and the Japanese cite the Americans. It also finds that latecomers' ability

J. Mahlich and W. Pascha (eds.), *Korean Science and Technology in an International Perspective*, DOI 10.1007/978-3-7908-2753-8_12,

to create local knowledge stock, namely the degree of intra-national and inter-firm knowledge diffusion, is proportional to the level of technological capability or order of entry, although organizational differences among the firms also comes into play. Thus, the findings will also let us examine whether Korea is still behind Japan in terms of degree of intra-national knowledge diffusion, and how it compares with Taiwan and China.

Another distinction between an NIE and an NAE would be the role of the government, with government or industrial policy having a smaller role in an NAE (Mahlich and Pascha 2007). In this regard, Frank et al. (2007) observes that government in Korea is still trying to play a key role in the telecommunication sector. However, Lee and Yoon (2010) find that compared to Taiwan, Korean firms relied less on the government and GRIs (government research institutes) as their knowledge sources, making fewer citations to GRI-owned patents. We will see if the same pattern holds between Taiwan and Korea in the mobile phone industry.

While concrete mechanisms for knowledge creation and diffusion in the two NAEs of Korea and Taiwan are the main focus, this paper deals with four Asian countries (Japan, Korea, Taiwan, and China) in a single framework using balanced weights. China's case is important for Korea as it presents severe competition (Frank et al. 2007). Thus, we will seek to answer how the four economies differ in terms of intra-national and inter-firm knowledge diffusion and how such differences can be related to their organizational differences.

Specifically, we will verify the proposition that the flow of knowledge follows the order of entry or catch-up in an industry, although different wireless standards might distort the order to some extent. These catch-up patterns imply that entrant firms are more likely to learn from leaders immediately ahead rather than those far ahead of them (Lee and Yoon 2010). While advanced economies tend to create most of this knowledge stock, the latecomer economies typically tap into the available stock. However, latecomer economies become constrained by the relatively fewer institutional channels for both international and intra-national knowledge diffusion, and that subsequently influences their abilities to absorb and create new knowledge (Lee and Yoon 2010). Thus, it can be reasoned that the more advanced country is more likely to engage in intra-national knowledge diffusion. Specifically, we will examine the second proposition that intra-national and inter-firm knowledge diffusion is proportional to the level of technological capability or order of entry: Japan is characterized by the highest degree, China the lowest, and Korea and Taiwan in between.

Furthermore, following Lee and Yoon (2010), this paper also investigates the organizational differences between Korean and Taiwanese firms in the mobile phone industry. It is well known that while the two economies are similar in their overall size, the Korean economy and its innovative advancements are spearheaded by a small number of large diversified business groups, whereas the Taiwanese economy is dominated by a large number of relatively smaller, more specialized firms (Mahmood and Singh 2003; Choung and Hwang 2000; Saxenian and Hsu 2001). This then carries into their organizational differences: the core competence of Korean firms or business groups is based on vertical or horizontal integration within their own brand marketing, whereas Taiwanese firms are largely based on specialized division of labor among relatively small firms (Chung et al. 2004). Hence, we put forward the third proposition that Korean firms tend to be less oriented towards diffusion across different firms compared to Taiwanese firms (P3).

The main empirical analysis in this paper uses the patent data of USPTO, and is an extension of Jin (2009) which follows the methodology of Lee and Yoon (2010). Section "Evolution of the Mobile Phone Industry and Catch-Up by Latecomers" provides a brief introduction to the development of the mobile industry and the catch up carried out by late comers. Section "International Knowledge Diffusion and Sources of Technological Advance" examines the patterns of knowledge diffusion among the four economies. Section "Intra-national and Inter-firm Knowledge Diffusion" compares Japan, Korea, Taiwan, and China in terms of intra-national and inter-firm knowledge diffusion, with a discussion of the different government policies. Section "Different Roles for Governments and GRIs" compares the roles of the governments and GRIs in Korea and Taiwan, and finally a section on summary and concluding remarks follows.

Evolution of the Mobile Phone Industry and Catch-Up by Latecomers

Evolution of the Industry

The invention of the mobile phone evolved from a car radiotelephone in the early 1970s. Engineers at Motorola, which used to be a major supplier of pre-cellular car telephones in the 1960s in the U.S., came out with the idea that "You park your car and leave; you can't use your car phone but you can take your portable with you." Due to Motorola's two-way radio and semiconductor experience, the company was able to manufacture much of the basic electronics needed for a portable phone system. In February of 1973, Motorola had produced a DynaTAC (DYNamic Adaptive Total Area Coverage) portable phone prototype. After 10 years of investment and strife, Motorola then announced the world's first commercial portable cell phone on September 21, 1983. This marked the arrival of the wireless communication age, which has changed the lives of people around the world.

Motorola's work was then quickly followed in Europe as telecommunications markets there were deregulated. Nokia, Ericsson and Siemens were the most successful representative European firms at that time. Later Ericsson became Sony Ericsson in 2001, and Siemens' mobile equipment business was 100% acquired by Taiwan's biggest handset maker BenQ in 2005. Japanese equipment manufacturers were prominent in global markets from the 1980s to the early 1990s (Kushida 2008), led by keiretsu companies (established business groups), like NEC, Fujitsu, Hitachi, and OKI, together with several new entrants such as Matsushita (later Panasonic), Sony, Toshiba, Sharp, etc.

Since the early 1990s, the mobile phone industry's growth has been full-fledged and worldwide. That was the moment Korea and Taiwan entered the industry. From the early 1990s, the Korean government and industry introduced the world's first CDMA network system, in cooperation with Qualcomm. Benefiting from their successful catch-up in many high-tech sectors, especially in the semiconductor industry, Samsung, LG and other Korean chaebols became the main force in developing mobile equipment. Samsung introduced its first CDMA handset in late 1991.

Taiwanese firms entered the mobile phone industry around the same time that Korean firms did. Different from Korea's reliance on large business groups, Taiwanese manufacturers are relatively small and specialized firms, primarily those that used to be successful in the PC sector. To enter mobile phone production, they leveraged their international production networks and their advantages in manufacturing and manpower. The mobile phone industry is comparable to the PC industry in terms of recent modularization trends. The main players in Taiwan's mobile phone industry – Benq, Compalcomm, Compal, Arimacomm, Lite-on, Inventec, Honhai ASUS and Acer – all used to be engaged in the PC industry, and started doing ODM (original design manufacturing) and EMS (electronic management system) for the advanced mobile phone manufacturers, such as Motorola, SonyEricsson, Siemens and so on. For example, Motorola brought up some main Taiwanese manufacturers by placing ODM orders since the early 1990s. Until 2006, Compalcomm has taken around 20–25% of Motorola's order to manufacture handsets.[1] Many advanced foreign companies have established ODM cooperation with Taiwanese vendors. Until 2004, the world's top six phone manufacturers have accounted for 65 ~ 72% of the orders placed with Taiwanese manufactures.

With the robust growth of China's economy and the emergence of every industrial sector after the Reform and Open policy, China has become a large and fast growing part of the mobile phone market. While it has been dominated by foreign brands like Nokia, Motorola and Ericsson for a long time, Chinese local manufacturers began to enter the business in the late 1990s. By leveraging advanced foreign technology and utilizing their advantages in satisfying local niche markets (low-end models with designs that fit Chinese tastes), local firms, such as Bird, TCL, Amoi, Konka, and Haier, soon expanded and started to become a threat to the incumbent foreign companies. For a short time during the early days of China's mobile handset industry, many Korean and Taiwanese firms thrived by selling designs, knock-down kits or finished handsets to Chinese makers (Ken and Shiu 2007).

The above story suggests a dynamic shake-up and catching up by the latecomers in the mobile phone industry. In 1990, the Japanese brand NEC was tied with Nokia in second place in global market shares, and five out of the nine top manufacturers

[1]http://www.ce.cn/cysc/communications/csdt/200608/18/t20060818_8181719.shtml.

were Japanese. However, by 2000, they became marginalized. By 2007, Samsung had edged out Motorola to come in second, with LG appearing in fifth place, with only slightly less than the estimated total of all Japanese manufacturers combined.

Benefiting from the increasing demand from world top equipment providers like Nokia, Motorola and SonyEricsson, Taiwanese-built handset shipments increased from 3.3% in 2001 to around 8% of world production in the mid-2000s, and it was estimated that Taiwanese handset exports would increase to 281 million units by 2008, 79% of that driven by the demand from ODMs.[2] When we look into China, driven by the growth of its robust local demand and rising indigenous firms, China's share of world mobile phone production has increased sharply to almost 55% in 2007. Also, there was a sharp rise in domestic market share for indigenous firms, with their stake growing dramatically from none in 1998 to 58% in 2003 (Lee and Yoon (2010)).

Timing of Successive Entry by Innovative Latecomers

Now we will examine patent data to show the successive entries of latecomers as innovators in the mobile phone industry. In our analysis of United States Patent and Trademark Office (USPTO)-registered patents, we refer to patents found by searching key words such as "mobile/cellular phone" and "mobile/wireless communication" in the fields of title, abstract, and specification code. More precisely, the data we are using should be referred to as "mobile phone" related patents.

Our data in Table 12.1 shows that the number of mobile phone patents increased rapidly from the 1980s to the early 1990s; the period when the industry began full-fledged growth worldwide. It grew from 47 patents in 1987 to 247 patents in 1993, more than a fivefold increase. Moreover, the share of mobile phone patents among total U.S. patents increased from around 0.1% in the late 1980s to higher than 1.5% in the late 1990s, which is much higher than that of semiconductor patents that accounted for 0.5% at that time. It has reached an even higher share of 5.5% on average in recent years. These numbers imply that the mobile phone industry is a high-technology industry characterized by rapid technological progress.

The number of patents granted to each country then shows a picture of successive entries and catch up. Both the United States and Japan were granted their first patents in 1972. It is not surprising that Japan filed for patents so early since our data covers all patents related to mobile phones and wireless technology in general; Japan began radio phone and cordless telephone manufacturing in the 1970s. Japan wanted to stand on its own feet in the development of wireless technology. In spite of its late entry compared to the U.S., Japan had a successfully developed telecommunication industry by the 1980s. Japanese firms were actually filing a similar

[2]http://news.driverchina.com/Html/news/tele/tele/101557440.html.

Table 12.1 Mobile phone patents by country and year (Data source: USPTO Granted Patent Database)

Year	U.S.	Japan	Korea	Taiwan	China	Others	Total
1976	9	2	0	0	0	3	14
1977	10	3	0	0	0	6	19
1978	8	2	0	0	0	8	18
1979	5	4	0	0	0	3	12
1980	12	5	0	0	0	9	26
1981	6	6	0	0	0	5	17
1982	13	11	0	0	0	7	31
1983	9	16	0	0	0	3	28
1984	12	9	0	0	0	4	25
1985	14	3	0	0	0	8	25
1986	18	8	0	0	0	7	33
1987	23	14	0	0	0	10	47
1988	28	18	0	0	0	14	60
1989	45	37	1	0	0	12	95
1990	49	31	1	0	0	20	101
1991	41	27	0	0	0	25	93
1992	73	54	2	0	0	36	165
1993	134	63	2	2	0	46	247
1994	207	104	1	2	0	76	390
1995	273	126	5	3	0	99	506
1996	501	206	7	2	0	170	886
1997	627	201	16	6	0	204	1,054
1998	1,070	365	43	13	0	370	1,861
1999	1,418	491	69	15	0	568	2,561
2000	1,742	568	102	38	0	703	3,153
2001	2,011	650	122	64	2	814	3,663
2002	2,287	929	185	104	2	945	4,452
2003	2,727	1,182	227	118	10	1,188	5,452
2004	3,314	1,455	342	221	8	1,331	6,671
2005	3,499	1,678	355	218	9	1,345	7,104
2006	5,487	2,664	625	365	34	2,057	11,232
2007	5,036	2,648	930	441	56	2,113	11,224
2008	3,244	1,480	713	235	68	958	6,698

number of patents to their U.S. counterparts from the 1980s to the early 1990s, the period when they entered the mobile phone industry and made advances. For example, patent numbers for the U.S. and Japan were both 6 in 1981 and increased to 45 and 37, respectively, in 1989.

As shown in Table 12.1, in terms of patents, Korea entered the industry a couple of years earlier than Taiwan, although far later than Japan. The first patents assigned to Korea and Taiwan came in 1989 and 1993, respectively, and to China in 2001. Since then, Korean and Taiwanese firms have experienced divergent paths. That is, Korean firms have become top global handset manufacturers, while Taiwanese firms are still ODM and OEM vendors. In terms of patents, Korea holds a far larger number than Taiwan.

Having been granted its first patent in 2001, Chinese patent filing in the U.S. begins to show rapid growth from 2006. It jumped dramatically from nine patents in 2005 to 34 patents in 2006 and then to 56 and 68, respectively, in 2007 and 2008. Our analysis finds that around 36% of the relevant patents are held by foreign firms in China (mainly Taiwanese firms, who account for 33.16% of China's total). Chinese indigenous firms hold 37.43%, Chinese government research institutes (GRIs) and universities 8.56%, and firms and research institutes in Hong Kong 16.58%. Taiwanese firms, like Shenzhen Futaihong Precision Industry, Shenzhen Hong Fu Jin Precision Industry, Inventec Electronics, and Benq, hold the bulk of the patents assigned to foreign companies in China. They are mainly wireless technology and mobile phone equipment providers that have had success in Taiwan.

From the early 2000s, with the fast growth of the mobile phone market in mainland China and the saturation of the market in Taiwan, Taiwanese mobile phone manufacturers and the component providers started to invest in mainland China. Entry constraints, namely mobile phone manufacturing and selling licenses, led Taiwanese firms entering the mainland's market to establish joint ventures with authorized local firms or to do those firms' ODM orders. Later, when some of the Taiwanese firms were granted licenses, they actually started to transfer their production base from Taiwan to mainland China. For example, Benq first entered the mainland market by establishing joint ventures with local manufacturer CETC, which had a license to produce handsets. Through entry into the local market and cooperation with local firms, Taiwanese firms have gained access to cheaper human resources and the much larger Chinese market. In the meantime, they have brought the production network and advanced technology into the local industry, as manifested by the large share of Chinese patents owned by Taiwanese firms.

International Knowledge Diffusion and Sources of Technological Advance

This section investigates the sources of technological knowledge among latecomers, namely Japanese, Korean, Taiwanese and Chinese firms. Lee and Yoon (2010) found, in the semiconductor industry, that the Taiwanese had a strong tendency to cite Korean patents, the Koreans Japanese patents, and the Japanese American. As mentioned above, our research is an endeavor to investigate whether the mobile phone industry shows the same pattern seen with semiconductors.

If a patent is granted, a public document is created containing extensive information about the inventor, his/her employer, and the technological antecedents of the invention, all of which can be accessed in computerized form. Among this information are "references" or "citations." The citations serve as the legal function for delimiting the scope of a patent's property right. A grant for a patent is essentially a legal statement confirming that the idea, which is embodied in the patent, represents a novel and useful contribution surpassing the previous state of knowledge. In principle, a citation of Patent X by Patent Y means that X represents a piece of previously existing knowledge upon which Y builds (Jaffe et al. 1993, 2000). It can be presumed that citations are informative links across patented inventions, and that knowledge flows have left a paper trail in the form of citations in patents (Jaffe et al. 1993).

In tracing latecomers' oversea knowledge sourcing during their entry period, our analysis will primarily focus on the first 5–10 years after the initial entry or filing of a patent. All USPTO-registered data can be found with full text or full images in their computerized forms from 1790 until the present. Full text refers to the patent number, assignee names and individual inventor records, citation information, year of application and grant, technological classifications, abstract, claim and description, and the figure image of a patent.

Table 12.2 shows the ratio of mobile phone patents citing a selected country's mobile phone patents in relation to the total citations made by all mobile phone patents worldwide. For example, the column on Japan in Table 12.2 shows that about 20% of the citations made by all mobile phone patents worldwide in 1993 cited ones filed by Japanese registrants. Japan's share has fluctuated around 20%, whereas the shares of Taiwan and Korea have steadily increased over the years. China holds the lowest share and is slowly increasing due to their late entry and current technical insufficiencies.

Now, to measure more precisely the relative importance of each country as a source of mobile phone-related knowledge, we will use a measure of the "relative citation propensity," that was developed in Lee and Yoon (2010). By relative

Table 12.2 Sources of knowledge for the world's mobile phone patents by source country

During the entry period for Korea and Taiwan (1993~2000) and China (2001~2006)

	U.S.	Japan	Korea	Taiwan	China	Others
1993	57.70%	20.49%	0.00%	0.50%	0.00%	21.30%
1994	60.53%	16.43%	0.23%	0.00%	0.00%	22.81%
1995	54.14%	21.44%	0.20%	0.08%	0.00%	24.15%
1996	60.99%	21.56%	0.09%	0.18%	0.00%	17.18%
1997	63.53%	16.66%	0.29%	0.15%	0.00%	19.37%
1993~1997	60.81%	19.00%	0.19%	0.15%	0.00%	19.84%
1998	60.28%	17.54%	0.33%	0.62%	0.00%	21.23%
1999	59.25%	17.84%	0.44%	0.36%	0.00%	22.11%
2000	59.66%	18.29%	0.57%	0.54%	0.00%	20.94%
2001	66.03%	13.48%	0.61%	0.48%	0.00%	19.40%
2002	57.79%	19.12%	0.81%	0.67%	0.02%	21.60%
2003	62.57%	14.48%	0.76%	0.54%	0.03%	21.61%
2004	58.33%	18.85%	1.58%	0.71%	0.05%	20.49%
2005	54.47%	21.53%	2.06%	0.68%	0.09%	21.17%
2006	51.69%	22.91%	2.49%	1.21%	0.46%	21.24%
2001~2006	59.36%	17.68%	1.29%	0.69%	0.10%	20.88%

citation propensity, we mean country X's tendency to cite country Y's patents, compared to other countries. The relative citation propensity can be expressed explicitly as follows:

Relative propensity of country X (e.g. Korea)'s patent citations to refer to those of country Y's patents =

$$\frac{n_{xyt}/n_{xt}}{n_{cyt}/n_{ct}}$$

where

n_{xyt} : number of citations made to country Y's patents in patents filed by country X in year t.

n_{xt} : number of all citations made by country X's patents granted in year t.

n_{cyt} : number of citations made to country Y's patents by all patents, except those of country X, granted in year t.

n_{ct} : number of all citations made by all patents, except those by country X in year t.

If this number is higher than 1, we can say, for example, that Korean patents are likely to be cited by Korean firms' patents rather than by other countries'; to some extent, we can say that Korean firms have the tendency to cite their local firms' patents.

Using this procedure, we have first estimated the citation propensity of Japanese firms in the mobile phone industry, the results of which are presented in Table 12.3A. The ratio of local citations (Japanese firms citing Japanese patents) starts off high, implying that Japan initially deployed its proprietary analog standard rather than purchasing the U.S. or European standards. Following the 1990s, the ratio of local-citation increased, which is consistent with the fact that while entering the 2G period, Japan chose to deploy its self-developed wireless standard PDC, rather than use GSM or CDMA that were adopted by the U.S. and the rest of the world.

Table 12.3B shows the case for Korean firms, starting from 1993. Although the first mobile phone patent was granted to a Korean firm in 1989, citation propensities of earlier years were of little significance since the number of citations made was too small. During the first 5 years (data drawn from 1993 to 1997) after initial entry into the industry, Japan was a more important source than the U.S. of mobile phone-related knowledge for Korean firms. It is consistent with the findings of Lee and Yoon (2010) in regard to semiconductor patents.

Table 12.3C shows the case for Taiwanese firms. Interestingly, we found that during their entry into this business, namely the first 5 years from 1993 to 1997, Taiwanese patents listed not a single citation to Korean patents. However, after the late 1990s, the table shows a significant increase in the ratio for citing Korean patents. This pattern seems to reflect that the main network standard adopted by Taiwan during the initial period was the GSM standard, and that during the later

Table 12.3 Relative citation propensity of countries

	U.S.	Japan	Korea	Taiwan	China	Others
Part A: Japan						
1971	1.17	0.00	NA	NA	NA	0.00
1972	0.82	1.58	NA	NA	NA	1.48
1973	0.64	NA	NA	NA	NA	2.49
1974	NA	NA	NA	NA	NA	NA
1975	0.90	0.00	NA	NA	NA	1.88
1976	0.12	7.70	NA	NA	NA	4.40
1977	0.55	3.19	NA	NA	NA	2.05
1978	1.06	0.00	NA	NA	NA	1.18
1979	1.04	0.00	NA	NA	NA	0.89
1980	0.86	3.57	NA	NA	NA	1.04
1971~1980	0.72	1.91	NA	NA	NA	2.28
1981	0.84	2.14	NA	NA	NA	1.07
1982	0.90	1.72	NA	NA	NA	1.03
1983	0.88	1.79	NA	NA	NA	0.76
1984	0.81	2.00	NA	NA	NA	1.33
1985	0.81	2.50	NA	NA	NA	0.94
1986	0.64	2.02	NA	NA	NA	1.67
1987	0.75	2.35	NA	NA	NA	1.31
1988	0.85	1.50	NA	NA	NA	1.08
1989	0.79	1.54	NA	0.00	0.00	1.38
1981~1989	0.76	2.33	NA	0.00	0.00	1.28
Part B: Korea						
	U.S.	Japan	Korea	Taiwan	China	Others
1993	1.30	0.61	NA	0.00	NA	0.59
1994	0.55	4.06	0.00	NA	NA	0.00
1995	0.92	1.36	0.00	0.00	NA	0.86
1996	1.03	1.31	23.07	0.00	NA	0.38
1997	1.00	1.47	7.03	7.03	NA	0.47
1993~1997	1.00	1.45	8.36	3.63	NA	0.49
1998	0.85	1.74	10.63	1.15	NA	0.65
1999	0.94	1.43	5.05	0.50	NA	0.73
2000	0.92	1.23	5.90	0.00	0.00	0.92
2001	0.76	1.83	6.26	1.37	NA	1.07
2002	0.94	0.90	11.02	1.66	3.27	0.85
2003	0.69	1.81	9.81	3.03	6.02	0.99
2004	0.66	1.44	7.84	3.39	3.44	0.95
Part C: Taiwan						
	U.S.	Japan	Korea	Taiwan	China	Others
1993	0.93	1.88	NA	0.00	NA	0.36
1994	0.47	2.61	0.00	NA	NA	1.25
1995	0.37	1.24	0.00	339.27	NA	1.10
1996	0.98	0.93	0.00	37.05	NA	0.78
1997	1.07	1.44	0.00	0.00	NA	0.41
1993~1997	0.83	1.47	0.00	44.81	NA	0.74

(continued)

Table 12.3 (continued)

	U.S.	Japan	Korea	Taiwan	China	Others
1998	1.12	0.52	3.89	18.99	NA	0.49
1999	1.14	0.79	2.03	12.17	NA	0.56
2000	0.86	1.36	2.43	7.65	177.57	0.86
2001	0.78	1.77	5.96	15.39	NA	0.71
2002	0.96	1.17	3.38	10.67	13.36	0.56
2003	0.75	1.47	8.51	17.26	20.97	0.72
2004	0.85	1.01	2.38	16.26	2.99	0.79
Part D: China (overall)						
	U.S.	Japan	Korea	Taiwan	China	Others
2001	1.04	1.95	0.00	0.00	NA	0.27
2002	0.93	0.80	0.00	0.00	338.63	1.07
2003	0.68	1.67	1.51	10.67	98.59	1.06
2004	0.49	1.70	3.12	5.25	50.91	1.39
2005	0.76	0.27	2.08	0.00	94.54	1.89
2006	0.83	0.65	2.94	2.22	11.20	1.29
2001~2006	0.70	0.99	4.35	4.04	49.95	1.31
Part D1: Foreign firms in China						
	U.S.	Japan	Korea	Taiwan	China	Others
2001	0.96	2.70	0.00	0.00	NA	0.00
2002	0.00	0.00	0.00	0.00	0.00	0.00
2003	0.50	1.64	3.46	29.32	0.00	0.24
2004	0.14	3.98	0.00	11.82	171.81	0.00
2005	0.00	0.00	0.00	0.00	0.00	0.00
2006	0.56	1.03	5.29	17.39	17.26	0.25
2001~2006	0.49	1.94	4.71	20.43	62.28	0.48
Part D2: Chinese indigenous firms						
	U.S.	Japan	Korea	Taiwan	China	Others
2001	0.00	0.00	0.00	0.00	NA	0.00
2002	1.73	0.00	0.00	0.00	0.00	0.00
2003	0.64	2.07	0.00	0.00	0.00	1.39
2004	0.58	1.69	4.31	6.45	0.00	1.11
2005	0.65	0.29	3.03	3.06	252.77	1.28
2006	0.72	0.49	3.11	0.00	16.94	1.69
2001~2006	0.63	0.82	5.35	1.75	91.53	1.49

period, Korean firms achieved a rapid growth with the increasingly global deployment of the CDMA standard.

Table 12.3D shows the Chinese case of international knowledge diffusion patterns. One of the important findings is the very high tendency of China to cite Taiwanese patents: 5.75%, compared to only 0.54% of all patents in the world citing Taiwanese patents in 2003. Given that foreign firms, local firms, and research institutes and universities in China all hold similar patent shares, we checked the pattern for each group and found that both foreign and local firms tend to cite

Taiwanese and Korean firms, while Chinese universities and research institutes tend to cite U.S. and certain European patents. As we have mentioned, Taiwanese firms hold the lion's share among those patents registered in the US by foreign firms located inside Chinas. Then it would be no doubt that, technologically, those foreign firms from Taiwan strongly rely on their parent companies and counterparties in Taiwan.

Moreover, Taiwanese firms entered the market of mainland China by selling SKD/CKD components or by doing the ODM for local companies. Local firms, which were in their infancy, benefitted greatly from this transfer of technology. The Chinese government's strict rules on entry into the mobile phone industry, has promoted foreign investments and the establishing of joint ventures. Nowadays, Taiwan's main mobile phone manufacturers, such as Benq, Inventec, Honhai and Dbtel, and component suppliers, like Jabil, Merry and Ichia, have all established subsidiaries or have moved their factories to the coastal cities in mainland China. That fostered the diffusion of technology from advanced Taiwanese firms to local firms in the early stages. On the other hand, benefiting from their advantages in CDMA technology and capacities in platform development, Korea's second-line companies, like Sewon, Pantech, and Teleson, cooperated with local Chinese firms through SKD/CKD. China's mobile handset makers have helped many Korean and Taiwanese firms thrive extravagantly by buying designs, knock-down kits or finished handsets (Ken and Shiu 2007).

While Chinese universities and GRIs do not hold a big share of the mobile phone patents, they have a stronger tendency to cite U.S. patents and those from European countries. This is consistent with the findings in Lee and Wang (2010) regarding the semiconductor industry, that is, patents from universities and research institutes are more likely to cite those from advanced nations like the U.S. and Japan, home to the most advanced and standard technologies that cater to universities and GRIs.

Other relevant observations can be made from the above-mentioned data, such as Japan's entry into the industry following the U.S.; Korea and Taiwan following Japan; and finally, China following Taiwan and utilizing Taiwanese patents with a larger number of citations from them compared to the number of those made by other countries. This pattern implies that latecomers have tendencies to cite the patents of frontrunners just ahead of them. Also, while latecomers prefer the newest technology, they are not likely to cite those with different technology standards.

Intra-national and Inter-firm Knowledge Diffusion

Measuring and Comparing Japan, Korea, Taiwan and China

In this section, we focus on intra-national knowledge diffusion, namely, the degree to which knowledge diffuses effectively among firms within a nation (Lee and Yoon 2010). This concept is important because the speed and degree that knowledge is diffused within a country does not only imply the effectiveness of using the foreign knowledge and local knowledge later generated, but also affects the further development of local knowledge stock. Thus, we tested the proposition that the more advanced a country is, the higher degree of intra-national knowledge diffusion it has, compared to its latecomer counterparts.

To evaluate the geographic localization of patent citations for each country, Jaffe et al. (1993) suggested an approach of comparing the probability of a patent matching the original patent by geographic area, which is conditional on the original patent, with the probability of a match that is not conditioned by a citation link. The non-citation-condition probability provides the baseline or reference value vis-à-vis the proportion for matched citations. The basic insight offered by the approach of Jaffe et al. (1993) is that the probability that a citation of a patent by another patent should be compared to a similar probability defined for reasonably comparable reference patents (Lee and Yoon 2010). We will explain this concept and our methodology using Table 12.4.

Part A reports the simple ratio by which country A's patents cite country A's patents (including self-citation) over the 1993–2004 period. For instance, in 1996, of all the citations made by Japanese mobile phone patents, 39.36% refer to other Japanese mobile phone patents, while the number is 2.17% and 6.67% for Korean and Taiwanese patents, respectively. However, we are not taking these numbers alone as representing a comparable degree of pure intra-national knowledge diffusion. From these numbers we are going to subtract the propensity y that Korean patents (for instance) are cited by non-Koreans. Part B of the Table shows that among all the patent citations by non-Koreans, 0.09% of them cited Korean patents in 1996.

As such, the degree of pure intra-national diffusion, which is the same as the measure of localization in Jaffe's and Lee and Yoon's work, can be expressed as follows:

$$\begin{aligned}&\text{Degree of (pure) intra-national diffusion in Japan} = \text{A} - \text{B}\\&= \text{(propensity of Japanese patents citing Japanese patents)}\\&- \text{(propensityof non-Japanese patents citing Japanese patents)}\end{aligned}$$

In the above formula, the tendency of non-Korean patents citing Korean patents can be interpreted as a measure of the global influence of Korean patents. Part B of Table 12.4 shows the difference in the global impacts of patents across countries, the most notable of which show Japanese patents having much higher impacts and thus are being cited more often by non-Japanese patents. The *t*-test results of the Japanese versus the two latecomers, Korea and Taiwan, substantiate this significant difference, which has been confirmed to exist in the semiconductor industry through Lee and Yoon's work.

However, there is a slight difference in the case of Korea and Taiwan. It should be noted that in the early years of the entry period, Korean patents and Taiwanese patents had a similar amount of global influence, that is, sometimes Korean patents

Table 12.4 Intra-national and inter-firm knowledge diffusion in Japan, Korea, Taiwan and China

Part A. Ratio of locally cited patents to total citations (e.g. Japanese patents citing Japanese patents)

	1993	1994	1995	1996	1997	1998	1999	2000	2001	2002	2003	2004
Japan (J1)	34.78%	36.99%	40.50%	39.36%	46.07%	34.50%	49.17%	38.92%	50.99%	55.27%	49.50%	51.79%
Korea + Taiwan (KT1)	0.00%	0.00%	5.13%	3.28%	2.44%	6.13%	2.89%	3.83%	6.32%	9.97%	11.11%	14.92%
Korea (K1)	0.00%	0.00%	0.00%	2.17%	2.04%	3.55%	2.20%	3.35%	3.79%	8.90%	7.47%	12.41%
Taiwan (T1)	0.00%	0.00%	26.67%	6.67%	0.00%	11.69%	4.42%	4.15%	7.45%	7.13%	9.29%	11.47%
China (C1)									0.00%	7.69%	3.45%	2.47%

Part B. Quality (Global Influence) of patents (other country's patents citing country A's patents)

	1993	1994	1995	1996	1997	1998	1999	2000	2001	2002	2003	2004
Non-Japan citing Japan (Jc)	16.17%	12.39%	15.49%	18.14%	12.54%	13.96%	13.87%	14.25%	10.21%	10.83%	9.98%	11.95%
Non-Korea citing Korea (Kc)	0.00%	0.23%	0.20%	0.09%	0.28%	0.29%	0.41%	0.51%	0.57%	0.63%	0.66%	1.19%
Non-Taiwan citing Taiwan (Tc)	0.51%	0.00%	0.00%	0.17%	0.15%	0.58%	0.35%	0.52%	0.45%	0.60%	0.48%	0.56%
Non-(Kor. + Tai.) citing Korea or Taiwan (KTc)	0.51%	0.23%	0.20%	0.26%	0.43%	0.87%	0.77%	1.05%	1.03%	1.25%	1.15%	1.79%
Non-China citing China (Cc)									0.00%	0.02%	0.03%	0.05%
Japan vs. (Kor. + Tai.) (Jc − KTc)	15.66%	12.16%	15.30%	17.88%	12.11%	13.09%	13.10%	13.20%	9.19%	9.59%	8.83%	10.16%
t-statistic	5.88	7.69	9.46	15.72	13.65	16.82	22.48	22.51	27.06	22.81	30.79	29.44
Korea vs. Taiwan (Kc − Tc)	−0.51%	0.23%	0.20%	−0.09%	0.13%	−0.28%	0.06%	−0.01%	0.12%	0.03%	0.18%	0.63%
t-statistic	−1.01	1.00	1.00	−0.58	0.78	−1.42	0.41	−0.07	1.08	0.20	1.84	4.88
(Kor. + Tai.) vs. China (KTc − Cc)									1.03%	1.23%	1.12%	1.74%
t-statistic									9.57	8.60	11.45	13.17

Part C. Total intra-national diffusion (including self-citations) after controlling quality difference

	1993	1994	1995	1996	1997	1998	1999	2000	2001	2002	2003	2004
Japan (J2 = J1 − Jc)	18.61%	24.60%	25.00%	21.22%	33.53%	20.54%	35.30%	24.68%	40.78%	44.43%	39.52%	39.84%
t-statistic	4.35	8.87	9.25	11.65	21.31	16.10	35.53	25.54	65.59	59.49	75.17	67.91
Korea (K2 = K1−Kc)	0.00%	−0.23%	−0.20%	2.09%	1.76%	3.26%	1.79%	2.84%	3.22%	8.28%	6.81%	11.22%
t-statistic	NA	−1.00	−1.00	4.80	4.44	7.80	6.85	9.11	14.77	21.98	27.47	32.81
Taiwan (T2 = T1 − Tc)	−0.51%	0.00%	26.67%	6.50%	−0.15%	11.11%	4.07%	3.63%	7.00%	6.53%	8.81%	10.91%
t-statistic	−1.01	NA	13.60	8.78	−1.45	15.53	11.64	10.62	24.28	19.09	32.74	33.84
Korea + Taiwan (KT2 = KT1 − KTc)	−0.51%	−0.23%	4.93%	3.02%	2.01%	5.25%	2.12%	2.78%	5.30%	8.72%	9.96%	13.13%

t-statistic	−1.01	−1.00	4.94	5.57	4.57	9.42	6.87	7.95	18.90	21.43	33.28	35.08
China (C2 = C1 − Cc)									0.00%	7.67%	3.42%	2.42%
t-statistic									NA	22.57	20.66	15.70
Comparing (Korea + Taiwan) and Japan (including self-citations)												
J2 − (KT2)	18.61%	24.83%	25.20%	19.13%	31.77%	17.28%	33.51%	21.84%	37.55%	36.16%	32.71%	28.63%
t-statistic	6.85	12.22	11.74	14.75	24.53	17.28	41.07	29.38	65.37	49.68	63.25	48.87
Comparing Korea and Taiwan(including self-citations)												
K2 − T2	0.51%	−0.23%	−26.86%	−4.41%	1.91%	−7.86%	−2.28%	−0.79%	−3.78%	1.74%	−2.00%	0.31%
t-statistic	NA	NA	−13.77	−5.28	5.76	−10.04	−5.86	−1.95	−11.44	3.69	−5.84	0.71
Comparing (Korea + Taiwan) and China (including self-citations)												
(KT2)−C2									5.30%	1.05%	6.54%	10.71%
t-statistic									22.23	2.12	20.72	29.31
Part D. Inter-firm diffusion (excluding self-citations)												
	1993	1994	1995	1996	1997	1998	1999	2000	2001	2002	2003	2004
Japan: inter-firm citations (J3)	26.83%	30.30%	37.39%	37.36%	43.53%	31.55%	44.85%	35.88%	47.82%	49.28%	41.22%	44.12%
Inter-firm diffusion tendency (J4 = J3 − Jc)	10.66%	17.92%	21.90%	19.22%	30.99%	17.59%	30.98%	21.63%	37.61%	38.45%	31.24%	32.17%
t-statistic	2.60	6.67	8.18	10.62	19.77	13.99	31.29	22.63	60.53	51.28	60.11	55.07
Korea: inter-firm citations (K3)	0.00%	0.00%	0.00%	0.00%	2.04%	0.74%	0.74%	2.34%	1.74%	5.47%	5.43%	8.11%
Inter-firm diffusion tendency (K4 = K3 − Kc)	0.00%	−0.23%	−0.20%	−0.09%	1.76%	0.45%	0.34%	1.83%	1.18%	4.84%	4.77%	6.92%
t-statistic	NA	−1.00	−1.00	−1.00	4.44	2.05	1.92	6.80	7.33	15.79	22.01	23.92
Taiwan: inter-firm citations (T3)	0.00%	0.00%	21.43%	6.67%	0.00%	8.11%	3.57%	2.80%	5.17%	3.92%	6.43%	8.68%
Inter-firm diffusion tendency (T4 = T3 − Tc)	−0.51%	0.00%	21.43%	6.50%	−0.15%	7.53%	3.22%	2.28%	4.72%	3.33%	5.95%	8.12%
t-statistic	−1.01	NA	11.78	8.78	−1.45	12.27	10.11	7.86	19.18	12.50	25.89	28.31
Korea + Taiwan: inter-firm citations (KT3)	0.00%	0.00%	7.89%	1.67%	1.63%	2.33%	1.23%	2.44%	2.71%	4.96%	5.73%	8.27%
KT4 = KT3 − KTc	−0.51%	−0.23%	7.70%	1.41%	1.20%	1.45%	0.47%	1.40%	1.68%	3.71%	4.58%	6.49%
t-statistic	−1.01	−1.00	6.35	3.49	3.21	3.78	2.02	4.69	8.28	11.95	19.87	21.54
China: inter-firm citations (C3)									0.00%	0.00%	3.45%	2.47%

(continued)

Table 12.4 (continued)

Part A. Ratio of locally cited patents to total citations (e.g. Japanese patents citing Japanese patents)												
	1993	1994	1995	1996	1997	1998	1999	2000	2001	2002	2003	2004
Inter-firm diffusion tendency												
(C4 = C3 − Cc)									0.00%	−0.02%	3.42%	0.68%
t-statistic									NA	−1.14	20.66	4.41
Comparing (Korea + Taiwan) and Japan (excluding self-citations)												
J4 − KT4	11.17%	18.14%	14.20%	17.82%	29.79%	16.14%	30.51%	20.24%	35.92%	34.74%	26.66%	25.68%
t-statistic	5.24	10.04	6.51	14.80	23.85	18.65	39.93	29.34	67.39	52.25	58.14	49.38
Comparing Korea and Taiwan (excluding self-citations)												
K4 − T4	0.51%	−0.23%	−21.63%	−6.58%	1.91%	−7.08%	−2.88%	−0.45%	−3.55%	1.51%	−1.18%	−1.20%
t-statistic	NA	NA	−11.96	−9.19	5.76	−12.00	−9.50	−1.39	−14.01	4.25	−4.10	−3.27
Comparing (Korea + Taiwan) and China (excluding self-citations)												
KT4 − C4									1.68%	3.73%	1.17%	5.80%
t-statistic									12.30	15.55	4.67	22.70
Part E: Self-citations	1995	1996	1997	1998	1999	2000	2001	2002	2003	2004		
Ratio of self-citations to total local citation												
Japan	12.24%	8.11%	9.76%	12.50%	15.94%	12.20%	11.91%	21.34%	28.45%	26.53%		
Korea	NA	100.00%	0.00%	85.71%	66.67%	30.77%	55.00%	40.83%	28.87%	37.67%		
Taiwan	25.00%	0.00%	NA	33.33%	20.00%	33.33%	32.26%	46.81%	32.89%	26.58%		
China							NA	100.00%	0.00%	0.00%		
Part F. Sample size: no. of mobile phone patents by country and no. of citations made by them												
	1993	1994	1995	1996	1997	1998	1999	2000	2001	2002	2003	2004
Japan	63	104	126	206	201	365	491	568	650	929	1,182	1,455
Citations	460	730	1,210	1,880	1,780	3,710	4,210	6,320	7,080	11,490	14,000	17,840
Korea	2	1	5	7	16	43	69	102	122	185	227	342
Citations	8	9	24	46	98	282	545	776	1,055	1,348	1,901	3,594
Taiwan	2	2	3	2	6	13	15	38	64	104	118	221
Citations	13	7	15	15	25	77	113	217	416	659	818	1,378
China									2	2	10	8
Citations									19	13	87	81

were cited more, while at other times Taiwanese patents were. Also, the *t*-test results of the global influence of the two countries' patents do not show a significant difference. However, in the later years, approximately from the year 2002, when Korean mobile phones started to take off globally, their patents also started to show a significant higher amount of influence than Taiwanese patents. As we have expected, Chinese patents significantly show the lowest influence, with a late entry in the year 2001 and small application numbers.

Part C of Table 12.4 presents the degree of intra-national knowledge diffusion, which is defined as total local citations (refer to Part A) minus global influence (refer to Part B). Based on these, Japan exhibits the highest degree of intra-national diffusion in the mobile phone industry, China the lowest, and Korea and Taiwan in between.

A comparison of Korea and Taiwan interestingly showed that Taiwan has higher degrees of intra-national diffusion compared to Korea across almost every year except for 1997 and 2002. The situation is similar in the semiconductor industry, as measured by Lee and Yoon (2010). Furthermore, following the methodology by Lee and Yoon (2010), we also try to control the self-citations (a phenomenon where firms cite their own previous patents) to distinguish diffusion occurrences among different firms from that within the same firm.

Lastly, Part D presents the degree of inter-firm knowledge diffusion, which is defined as total local citations (excluding self-citations) minus quality (as defined in Table 12.4, Part B).As shown in Part E, Korea has a higher ratio of the simple share of self-citations than Taiwan, while Japan commands the lowest ratio. The numbers in Part D further show that the degree of inter-firm diffusion has been increased from around 10–30% for Japan. Meanwhile, t-statistics indicate that the results have all been statistically significant at the 5% level since 1993. It shows positive and statistically significant results for Korean and Taiwanese firms since the mid 1990s and that for Chinese firms since 2003. When we conducted the t-tests across countries, inter-firm diffusion degrees of Japanese firms stand significantly higher than the joint degrees of Korean and Taiwanese firms across the years. Between Korea and Taiwan, except for years 1997 and 2002, the Taiwanese showed higher degrees of diffusion, with significant values from 1993 to 2004. Furthermore, the joint degrees of Korean and Taiwanese firms show significantly higher values than that of Chinese firms.

With similar results as those drawn from the semiconductor industry, this section proves that while the global impacts of the mobile phone patents follow the order of entry into and the associated strength of the industries in Japan, Korea, Taiwan, and China, the degrees of inter-firm and intra-national knowledge diffusion do not necessarily follow the same order but reflect the differences in industrial organization and firm types. The proposition of a higher degree of intra-national knowledge in a country with superior technological development is confirmed only when Japan is compared with Korea and Taiwan jointly. Comparing Korea and Taiwan or comparing them with China yet show a different story. As explained by Lee and Yoon (2010), inter-firm or intra-national knowledge diffusion is not only affected by the level of technological capability but also by the organizational characteristics of firms constituting the national economy of each country.

The fact that Taiwanese firms have higher degrees of inter-firm diffusion than Korean firms confirms our initial proposition about the organizational difference: similar to the semiconductor industry, the Korean mobile phone industry is dominated by large business groups, chaebols like Samsung and LG, while the Taiwanese mobile industry is dominated by relatively small and specialized firms, like Benq and Compalcomm. Thus, in the mobile phone industry, it is likewise proved that Korean hierarchy-oriented firms are stronger in internalization and combination, while Taiwanese network-based firms do better in externalization and socialization.

Different Roles for Governments and GRIs

Being a core block of any national innovation system, the wireless communication industry is always in the list of many government's top policy agenda. In order to promote national wireless communication industries and to bring up indigenous firms, some government tends to enact policies and activities that promote technology diffusion from abroad and its localization.

In Korea, a notable example is the CDMA mobile phones, and the existing analyses tend to credit the government for its strategy and ability to take a potentially risky gamble (Chung and Lee 1999; Yang et al. 2003). By introducing a standard that has not been commercialized in other countries, it helped in alleviating Korean firms' weakness in technology abilities as newcomers, since Motorola and the other foreign firms that dominated Korea's equipment markets lacked expertise in this technology (Kushida 2008). The government then favored local firms by stimulating them to develop their own knowledge base and become technologically self-sufficient. In the first phase, the Ministry of Commerce (MOC) gave the Electronics and Telecommunications Research Institute (placed under MOC control in 1992) the task of developing digital technology, and it brought together researchers from the government, scientists, and large equipment manufacturers, such as Samsung, Hyundai, LG, and Maxon (Lee and Lim 2001). In exchange for significant royalties to Qualcomm (owner of CDMA core intellectual property), Korean manufactures were given the rights to distribute CDMA handsets worldwide. With a well-planned start and sustained efforts at rolling out handsets and participating in the building of network systems, Samsung, LG and other Korean firms have gained comprehensive technological development abilities and have now become the top CDMA brands in the world.

In Taiwan, from the early 1990s, the Ministry of Economic Affairs set out policies like the "Four-Year Plan of Telecommunication and Electronic Technology Development" and the "Five-Year Plan of Telecommunication and Electronic Technology Development" in order to promote the development of telecommunication technology. The development of the wireless communication industry used to be included in policies that promoted the development of wire communication. Aware of the high-growth and promising future of the wireless communication industry, the Taiwanese government then enacted separate plans and provided special financial support for the development of wireless technology. The Industrial Technology Research Institute (ITRI), the Chung-Shan Institute of Science Technology, Telecommunication Laboratories of Chunghwa Telecom, and other GRIs were designated as the main research arms to conduct these projects. Like their successful experience in supporting the semiconductor industry by absorbing foreign knowledge and its localization, Taiwanese GRIs continued to be the main strength in the early development of the wireless communication industry. Our examination of the patent data indicates that although the patent numbers are relatively small during the early days, there are more patents by GRIs, like ITRI and NSC, compared with the much smaller number of patents, namely one or two, by private firms. Besides directly implementing the R&D projects, ITRI was also authorized by the Ministry of Economic Affairs to hold training courses to cultivate high-tech talent in wireless technology.

In the early stages, the Chinese government maintained policies promoting foreign investment in this industry. Diverse forms were adopted, including completely foreign owned enterprises, joint venture enterprises, joint cooperation enterprises, etc. The presence of many joint venture handset manufacturers in China fostered the diffusion of technology expertise across the country (Ding et al. 2006). Entry regulations set by the Chinese government granted indigenous makers more access to foreign knowledge Lee and Yoon (2010). The Chinese government or the ministry in charge, the Ministry of Industry and Information Technology (MII), maintained a tight control over entries by selectively issuing mobile phone manufacturing and selling licenses. In 1999, when the first round of licenses were granted, only Chinese enterprises were eligible, including Kejian, Xiahua, Eastcom, TCL, Haier, Konka, ZTE, BIRD and Soutec. The number of licenses was quite limited for a while although it gradually loosened over time. Thus, by the end of 2003, 30 GSM and 19 CDMA handset manufacturing licenses had been issued to 37 enterprises, among them 13 foreign and 24 local companies. Until this time, there were many foreign companies who wanted to produce and sell in the Chinese market but were not permitted to do so due to regulations. Thus, many foreign companies, primarily Sewon, Telson and other medium-sized Korean companies, had to collaborate with licensed Chinese companies after providing the platforms and other key software to Chinese makers (Zhu et al. 2006).

The State Council adopted policies designed to allow Chinese firms to increase technological capacity and occupy the domestic market in order to attract more investment and to collect special program funds to improve the technological competency of the domestic mobile communications equipment producers (Network Weekly 2004). The State Council financially supported R&D for mobile technology by (1) transferring 5% of fixed-line telephone installation fees as a special grant from 1999 to 2003, and (2) having the MII invest 1.4 billion Yuan ($169.7 million) from mobile connection fees. In addition, it stopped issuing licenses for joint ventures in mobile handset manufacturing after 1999 (Ding et al. 2006).

In leveraging foreign knowledge and creating local knowledge stock, GRIs are usually the key players. Specifically, when we refer to patent citations, we find that they have played different roles in each country. That is, Taiwanese firms relied greatly on GRIs, while Korean firms did not much. When we examine who filed more patents in Korea and Taiwan, we find that in Korea, most of the mobile phone patents belong to chaebols. In Taiwan, patents are more often collectively registered by GRIs than by private firms. This pattern is consistent with Lee and Yoon (2010)'s findings on governments' different roles in the semiconductor industry. Existing studies likewise argue that the success of Taiwan SMEs is propelled by the efforts of ITRI and other GRIs, whose laboratories have acted as the prime vehicle for leveraging and modifying advanced technology from abroad. This technology was then rapidly diffused among various Taiwanese firms (Amsden and Chu 2003; Mathews 2002).

Taiwanese firms' heavy reliance on GRIs has to do with their disadvantage in scale and capacities for technology development. With regard to the large scale R&D budget needed and technological uncertainty faced by the development of say 3G technology as 2G technology becomes obsolete, small Taiwanese firms are unlikely to take the risk, given their financial and technological limitations. These deficiencies can be overcome by utilizing the relative abundant R&D resources of the government and its GRIs, together with financial support and preferential policies. As mentioned above, in the first phase, the Taiwanese government carried out specific projects on wireless technology development. Then ITRI, Chung-Shan Institute of Science and Technology and other GRIs were designated to be the main research arms to conduct projects. ITRI has also held training courses to build up the manpower pool and provide specialists for the firms. Newcomers at that time tend to scout from the GRIs. ASUS is now a famous mobile phone brand in Taiwan, which has roots in the PC business. When they entered the wireless industry, their main technology development force came from those who were transferred from the wireless communication team of Chung-Shan Institute of Science and Technology (Yang 2004).

Summary and Concluding Remarks

Spearheaded by U.S. firms in the 1970s, mobile telecommunications technology was further commercialized in the 1980s with the entering of Japanese firms. It then boomed globally from the late 1980s and early 1990s with the entrance of Korean and Taiwanese firms, and most recently Chinese firms. Focusing on the building up of technological ability, this paper has examined the process of technology acquisition and catch-up carried out by latecomer countries, with a focus on Korea, compared in terms of degrees of inter-national, intra-national, and inter-firm knowledge diffusion.

First, in terms of inter-national diffusion, we find, similarly to Lee and Yoon (2010), that the flow of knowledge seems to follow the order of entry into the industry, that is, Japanese patents tend to cite American patents, Korean patents tend to cite Japanese patents, Taiwanese patents tend to cite Korean patents, and Chinese tend to cite Taiwanese. We also find that in the mobile phone industry, technology standards are vital factors affecting latecomers' citation patterns. That is, if different wireless technology standards are adopted by the late entrants, it tends to reduce the latecomers' propensity to cite the work of its forerunners.

Second, we find that intra-national and inter-firm knowledge diffusion is also closely related to the level of technological capability or order of entry; Japan is characterized by the highest degree, China the lowest, and Korea and Taiwan in between. It is also shown that Taiwan has a higher degree than Korea both in intra-national and inter-firm knowledge diffusion, which can be explained by the organizational differences of the firms in the two countries; Korea consists of big group firms, compared with Taiwanese SMEs, as manifested in the semiconductor industry study by Lee and Yoon (2010). This paper also discussed the role of the government in latecomer entries, with some focus on the GRIs. We find that Taiwanese firms rely greatly on GRIs, while Korean firms do not.

Overall, the analysis in the paper shows both the achievements and limitations of Korea as an NAE. In terms of knowledge creation within its territory, Korea has yet to catch up with other advanced countries, like Japan, in terms of intra-national knowledge creation and diffusion. However, Korea's big businesses have now become quite independent in their R&D with less reliance on the government or GRIs than before and Taiwanese firms. This is consistent with Mani (2007)'s finding that Korea's innovative capabilities have shifted from the GRIs to private sector firms.

References

Amsden A, Chu W (2003) Beyond late development. MIT Press, Cambridge

"China," Network Weekly (2004) Implications of bird's annual report: intention of a soft-landing. http://www.chinabird.com/birdnews/news-detail.asp?newsid=206. Accessed dates October 15, 2008

Choung J-Y, Hwang H-R (2000) National systems of innovation: institutional linkages and performance in the case of Korea and Taiwan. Scientometrics 48(3):413–426

Chung C, Tsai P, Wang S-Y (2004) US semiconductor patents granted to Taiwan, South Korea and Japan. In: Chen T, Lee J (eds) The new knowledge economy of Taiwan. Edward Elgar, Glos, pp 188–227

Chung K, Lee K (1999) Mid-entry technology strategy: the Korean experience with CDMA. R&D Manage 29(4):353–363

Ding L, Haynes KE (2006) Technology, innovation and latecomer strategies: evidence from the mobile handset manufacturing sector in China European Regional Science Association in its series ERSA conference papers with number ersa06: 17, http://www.ersa.org/ersaconfs/ersa06/papers/17.pdf. Accessed dates October 1, 2008

Frank R (2007) Korean's telecommunication industry. In: Mahlich J, Pascha W (eds) Innovation and Technology in Korea. New York: Physica-Verlag

Imai K, Jingming S (2007) A divergent path of industrial upgrading: emergence and evolution of the mobile handset industry in China IDE discussion paper no.125

Jaffe AB, Trajtenberg M, Forgaty MS (2000) Knowledge spillovers and patent citations: evidence from a survey of inventors. American Economic Review, Papers on proceedings, pp 215–218
Jaffe AB, Trajtenberg M, Henderson R (1993) Geographic localization of knowledge spillovers as evidenced by patent citations. Q J Econ 108(3):577–598
Jin J (2009) Knowledge diffusion in mobile phone industry. MA thesis, Economics Department, Seoul National University
Kushida KE (2008) Wireless bound and unbound: the politics shaping cellular markets in Japan and South Korea, is forthcoming. J Inf Technol Polit/Telecommun Policy Rev (in Korean) 13(3):91–119
Lee K, Wang L (2010) Science and technology institutions and performance in China: the semiconductor industry. In: Xiaoming H (ed) Institutional dynamics of China's great transformation. Routledge, New York/London
Lee K, Yoon M (2010) International, intra-national, and inter-firm knowledge diffusion and technological catch-up: the US, Japan, Korea, and Taiwan in the memory chip industry. Technol Anal Strateg Manage 22(5):553–570
Lee K, Cho S-J, Jin J (2009) Dynamics of catch-up in mobile phones and automobiles in China: a sectoral innovation system perspective. China Econ J 2(1):25–53
Lee K, Lim C (2001) Technological regimes, catch-up and leapfrogging: findings from the six industries in Korea. Res Policy 30(3):459–483
Mahlich J, Pascha W (2007) Innovation and technology in Korea. Physica, New York
Mahmood IP, Singh J (2003) Technological dynamism in Asia. Res Policy 32:1031–1054
Mani S (2007) Keeping pace with globalization: innovation capability in Korea's telecommunication equipment industry. In: Mahlich J, Pascha W (eds) Innovation and Technology in Korea. New York: Physica-Verlag
Mathews JA (2002) The origins and dynamics of Taiwan's R&D consortia. Res Policy 31:633–651
Saxenian A, Hsu JY (2001) The Silicon Valley-Hsinchu connection: technical communities and industrial upgrading. Ind Corpor Change 10(4):893–920
Yang H, Yoo Y, Lyytinen K, Joong-Ho A (2003) Diffusion of broadband mobile services in Korea: the role of standards and its impact on diffusion of complex technology system. Case Western Reserve University, Cleveland
Yang T-P (2004) Exploration of business strategies of mobile phone industry in Taiwan (in Chinese), Master's thesis, Department of Economics, Mingchuan University
Zhu HY, Yang Y, Tintchev MT, Wu GSH (2006) The interaction between regulation and market and technology opportunities: a case study of the Chinese mobile phone industry, innovation management. Policy Pract 8(1–2):102–112

Chapter 13
Comment On: From Learning Knowledge Outside to Creating Knowledge Within: Korean Mobile Phone Industry Compared with Those of Japan, Taiwan and China, by Lee and Jin

Bernhard Dachs

How were firms from East Asian countries able to become technology leaders in a number of important fields? What sources of knowledge did they utilize in the early phases of their expansion? Many readers would assume that these firms learned from the industry leaders at the time, in particular from the US and European firms. However, Lee Keun and Jin Jia come to a different, somewhat surprising answer. They use patent portfolios to demonstrate that in the mobile phone industry, latecomer firms have a tendency to draw on knowledge not from the industry leaders but from their immediate predecessors.

Lee and Jin employ patent data from the US Patent and Trademark Office for their analysis. Based on patent citations, they show that during their entry in the mobile phone industry, Japanese firms tended to cite US patents, while Korean firms had a tendency to cite Japanese patents, and, only to a lesser degree, US patents. This pattern continues with Taiwanese firms who showed a tendency to learn from Korean firms, while Chinese firms tended to cite Taiwanese firms during their industry entry phase.

This 'leapfrogging' pattern is not new in the literature. We can already find it in previous studies on international technology diffusion, for example in the work of Hobday (1995), or Peréz and Soete (1988). Moreover, the idea of leapfrogging should not be confused with the idea of the 'fast second' (Geroski and Markides 2005). The story Lee and Jin tell is not about radical innovation and subsequent commercialisation, but about incremental development of new technology.

The paper demonstrates that patent data holds some strengths for the analysis of technological change: Lee and Jin combine the analysis of aggregate behavior and a sound knowledge of the evolution of the mobile phone industry with very detailed information on the behavior of individual actors, such as universities, governmental research centers or individual firms. This allows them to relate patterns of knowledge diffusion to differences in the industrial structure and institutional set-up of countries. The lower degree of inter-firm knowledge diffusion in Korea compared to Taiwan, for example, is explained by the role of large conglomerates in the Korean economy, which have no equivalent in the Taiwanese economy. Public research centers were more important patent holders in Taiwan than in Korea

J. Mahlich and W. Pascha (eds.), *Korean Science and Technology in an International Perspective*, DOI 10.1007/978-3-7908-2753-8_13,

during entry due to size disadvantages, while Korea had critical masses of R&D in their large firms.

The paper also delivers some interesting sub-plots besides the main story. We learn about the role of standards in catch-up: For 2G mobile communications, Japan chose to adapt a self-developed standard (PDC), rather than use the international GSM or CDMA standards. This 'go-it-alone' strategy led to high domestic citations and a strong domestic market presence, but Japanese firms lost international markets after 2000. We also get some insights into technological regimes and the interplay of technological trajectories, the cumulativeness of the knowledge base, and market entry: a shift in the technological trajectory often serves as a window of opportunity for latecomers. Korean, Taiwanese and Chinese firms could enter the market more easily because new generations of mobile communication and new standards eroded the advantages of incumbents.

This finding, however, also points to some further questions: how general are the results of this paper? Can we find similar patterns in other industries, or are the mobile phone and semiconductor industry cases special? One may assume that the pattern revealed by Lee and Jin depends very much on the propensities of the technological regime involved, such as the degree of cumulativeness of the knowledge base, the relative importance of tacit knowlege, the breadth of competences required, or the need to link industrial R&D to academic research (Marsili 2001). Industries with different knowlege bases may therefore reveal different patterns (Lee and Lim 2001). Leapfrogging may be easier in the mobile phone industry than in pharmaceuticals or in industries with a high degree of more tacit process knowledge.

Some readers may also be concerned by the data employed in the paper. Lee and Jin use patent citation data from the US Patent and Trademark Office to trace knowledge flows between patenting firms. Patent citations, however, are not only added by the inventor, but also by the patent office examiner (Criscuolo and Verspagen 2008; Alcácer et al. 2009). In some cases, patent citations may therefore reflect a legal requirement (distinguishing a new patent application from prior art) rather than provide evidence of knowledge flows. The debate over how severely this feature depletes the usefulness of patent data for the analysis of technological change has not yet been settled. Many authors (including myself) see more advantages than disadvantages in patent data. It would, however, be very useful to get complementary information on knowledge flows that support the leapfrogging pattern.

The outcomes of patent analysis may also depend very much on the choice of the patent office (Dernis and Guellec 2001). European firms tend to patent at the European Patent Office, while US firms tend to go to the US Patent and Trademark Office first. Would the results look differently with data from the Japanese, Korean or European Patent offices?

It would also be very interesting to learn more about the role of European firms in the process of leapfrogging. Nokia is currently the world's largest producer of mobile phones and its market share was already high in the 1990s and the years after 2000. Would the story be different if the authors included Nokia, Ericsson, or

Siemens in their analysis? Table 3 in Lee's and Jin's contribution indicates that Japanese firms indeed cited European ('Others') patents more frequently than US patents. The same is true for Chinese firms after 2000.

Finally, is China's leapfrogging the end of the story? Or is there another emerging country after China which may benefit from China's knowledge? It is, of course, impossible to give a definite statement on this question based on present data. The fast rate of technological change in the mobile phone industry, convergence between different types of media, and the high rate of market entrants, however, suggests that we have not seen the end of the story.

References

Alcácer J, Gittelman M, Sampat B (2009) Applicant and examiner citations in U.S. patents: an overview and analysis. Res Policy 38(2):415–427

Criscuolo P, Verspagen B (2008) Does it matter where patent citations come from? inventor vs. examiner citations in european patents. Res Policy 37(10):1892–1908

Dernis H, Guellec D (2001) Using patent counts for cross-country comparisons of technology output. STI Rev 2001(27):129–147

Geroski PA, Markides C (2005) Fast second: how smart companies bypass radical innovation to enter and dominate new markets. Jossey-Bass, San Francisco

Hobday M (1995) East asian latecomer firms: learning the technology of electronics. World Dev 23(7):1171–1193

Lee K, Lim C (2001) Technological regimes, catching-up and leapfrogging: findings from the korean industries. Res Policy 30(3):459–483

Marsili O (2001) The anatomy and evolution of industries: technological change and industrial dynamics. Edward Elgar, Cheltenham/Northampton

Peréz C, Soete L (1988) Catching up in technology: entry barriers and windows of opportunity. In: Dosi G, Freeman C, Nelson R (eds) Technical change and economic theory. Pinter, London

Chapter 14
Relational Capital, Knowledge Transfer and Performance in International Joint Ventures (IJVs) in Korea

Chol Lee, Chan-Soo Park*, and Ilan Vertinsky

Introduction

Forming international joint ventures has become a fashionable strategy through which Korean business firms add value by augmenting their resources and exploiting synergies in a variety of activities such as R&D, procurement, distribution and marketing. Joint ventures offer better ways to cope with the competitive challenges of rapid technological change and internationalization of markets (Harrigan 1986). An important dimension of the benefits that accrue from IJV formation is the transfer of knowledge from the foreign parents to the IJV. Some studies have shown, however, that such alliances experience high failure rates, ranging from 50% to 70% (e.g. Park and Ungson 1997). Other studies indicate that most alliances have failed to fully accomplish their goals (Das and Rahman 2001).

Alliances are complex and may form unstable organizations (Inkpen and Ross 2001). Key reasons for the high mortality rates include opportunistic behavior by partners (Hamel 1991) and inattention to relationship issues such as trust and commitment (Kauser 2007). Indeed, relationship issues seem to be forgotten when launching international joint venture agreements (Kauser 2007). Despite the recognition that social relationships are critical to the survival and success of IJVs, only a few studies have adequately examined how IJV partners can successfully manage alliance relationships and how relationships affect IJV success (Choi 2004).

Parkhe (1991) argued that the success of an alliance between IJV partners is related to the partners' characteristics. Yet prior studies have failed to identify and adequately assess the dimensions of partners' characteristics that affect performance. Studies on relational capital, which are still at an early stage, and studies regarding relationship management leave important gaps in the IJV literature (Robson et al. 2006). Robson et al. (2006) argued that the study of relational capital is characterized by insufficient conceptualization, research designs and analytical methods, which generate inconsistent findings.

*Corresponding author

J. Mahlich and W. Pascha (eds.), *Korean Science and Technology in an International Perspective*, DOI 10.1007/978-3-7908-2753-8_14,

In particular, there are inconsistent findings on the influence of relational capital on knowledge transfer and performance. For example, several studies examining performance in the context of relational capital found trust to be positively associated with an alliance's performance (Ramaseshan and Loo 1998; Lane et al. 2001) but other studies (Sarkar et al. 2001; Inkpen and Birkenshaw 1994; Currall and Inkpen 2002) found no significant relationship. Several studies revealed that commitment was positively associated with alliance performance (Cullen et al. 2000; Hyder and Ghauri 2000; Tsang et al. 2004), whereas others indicated no such linkage (Lee and Beamish 1995; Demirbag and Mirza 2000; Zeybek et al. 2003). With regard to communication, some studies (Hyder and Ghauri 2000; Zeybek et al. 2003; Ramaseshan and Loo 1998) supported a positive association with alliance performance, yet other studies revealed no significant association (Sarkar et al. 2001; Aulakh et al. 1996; Fryxell et al. 2002) or found a negative association.

The IJV literature emphasizes that knowledge acquisition is an important rationale for forming an alliance, reflecting an expectation of a positive relationship between knowledge acquisition and IJV performance (Lyles and Salk 2007; Hamel 1991). Alliances are often considered the most effective mechanism with which to transfer knowledge to other firms (Inkpen 2005; Kogut 1988). However, knowledge transfer, through alliances, is very difficult because social relational aspects such as trust, commitment and common goals play an important role in knowledge transfer and require significant investment and time to develop (Minbaeva 2007).

Our study contributes to the IJV literature in several ways. First, it develops a theoretical model that explores both the direct effects of relational capital on performance as well as the effects of relational capital on performance that flows through knowledge transfer as a mediator. The partial mediation model may resolve ambiguities in prior findings in the literature. Previous studies of IJV have emphasized the importance of relational capital but have not attempted an integrated framework involving the impact of relational capital on knowledge transfer and IJV performance as well as the impact of knowledge transfer on performance (Robson et al. 2006).

Second, while some theories have explained how the characteristics of alliance partners can critically influence performance, few identified which dimension of alliance relational capital has more impact on IJV performance. We contribute empirically to the literature by examining the significance of the impact that the key components of relational capital have on the transfer of knowledge and IJV performance. This study confirms the premise that relational capital has an impact on knowledge transfer and the success of an IJV (Kauser and Shaw 2004; Dhanaraj et al. 2004; Robson et al. 2006). We identify parental commitment and strength of the communication system as the key relational attributes that explain IJV performance. Surprisingly, trust did not contribute significantly to explaining variations in performance.

The structure of the remainder of this paper is as follows: firstly, we provide a literature review on alliance relational capital and knowledge transfer. Secondly, we offer an integrated model of partner relational capital, knowledge transfer and performance. Thirdly, we present the results of testing our hypotheses and discuss their theoretical implications. Finally, we conclude by providing managerial implications.

Theory

The Role of Relational Capital

Relational capital is defined as "the long and sticky nature of the relationship that constitutes a realistic alternative to egotistical power and control strategies" (Robson et al. 2006). It provides "the foundation for partners' investments in time and effort toward building positive feelings and interaction patterns in the IJV" (Robson et al. 2006). The dominant theories of alliance behavior such as transaction cost theory, resource based view theory and knowledge based view theory emphasize key elements of relational capitals such as trust and commitment as factors which differentiate a relationship based alliance from an arm's-length exchange.

A successful relationship requires the softer style of alliance management based on trust and commitment, which are the essential elements for a successful relationship within an IJV (Morgan and Hunt 1994; Cullen et al. 2000). Relational capital becomes more important for IJV success because not all issues arising in the IJV can be fully covered within IJV contracts (Cullen et al. 2000). The soft side of an alliance "relationship" consists of the socio-psychological attributes of the alliance that are positive to the alliance's success (Cullen et al. 2000). Dhanaraj et al. (2004) asserted that a "socially embedded relationship" goes beyond formal contracts and yet formal contracts define clear boundaries in terms of a partner's loyalty. The success of an IJV depends more upon informal processes (e.g. relational capital) than formal processes (e.g. alliance structure) (Kauser and Shaw 2004; Robson et al. 2006). An informal process underlines the development of relational capital which is important to the relationship (Morgan and Hunt 1994).

How firms learn and develop knowledge and skills has been considered for three levels of exchange: markets, hybrids and hierarchies (Powell 1990). Trust is the most critical factor in hybrids because of potential opportunism and market uncertainty (Luo 2002). Within hierarchies, the maintenance of common goals and systems is important for transferring knowledge (Brown and Duguid 2001).

In this study, we suggest three indicators for relational capital in IJVs (see Table 14.1): trust, commitment and communication (Sarkar et al. 2001). Extending previous work in the literature, we define relational capital in the IJV as the degree to which 'the relationships are embedded in social attachments' between the alliance partners (Robson et al. 2006; Uzzi 1997; Dhanaraj et al. 2004).

Knowledge Transfer

Several theoretical approaches such as the resource-based view of the firm, organizational knowledge transfer theory and the knowledge-based view have made contributions to the study of knowledge transfer (Minbaeva 2007). Knowledge is more easily transferred in alliances than through market transactions (Shenkar and

Table 14.1 Definitions of relational capital

Relational capital	Definitions	Sources
Trust	The degree of confidence shared by the partners regarding the other's integrity	Uzzi (1997), Cullen et al. (2000), Moorman et al. (1993), Sarkar et al. (2001)
Commitment	An implicit and explicit pledge of relational continuity between exchange partners to maintain the alliance relationship	Sarkar et al. (2001), Dwyer et al. (1987)
Communication	The existence of effective formal and informal mechanisms to share information between partners and the willingness of partners to share meaningful information	Anderson and Narus (1990)

Li 1999). An alliance is therefore an efficient mode to compensate for deficiencies in a firm's knowledge.

It is important for global firms to know how new organizational knowledge is acquired (Inkpen 1998). Knowledge has been considered a key resource by firms seeking competitive advantage (Teece et al. 1997). Knowledge provides the foundation for new skills, leading to competitiveness (Inkpen 1998). Arguably, knowledge acquisition is a primary motivation for engaging in an IJV. IJV success is determined by the extent to which the firm is able to increase its knowledge drawing on its parent's knowledge resources (Kogut 1988; Hamel 1991; Suseno and Ratten 2007).

Knowledge transfer is defined as a knowledge combination and knowledge creation (Bartlett and Ghoshal 1989) of each partner's "know-how" and "know-what" to generate IJV performance. Knowledge transfer depends on characteristics of the knowledge itself, its senders and receivers and the relationship between the senders and receivers (Minbaeva 2007). Knowledge is more effectively transferred when close relationships are established between partners (Minbaeva 2007) because attributes of socio psychological behavior may play a critical role in knowledge transfer. Effective knowledge transfer often requires informal processes (e.g. 'give and take', communication, etc.) because a relationship built on a formal contract in this case may not be enough as asymmetries of knowledge imply that in most cases contracts are incomplete (Dhanaraj et al. 2004).

Knowledge acquisition through alliances occurs in multiple stages. The first stage is the formation of the alliance and interactions between individuals. The second stage is the internalization process, which means the transfer of individually acquired knowledge to an organization's knowledge database. Despite a number of academic papers investigating knowledge transfer as a precursor to a successful alliance, there is still little understanding how knowledge transfer is facilitated (Suseno and Ratten 2007). In this paper we focus on the key behavioral attributes (relational capital) that enhance knowledge transfer. In our empirical analysis we control for a variety of factors that relate to other important elements of the transfer process such as the absorptive capacity of the receiver (the IJV) and the dissemination capacity of the transmitter (the foreign parent). We also control for other factors that may affect performance.

Hypotheses

A conceptual framework presented in Fig. 14.1 identifies the role of relational capital and how it is assumed to positively influence knowledge transfer and facilitate a more effective IJV performance. The key indicators of relationship capital (trust, commitment, communication) are adapted from Sarkar et al. (2001). Previous studies of IJVs found that relational capital has a positive impact on knowledge transfer and performance because relational capital increases the quality of an inter-partner relationship.

Relational Capital, Knowledge Transfer, and Performance

The sociopsychological aspects embedded in relational capital are important since they determine the quality of the relationship in the collaboration (Sarkar et al. 2001). One very important factor found to contribute to the degree of knowledge transfer was the quality of the relationship between partners (Argote 1999).

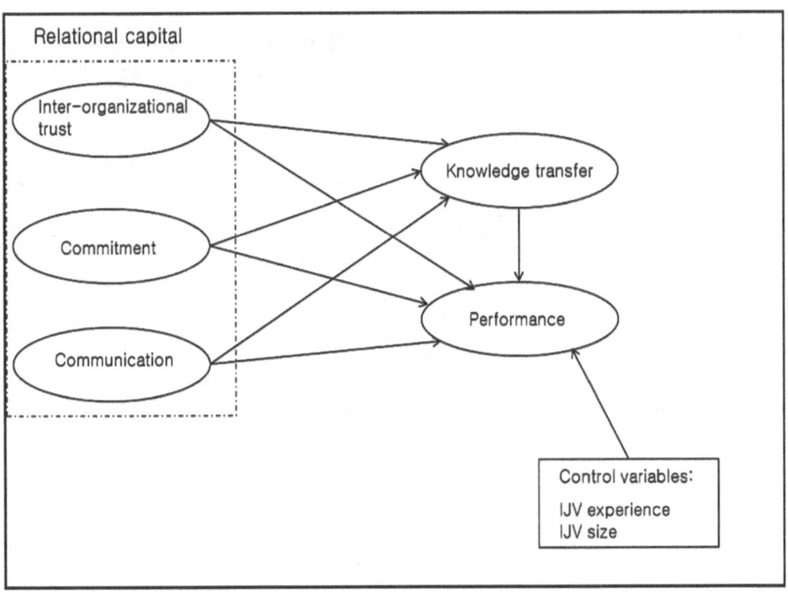

Fig. 14.1 Conceptual framework

Trust is one of the more frequently cited elements in analyzing relational (or social) capital. The more trust present between the IJV partners, the lower their transaction costs. Trust is a critical factor in the IJV because of potential opportunism and market uncertainty (Luo 2002). Trust between partners can reduce transaction costs by preventing opportunism and market uncertainty thus reducing monitoring and enforcement costs as well as other costs involved in guarding against opportunistic behavior. Lower risk of opportunism facilitates knowledge exchange as those with knowledge resources do not need to constrain knowledge flows to guard against undesirable spillovers (Luo 2002).

Commitment refers to the extent to which each party is willing to exert effort, and demonstrate its commitment through its willingness to maintain the relationship (Kauser 2007; Morgan and Hunt 1994). Commitment reflects the fact an alliance partner believes the relationship is so important as to warrant maximum effort (Sarkar et al. 2001).

Communication is a very important factor because the knowledge needs to be effectively gathered, transferred and learned from the alliance partners (Nonaka 1991) to create a competitive advantage. Communication reflects the ability of the partners to share information regarding organization skills and resources (Inkpen and Birkenshaw 1994) and is critical to knowledge transfer. Knowledge is effectively transferred in the IJV with open communication.

Drawing from the above arguments, it can be expected that higher accumulation of relationship capital with respect to the alliance is associated with higher levels of knowledge transfer.

Hypothesis 1: Relational Capital Will Have a Positive Impact on Knowledge transfer from the Foreign Parent to the IJV

Prior studies have found also that relational capital variables influence alliance performance directly (Aulakh et al. 1996; Bradrach and Eccles 1989). Some studies have revealed that such relational attributes positively influence IJV performance (Morgan and Hunt 1994; Dwyer et al. 1987) while others have indicated no significant influence. Robson et al. (2006) indicated that even though there is a direct association between relational capital and alliance performance, the effect of each relational capital component (behavioral attribute) may vary.

Relational capital is considered to be a substitute for hierarchical governance, thus accomplishing organizational objectives in inter-firm partnerships. As conditions present opportunities for the partner firm to exploit each other's vulnerabilities, relational capital becomes particularly important to explain performance. For example, when a firm has invested in important assets specific to its relationship with the foreign partner, relational capital is needed to offset the possibility of opportunistic behavior, having a positive influence over alliance performance (Aulakh et al. 1996). Relational capital plays a significant role in reducing transaction, monitoring and enforcement costs thus contributing to higher performance in IJVs.

Hypothesis 2: Relational Capital Will Have a Positive Impact on Performance

The Influence of Knowledge Transfer on Performance

Knowledge is essential to create competitive advantages (Kogut and Zander 1992). Successful knowledge transfer is a precursor for positive IJV performance, which is an important prerequisite for an alliance's success (Bresman et al. 1999). IJV performance is measured by the degree of satisfaction, the extent to which mutual desired goals are achieved in terms of increasing sales revenue, increasing market share, achieving planned goals and making profit (Dhanaraj et al. 2004).

Prior studies support a positive relationship between knowledge transfer and IJV performance (Lyles and Salk 2007; Dhanaraj et al. 2004; Inkpen and Beamish 1997). The relationship between knowledge transfer and performance is very significant throughout the acquisition of a partner's knowledge (Lyles and Salk 2007). The knowledge transfer builds the organizational capabilities leading to enhanced performance (Dhanaraj et al. 2004). The skills and know-how that improve the quality of products and organizational efficiency result in better performance.

Hypothesis 3: Knowledge Transfer from Parents Will Have a Positive Impact on the Performance

Methodology

Sample Selection

We collected data by fax, web surveys and emails from managers involved in IJV operations in Korea. We sent a fax and email plus a cover letter to firms who had agreed to participate after a telephone call and took steps to increase the response rate while conducting the survey. The sample frame was derived from data in the *Foreigner Investment Statistical Yearbook* published by the Republic of Korea's Ministry of Knowledge Economy. To increase the number of observations, we participated in the 2008 foreign investment companies job fair (www.2008jobfair.org) to collect additional data from IJV managers.

South Korea represents an excellent empirical context to study the impacts of relational capital on knowledge transfer and knowledge transfer on performance. South Korea is a newly industrialized country where foreign knowledge transfers have contributed and continue to contribute significantly to rapid economic development. The country has attracted a large number of foreign IJV investors from geographically, culturally, and economically diverse countries. Studying relationships between IJVs and foreign firms in Korea ensures diversity of foreign parent origins and is likely to produce more robust findings.

The Korean environment, however, best represents the context of a young industrialized country with a highly collectivistic culture. This context can provide insights, for example, about the future of rapidly industrializing countries such as China. To the extent that our findings are culture bound, we expect that findings concerning the importance of relational capital are likely to be more significant compared to those that can be found in IJVs in individualistic cultures with parents from similar cultures.

Sample Characteristics

Overall, a total of 337 surveys were returned for a response rate of 16.1% (337/2094). After deleting the unusable responses, there were 334 IJVs in the final data sample. We included only IJVs that had been in business more than 3 years to allow sufficient time for relational capital to be built and influence behavioral attributes in terms of knowledge transfer. We also limited the target sample to IJVs having more than 30% and less than 70% of foreign equity because we wanted to examine knowledge transfer from the foreign parents to the Korean IJV. The average years of operation of the IJVs (N = 334) is 14.05. The minimum number of employees is 1 and the maximum is 2,000, with 140 as the average. Of the 334 firms, 179 (56.6%) are of Asian origin. Sixty-eight (20.4%) firms are of European origin, 54 of North American (16.2%) origin and three in Oceania (0.9%) (see Table 14.2). The distribution by industrial sector of the firms in our sample highlights the prevalence of IJVs in sectors where transfers of technology are important such as electronics, chemicals and machinery, but also indicates the large number of IJVs occurring in low technology sectors such as retail and distribution (see Table 14.3).

Measures

Sarkar et al. (2001) identified three dimensions of relational capital; trust, commitment and communication. These have been highlighted in prior studies as factors which differentiate a relationship based from an arm's length exchange (Sarkar et al. 2001; Heide and John 1992; Morgan and Hunt 1994). Table 14.4 identifies the items used in each construct. All scales measuring the latent variables were validated and refined using Confirmatory Factor Analysis.

Table 14.2 National origin of partner firms

Continent	Country	Number	%
	Netherlands	7	2.1
	Germany	24	7.2
	UK	6	1.8
	France	16	4.8
	Switzerland	4	1.2
	Luxemburg	3	0.9
	Italy	2	0.6
	Norway	4	1.2
Europe	Austria	2	0.6
	Japan	149	44.6
	China	15	4.5
	Pakistan	2	0.6
	Hong Kong	4	1.2
	Singapore	13	3.9
Asia	Taiwan	6	1.8
	U.S.	53	15.9
North America	Canada	1	0.3
Oceania	Australia	3	0.9
Others	Others	20	6.0
Total		334	100.0

Table 14.3 Industry classification

Industry	Number	Percent
Metal	22	6.6
Finance and insurance	7	2.1
Machinery	32	9.6
Manufacturing	5	1.5
Retail and distribution	61	18.3
Real Estate	2	0.6
Nonmetal	5	1.5
Business services (public service)	33 (2)	9.9
Textile	4	1.2
Food	10	3.0
Construction	2	0.6
Transportation machinery	28	8.4
Logistics	25	7.5
Lodging	4	1.2
Chemical (medical)	46 (1)	13.8
Electronics	43	12.9
Paper	3	0.9
Telecommunications	2	0.6
Total	334	100.0

Table 14.4 Measurement

		Alpha	Item-total correlation
Trust		.939	
TRUST_1	1. Understand each other		.734
TRUST_2	2. Have never had the feeling of being misled		.684
TRUST_3	3. Avoid damaging demands		.732
TRUST_4	4. Expect not to pursue its interest		.690
TRUST_5	5. Accept informal agreements as formal contracts		.752
TRUST_6	6. Do not take advantage of the other		.752
Commitment		.952	
COMMT_1	1. Dedicate whatever people and resources are needed		.799
COMMT_1	2. Provide experienced and capable people to IJV		.805
COMMT_1	3. Committed to making IJV a success		.804
Communication		.930	
COMMU_1	1. Communication is inaccurate/accurate		.831
COMMU_2	2. Communication is not credible/credible		.846
COMMU_3	3. Ask for advice when making decisions		.848
COMMU_4	4. Inform in advance of changing needs		.858
COMMU_5	5. Share important information with partner		.738
COMMU_6	6. Participate in regular meetings		.666
Knowledge transfer		.926	
KT_1	1. Written knowledge about the technology		.755
KT_2	2. Procedural manuals or technical manuals		.786
KT_3	3. Written knowledge about management techniques		.780
KT_4	4. New marketing expertise		.806
KT_5	5. Knowledge about foreign cultures and tastes		.800
KT_6	6. Managerial techniques		.796
Performance		.933	
PERFOR_1	1. Key managers in the Korean parent would rate the JV's performance as		.826
PERFOR_2	2. Key managers in the foreign parent would rate the JV's performance as		.898
PERFOR_3	3. You would rate the JV's performance as		.867

Structural Equation Modeling

Structural equation modeling (SEM) is particularly appropriate when testing models constructed from path analytic frameworks containing several latent variables (Fornell and Larcker 1981; Dhanaraj et al. 2004). SEM is a multivariate technique combining several aspects of factor analysis and multiple regressions enabling the researcher to simultaneously examine a series of interrelated dependent relationships among the measured variables and latent constructs as well as those between several latent constructs (Hair et al. 2006). We have tested the research model in Fig. 14.1 using the AMOS version 17.0. The unit of observation was the IJV.

Results

Table 14.5 shows the structural model with unstandardized coefficients. The relative effect of relational capital on knowledge transfer is shown in Table 14.5, which also provides the path coefficients.

Hypothesis 1 suggests relational capital has a positive effect on knowledge transfer. The coefficients representing commitment and communication in relation to knowledge transfer are positive and significant, suggesting commitment and communication are important for knowledge transfer in IJVs. However, trust does not influence knowledge transfer. Thus Hypothesis 1 is not fully supported.

Hypothesis 2 predicts a positive relationship between relational capital and performance. The links between commitment and performance, and communication and performance are significant. The link between trust and performance is insignificant. Hypothesis 2 is not fully supported.

Hypothesis 3 suggests knowledge transfer has a positive impact on performance. The coefficient for knowledge transfer on performance is positive and significant ($p < 0.001$). Hypothesis 3 is fully supported.

Discussion

Transfer of knowledge from foreign parents to an IJV was identified by many researchers as a prime motive for formation of IJVs and a key to their success. The foci of this study are the direct impacts of relational capital on transfer of knowledge to IJVs and their performance as well as the mediated impact of relational capital on IJVs performance through knowledge transfer. We have developed and tested an integrated model which accounts for both the direct impact of relational

Table 14.5 Path analysis: structural model coefficients

		Path	Estimate	S.E.	C.R.	P
Knowledge	<--	Commitment	0.252	0.068	3.689	***
Knowledge	<--	Communication	0.507	0.118	4.29	***
Knowledge	<--	Trust	−0.012	0.101	−0.12	0.905
Performance	<--	Commitment	0.248	0.07	3.553	***
Performance	<--	Trust	0.048	0.102	0.467	0.641
Performance	<--	Knowledge	0.211	0.064	3.273	0.001
Performance	<--	Communication	0.255	0.119	2.146	0.032

The C.R. (critical ratio) is equivalent to the t-statistic in a regression model to compute the statistical significance of the coefficient estimates capital on performance as well as its mediated impact through facilitation of knowledge transfer. We contend that the partial mediation model developed provides a more accurate account of the total impact of relational capital on performance. It also highlights the role relational capital plays in improving knowledge transfer and providing empirical assessment of the importance of knowledge transfer to the success of the IJV.

Our empirical results suggest that not all the elements identified in the literature as key ingredients to the development of relational capital have a significant impact on an IJV's performance or transfer of knowledge from its parents. While, as hypothesized, commitment and communication were shown to significant impacts on both the transfer of knowledge and performance, trust did not contribute significantly to the explanatory powers of the model. This result is consistent with the findings of Lane et al. (2001) who suggested that perhaps trust between the parents of an IJV may not reflect their relationship to the IJV. In our sample, almost all top managers of the IJV were former employees of the Korean parent thus it was unlikely that trust relationships could exist with the IJV and not with its Korean parent. There are several other possible explanations why the links between trust and knowledge transfer, and trust and performance were not significant. We have ruled out the possibility that this resulted from low variation in trust levels in our sample. An alternative explanation may lie in the nature of trust.

Trust is a measure of the willingness of the partners in an alliance to take risk by taking actions which may make them vulnerable to the opportunistic behavior of other partners. In an environment where the risk of opportunistic behavior is low, trust is less important to the formation of effective relationships. High levels of alignment in the objectives of partners with respect to the IJV, clear contractual arrangements and strong enforcement of contracts and the ability of betrayed partners to retaliate in future transactions, reduce the risk of opportunism and thus the importance of non-calculative trust.

It is likely, however, that trust is an important pre-condition to the formation of a commitment to the joint venture and to the establishment of an effective communication system, both key elements which we found to positively facilitate knowledge transfer and improve performance. The formation of commitment and the establishment of effective communication systems are especially important at the early stages of IJV life and less important for mature IJVs such as those in our sample.

The theoretical implication of our findings suggest that modeling explicitly the interrelation between the different elements of relational capital is required to understand more fully the role that trust plays in the formation of relational capital.

Our study also contributes to empirical knowledge by confirming the significant contribution transfers of knowledge make to IJV performance. The managerial implications of our study are obvious. The development of relational capital is an important precondition to IJV success. Fostering an effective credible communication system, alignment of objectives and the commitment of parents to the IJV are key to strong performance.

Conclusion

Prior studies highlighted the importance of relational capital but have not accounted fully for its direct and indirect impacts. Our study highlights the importance of partially mediated impacts of relational capital on performance through their impacts on knowledge transfer.

Our findings suggest that IJV partners in Korea can improve performance by better understanding the roles that relational capital plays and taking steps to facilitate its formation. In particular, steps should be taken to improve communication and ensure the parents' commitment to the IJV.

Acknowledgements We acknowledge the assistance and support of Jeremy Kim, Diana Lam, Irene Kang, and Angela Ahn. This work was supported by the National Research Foundation of Korea Grant funded by the Korean Government (NRF-2009-327-B00306).

References

Anderson J, Narus J (1990) A model of distributor firm and manufacturer firm working partnerships. J Mark 54:42–58

Argote L (1999) Organizational knowledge transfer: creating, retaining, and transferring knowledge. Kluwer, Boston

Aulakh PS, Kotabe M, Sahay A (1996) Trust and performance in cross-border marketing partnerships: a behavioral approach. J Int Bus Stud 27:1005–1032

Bartlett C, Ghoshal S (1989) Managing across borders. Harvard Business School Press, Boston

Bradrach JL, Eccles RG (1989) 'Price, authority, and trust: from ideal types to plural forms. Annu Rev Sociol 15:97–118

Bresman H, Birkinshaw J, Nobel R (1999) Knowledge transfer in international acquisitions. J Int Bus Stud 30(3):439–462

Brown JS, Duguid P (2001) Knowledge and organization: a social-practive perspective. Organ Sci 12:198–213

Choi YT (2004) Alliance coordination effectiveness and the performance of international joint ventures: development of the partnership and moderating role of market environment turbulence. Doctoral dissertation, Texas A&M University

Cullen JB, Johnson JL, Sakano T (2000) Success through commitment and trust: the soft side of strategic alliance management. J World Bus 35(3):223–240

Currall SC, Inkpen AC (2002) A multilevel approach to trust in joint ventures. J Int Bus Stud 33(3):479–495

Das TK, Rahman N (2001) Partner misbehaviour in strategic alliances: guidelines for effective deterrence. J Gen Manage 27(1):43–70

Demirbag M, Mirza H (2000) Factors affecting international joint ventures success: an empirical analysis of foreign-local partner relationship and performance in joint ventures in Turkey. Int Bus Rev 9:1–35

Dhanaraj C, Lyles MA, Steensma HK, Tihanyi L (2004) Managing tacit and explicit knowledge transfer in IJVs: the role of relational embeddedness and the impact on performance. J Int Bus Stud 35:428–442

Dwyer FR, Schurr PH, Oh S (1987) Developing buyer-seller relationships. J Mark 51:11–28

Fornell C, Larcker DF (1981) Structural equation models with unobservable variables and measurement error: algebra and statistics. J Marketing Res 18:382–388

Fryxell GE, Dooley RS, Vryza M (2002) After the ink dries: the interaction of trust and controlling US-based IJV. J Manage Stud 39(6):865–886

Hair JF, Black WC, Babin BJ, Anderson RE, Tatham RL (2006) Multivariate data analysis. Pearson Prentice Hall, New Jersey

Hamel G (1991) Competition for competence and inter-partner knowledge transfer within international joint ventures. Strateg Manag J 12:83–103

Harrigan KR (1986) Managing for joint ventures success. D.C. Heath, Lexington

Heide JB, John G (1992) Do norms matter in marketing relationships? J Mark 56:32–44

Hyder AS, Ghauri PN (2000) Managing international joint venture relationships. Ind Mark Manage 29(3):205–218

Inkpen AC (1998) Knowledge transfer and knowledge acquisition through international joint ventures. Acad Manag Exec 12(4):69–80

Inkpen AC (2005) Knowledge transfer through alliance: general motors and nummi. Calif Manage Rev 47(4):114–136

Inkpen AC, Beamish PW (1997) Knowledge, bargaining power, and the instability of international joint ventures. Acad Manage Rev 22(1):177–202

Inkpen AC, Birkenshaw J (1994) International joint ventures and performance: an interorganizational perspective. Int Bus Rev 3(3):201–217

Inkpen AC, Ross J (2001) Why do some strategic alliances persist beyond their useful life? Calif Manage Rev 44(1):132–148

Kauser S (2007) Alliance relationship dynamics: conflict, structure and control. J Euro-Mark 16(3):5–25

Kauser S, Shaw V (2004) The influence of behavioral and organizational characteristics on the success of international joint ventures. Int Mark Rev 21(1):17–52

Kogut B (1988) Joint ventures: theoretical and empirical perspectives. Strategic Manage J 9(4):319–332

Kogut B, Zander U (1992) Knowledge of the firm, combinative capabilities, and the replication of technology. Organ Sci 3(3):383–397

Lane PJ, Salk JE, Lyles MA (2001) Absorptive capacity, learning, and performance in international joint ventures. Strateg Manag J 22(12):1139–1161

Lee C, Beamish PW (1995) The characteristics and performance of korean joint ventures in LDCs. J Int Bus Stud 26(3):637–654

Luo Y (2002) Contact, cooperation, and performance in international joint ventures. Strateg Manag J 23(10):903–919

Lyles MA, Salk JE (2007) Knowledge acquisition from foreign parents in international joint ventures: an empirical examination in the Hungarian context. J Int Bus Stud 38(1):3–18

Minbaeva DB (2007) Knowledge transfer in multinational corporations. Manag Int Rev 47(4):567–593

Moorman RH, Niehoff BP, Organ DW (1993) Treating employees fairly and organizational citizenship behaviors: sorting the effects of job satisfaction, organizational commitment, and procedural justice. Empl Responsibilities Rights J 6:209–225

Morgan RM, Hunt SD (1994) The commitment-trust theory of relationship marketing. J Mark 58(3):20–39

Nonaka I (1991) The knowledge creating company. Harv Bus Rev 69(6):96–104

Park SH, Ungson GR (1997) The effect of national culture, organizational complementarity, and economic motivation on joint venture dissolution. Acad Manage J 40(2):279–307

Parkhe A (1991) Interfirm diversity, organizational knowledge transfer, and longevity. J Int Bus Stud 22(4):579–601

Powell W (1990) Neither market nor hierarchy: network forms of organizations. In: Cummings LL, Staw B (eds) Research in organizational behavior, vol 12., pp 295–336

Ramaseshan B, Loo PC (1998) Factors affecting a partner's perceived effectiveness of strategic business alliance: some Singaporean evidence. Int Bus Rev 7:443–458

Robson MJ, Skarmeas D, Spyropoulou S (2006) Relational capital and performance in international joint ventures: review and future directions. Int Marketing Rev 23(6):585–609

Sarkar MB, Echambadi R, Cavusgil ST, Aulakh PS (2001) The influence of complementarity, compatibility, and relationship capital on alliance performance. J Acad Market Sci 29(4): 358–373

Shenkar O, Li J (1999) Knowledge search in international cooperative ventures. Organ Sci 10(2): 134–143

Suseno Y, Ratten V (2007) A theoretical framework of alliance performance: the role of trust, social capital and knowledge development. J Manag Organ 13(1):4–23

Teece D, Pisano G, Shuen A (1997) Dynamic capabilities and strategic management. Strategic Manag J 7(1):37–51

Tsang EW, Nguyen DT, Erramilli MK (2004) Knowledge acquisition and performance of international joint ventures in the transition economy of Vietnam. J Int Marketing 12(2):82–103

Uzzi B (1997) Social structure and competition in interfirm networks: the paradox of embeddedness. Adm Sci Q 42:35–67

Zeybek AY, O'Brien M, Griffith DA (2003) Perceived cultural congruence's influence on employed communication strategies and resultant performance: a transitional economy international joint venture illustration. Int Bus Rev 12(4):499–522

Chapter 15
Low Carbon, Green Growth Korea

Greg Scarlatoiu

Korea's Green Pledge

During the focus on rapid industrialization and accelerated economic growth, Korea had little leisure to worry about environmentally sustainable growth, with severe and lingering consequences: Korea is one of the top ten carbon dioxide emitters and ranks 94th on the 2010 Yale Environmental Performance Index (EPI), a measure of environmental public health and ecosystem vitality in 163 countries. Korea's current commitment to Green Growth relates to the concern for the environment which Koreans have always fundamentally shared but had chosen to put on the back burner for decades. This green pledge was articulated in President Lee Myung-bak's New Green Growth Formula, summarized in his August 15, 2008, address on the 60th Anniversary of the Republic of Korea. Noting that, after the agricultural, industrial, and information revolution, the world is about to enter the age of an environmental revolution, President Lee presented Low Carbon, Green Growth as a "new national development paradigm" that will create new growth engines and jobs in the twenty-first century, with the advent of green technology and clean energy.[1]

During the recent global financial turmoil, in particular, the vulnerability created by Korea's fossil-fuel dependent growth became even more obvious. Korea imports 97% of its total energy requirements and is the world's sixth largest petroleum importer. In addition to concern for the environment, several other considerations are driving Korea's commitment to Green Growth: a recognition that, as a "Newly Advanced Economy," the country must curb its reliance on energy imports; a corresponding desire to reduce dependence on fossil fuels, increase reliance on renewable energy and generate "green" jobs; and a sustained effort to assume an international leadership role and enhance its transformation from a recipient to a

[1]*Road to Our Future: Green Growth-National Strategy and the Five-Year Plan (2009–2013).* Presidential Commission on Green Growth, Republic of Korea.

J. Mahlich and W. Pascha (eds.), *Korean Science and Technology in an International Perspective*, DOI 10.1007/978-3-7908-2753-8_15,

provider of international development assistance. As Korea emerges as a middle power, as acknowledged by it hosting the G-20 summit in November 2010 and other developments, Green Growth presents an area where Korea can assume international leadership and bridge the gap between developing and developed nations, a role that it appears willing and ready to assume.

Korea's Low-Carbon Green Growth (LCGG) Initiative

President Lee's commitment to "Low Carbon, Green Growth" (LCGG) aims to decouple economic growth from environmental degradation and to convert a vicious cycle of development and environmental degradation into positive synergic relations among energy, economy, climate change, and environment (Lee 2008). LCGG results from concern over the effects of climate change as well as a variety of other internal and external considerations.

According to the United Nations Environmental Program (UNEP), the average surface temperature rose 1.74°C in South Korea between 1912 and 2008, and in the past 40 years, the sea level around Korea has risen by 22 cm. LCGG is an attempt to decouple economic growth from environmental degradation by reducing greenhouse gas emissions and fossil-fuel dependency.

Korea has maintained developing country status within the Kyoto protocol, but Korea's greenhouse gas emissions hardly belong in the developing nation category. In 2007, Korea's CO2 gas emissions from fuel combustion reached 489 megatons, 1.7% of the world total and the ninth highest in the world.[2] Amid growing complaints regarding its greenhouse gas emissions, Korea also has reason to worry that its exports may be subject to punitive measures from developed countries unless it pursues voluntary greenhouse reduction goals. Initially, the centerpiece of the Lee Myung-bak administration's domestic economic revitalization vision was the 747 Plan, aiming to achieve a 7% economic growth rate and to complete the Grand Canal Project (Cho 2009). Due to the global recession and President Lee's controversial handling of the issue of U.S. beef imports in 2008, it became difficult for the Lee administration to move ahead with previous commitments, and LCGG was suggested as a political breakthrough aimed to enhance the government's legitimacy and consolidate its support base, while coping with the global challenges of climate change and energy. Under difficult economic circumstances worldwide, President Lee's LCGG vision also provided an opportunity to create jobs and curb unemployment.

---

[2] *Road to Our Future: Green Growth-National Strategy and the Five-Year Plan (2009–2013).* Presidential Commission on Green Growth, Republic of Korea.

The Green New Deal

"The Green New Deal," Korea's grand stimulus package announced in early January 2009 in conjunction with tax revisions, called for the injection of KRW50 trillion (USD38.1 billion) to create 960,000 new jobs: KRW39 trillion for 9 large-scale development projects, expected to generate 690,000 jobs, and KRW11 trillion for 21 small-scale projects, set to produce 270,000 jobs. A United Nations Environment Programme (UNEP) report "Global Green New Deal: An Update for the G20 Pittsburgh Summit" noted that 80% of Korea's stimulus package addresses renewable energy, low-carbon vehicles, water management and other environmental themes.

The nine key projects proposed in the Green New Deal include:

- Revitalizing Korea's four major rivers by building dams, reservoirs and other water management facilities, while creating 280,000 jobs[3]
- Building a green transportation network. The plan includes the creation of low-carbon emitting railways, other "green" transportation systems and the most visible of all "green" projects, a bicycle path around the country. Such "green" transportation networks will eventually create an estimated 160,000 jobs
- Building a database on national territory and resources to integrate territory management
- Building a better water resource management system
- Creating a cleaner energy program, including the development of greener cars
- Building a waste resource recycling program, managing forestry and infrastructure
- Improving forest management
- Building green homes and schools
- Cleaning up rivers to turn them into "ecological rivers"

As endorsed by the National Committee on Science & Technology, the stimulus plan also called for a twofold increase in R&D spending on 27 key technology areas by 2012, up from USD769 million in 2008. Climate change prediction and modeling, photovoltaic solar panels and LED technology were among the areas targeted.

Nuclear energy has also been targeted as a core component of low-carbon growth in Korea. Nuclear power accounts for 24% of the country's power supply, but it is envisioned that nuclear energy will account for 32% of all electricity generated nationally by 2020. There are plans to build ten more nuclear power plants, with land acquisition for the new reactors to be finalized by 2010.

---

[3]The project involves revitalizing Korea's Han, Nakdong, Geum and Yeongsan rivers.

The Five-Year Green Growth Plan

In July 2009, the Korean government introduced a five-year green growth plan to translate some of its major green growth strategies into operational plans. The plan – reminiscent of the nostalgia- and controversy-inducing five-year economic development plans initiated in the 1960s by authoritarian governments to rebuild Korea – integrated projects announced as part of the Green New Deal with new initiatives, as part of a long-term economic development strategy. A total of KRW107.4 trillion (about USD80 billion), including the original KRW50 trillion that was part of the Green New Deal, was allocated for the five-year (2009–2013) Green Growth plan.

According to the UNEP's "Overview of the ROK's Green Growth National Vision," published in August 2009, the five-year plan is expected to create 1.56–1.81 million jobs. It consists of three strategies, ten policy directions, and 50 core projects.

The National Strategy for Green Growth: Ten Policy Directions

In order to implement the Green Growth vision more effectively and ensure ample citizen participation, the government is integrating several projects, including New Growth Dynamics, Green New Deal, and Green Technology R&D, into one unified package, under LCGG.[4] On the policy-planning level, the National Strategy for Green Growth was adopted along with the Five-Year Plan for Green Growth. Together, they contain ten primary policy directions:

1. Effectively mitigating greenhouse emissions, by adopting a Carbon Visibility Index, a Carbon Emissions Trading System, greenhouse gas emission standards, and expanding green areas.
2. Reducing the use of fossil fuels and the enhancement of energy independence through increased R&D investment in energy efficiency and the adoption of a Smart Power Network (New and renewable energy) supplies will be increased from 2.7% in 2009 to 3.78% in 2013 and 6.08% in 2020. The use of nuclear energy for power generation will be enlarged to reduce CO2 output. A Renewable Energy Portfolio Standard (RPS0 and Renewable Energy Fuel Standard (RFS) will be introduced.)
3. Strengthening the capacity to adapt to climate change, through the Four River Restoration Project, expanding climate change monitoring systems and increasing the share of environment-friendly agricultural products up to 18% by 2020.

---

[4]*Road to Our Future: Green Growth-National Strategy and the Five-Year Plan (2009–2013).* Presidential Commission on Green Growth, Republic of Korea.

4. Developing green technologies, including silicon-based solar cells, bio-energy, advanced LWRs, high-efficiency fuel cells, integrated coal gasification combined cycle technology, and smart grids for electric vehicles.
5. The "greening" of existing industries and promotion of green industries. (The steel, automobile, shipbuilding and the chemical industries will be encouraged to increase their green expenditures in R&D and plant equipment investments.)
6. Advancing industrial structure by developing healthcare services, education, financing and banking, the contents industry, software, and the tourism industry. (Also, state-of the-art technology will be promoted within the broadcasting and telecommunications, IT, robotics, new materials, nano-materials, bio resources, medical appliances, and green food industries.)
7. Engineering a structural basis for the green economy by expanding financial assistance to green industries and encouraging the private sector to disclose carbon information, developing a green stock index, and increasing investment in green financing.
8. Greening the land and water in part by building a green transportation infrastructure.
9. Bringing the green revolution into daily lives, or inducing green behavior changes through the development of a green lifestyle index and a nationwide green life movement. (Carbon footprint labeling and certifying systems for goods will also be developed.).
10. Becoming a role-model for the international community as a green growth leader. (Korea will raise the ratio of Green Overseas Development Assistance (ODA) to 30% of total ODA by 2020.)

The Framework Act on Low Carbon, Green Growth

Submitted by the government to the National Assembly on February 25, 2009, the bill was reviewed by the Special Committee of Climate Change Measures (SCCCM) in the National Assembly, passed in November 2009, and entered into force on January 13, 2010. The bill aims to establish a comprehensive legal basis for the diverse plans to transform Korea into a green economy and a green society that is much more focused on green industry than before. Between the submission and the passing of the Green Growth Act, the bill was a source of political dispute between the ruling and opposition parties, as the Lee Myung-bak administration pushed LGGC as one of its core policy objectives.

LCGG: Dissenting Opinions

President Lee's Green Growth paradigm created enthusiasm but was also met with skepticism. Several Korean businesses showed reluctance to implement pervasive Green Growth standards without proper preparations, and skeptics criticized the new commitment to Green Growth as populist, excessively focused on infrastructure projects, and a mere reaction to the new U.S. commitment to Green Growth. One skeptic, Rep. KIM Yong-goo of the Freedom Forward Party (FFP), stated that only 17 out of the 91 Green Growth projects were actually new and expenditures on these accounted for only 1.7% of total annual Green Growth expenditures for 2009, or KRW7.72 billion out of KRW4.4 trillion. In other words, opponents contend that the LCGG paradigm is simply the greening of existing projects.

Opponents also view the Four Rivers Restoration Project as a crude revival of the widely unpopular Grand Canal Project. Opposition Democratic Party (DP) lawmakers have maintained that nearly half of the Green New Deal's KRW50 trillion budget is to be used for the Four Rivers Restoration Project earthworks and the Gyeongin Canal (Choi 2009).

Skeptics from ruling and opposition parties doubt the need to invest KRW292 billion to build 16 small hydro power facilities at 16 reservoirs on the major rivers, without concrete power production schemes (Park 2009a). They question the LCGG's relevance to renewable energy and its commercial and employment stimulus effect, claiming that LCGG is simply an aggregation of existing state projects. Skeptics also argue that the utility and marketability of renewable energy have been exaggerated. On one occasion, Rep. IM Dong-kyu of the ruling Grand National Party (GNP) pointed out that it would take approximately 40–77 years to retrieve the investment in three recently constructed government buildings that incorporate building-integrated photovoltaic (BIPV) systems (Lee 2009a). The opposition DP has also criticized LCGC's seeming contradictions: the Four Rivers Restoration Project, deregulation of metropolitan areas, and expansion of nuclear power facilities (Park 2009b).

Some financial analysts warn that, while short-term economic recovery may be achieved, abundance in liquidity may cause the price of raw materials to soar when the economy stabilizes and bounces back. It has also been argued that over stimulus of the economy will result in high inflation. There are also concerns over the lack of adequate LCGG planning and the affordability of the KRW107.4 trillion Five-Year Green Growth Plan, but the Ministry of Strategy and Finance has insisted that all green initiatives are built on long-term growth potential and fiscal soundness. While ensuring the fiscal soundness of Green Growth initiatives, the government set the goal to increase the proportion of distributing renewable energy up to 11% by 2030 and 20% by 2050.

The environmental movement in Korea has also often been at odds with the Lee Myung-bak administration's LGGC vision. The root of the problem may be that the beginnings of a genuine grassroots environmental movement in Korea were hampered by the confrontational culture fostered by many environmental activist groups. For years, numerous environmental groups in Korea were so entrenched in anti-establishment protests that they failed to connect with the scientific community and policymakers. Nonetheless, signs of a constructive, creative grassroots environmental movement are surfacing. The Daejeon Green Forum is testament to such development. Begun in early 2008 as a small gathering of researchers and scholars concerned about the environment, the group has expanded its membership and gained visibility by networking internationally and spearheading an effort to establish an association of eco-friendly cities that may team Daejeon up with Tsukuba, Japan, and Palo Alto, California.

Korea's International Commitment

South Korea has become the first developing country, according to the UN Framework Convention on Climate Change (UNFCCC), to adopt an emissions cap. The Presidential Committee on Green Growth, the council established to coordinate South Korea's green growth strategies, announced that the country could lower greenhouse gas emissions by 21–30% by 2020. According to the International Energy Agency, Korea's greenhouse gas emissions reached 538 million tons of CO_2 in 2005, making it the world's 16th largest emitter. Korea's commitment could encourage other developing countries to adopt voluntary goals in emission targets. The Framework Act on Low Carbon, Green Growth calls for the establishment of a carbon market[5] and the introduction of a cap and trade system in Korea.[6]

While implementation of necessary policy changes at the local level and other areas may need more time on the drawing board, criticizing this model as rhetorical, excessively focused on infrastructure, or a response to the projected greening of the U.S. economy fails to account for Korea's commitment to Green Growth, firmly stated in Seoul as early as 2005, when Korea affirmed its regional leadership on environmental matters, at the fifth Ministerial Conference on Environment and Development in Asia and the Pacific.

Korea's commitment to green growth actually goes as far back as 2002, when it participated in the World Summit on Sustainable Development in South Africa. Korea's resolve became indisputable when the 2005 Ministerial Conference on Environment and Development coined the term "Green Growth" and launched the Seoul Initiative, proposed by the Korean Ministry of Environment, and endorsed by the 61st Commission Session of the United Nations Economic and Social Commission for Asia and the Pacific (UNESCAP). The Initiative addressed the sources of the load on environmental carrying capacity in the Asia and Pacific region, including: the exponential growth in industrial production, agricultural expansion and intensification; the rising demand for water, energy and raw materials; rapid and unplanned urbanization; and growing and changing consumption patterns. The Seoul Initiative vision accounted for the need for further economic growth, as 670 million people were still living on less than $1 a day (PPP adjusted) in the ESCAP region in 2004. The Seoul Initiative identified three targets of environmentally sustainable economic growth in the region: improving eco-efficiency for

[5]Framework act on low carbon. Green Growth 28(5). http://eng.me.go.kr/board.do?method=list&bbsCode=law_law&categoryCode=00. Accessed 3 Sep 2010.

[6]Ibid, Article 46.

environmental sustainability, enhancing environmental performance, and promoting the environment as an opportunity for economic growth and development.

In late 2009, World Bank chief economist Kirk HAMILTON recognized Korea as a world leader in "green stimulus" and a role model for countries trying to address climate change. Those remarks came right as South Korea announced its first greenhouse gas reduction target. Since South Korea is not among countries that must cut emissions pursuant to the current Kyoto Protocol, such voluntary target-setting could put more pressure on developed nations, and encourage them to address global warming more aggressively.

The Korea-Africa Green Growth Initiative, aimed at strengthening cooperation in water supply, waste and air pollution management, was adopted at the Korea-Africa Forum in October 2010. Pursuant to that initiative, Korea and African nations agreed to proceed with the transfer of green technologies and the launching of joint eco-tourism and biomass energy projects to stimulate nascent green markets.

At the September 2009 UN Summit on Climate Change, President Lee Myung-bak proposed the establishment of a UN "registry" to monitor gas emissions by developing countries, emphasizing the need for coordination between developing and developed countries, in order to work towards a future climate change treaty.

While dissenting opinions questioning the motives and feasibility of LGGC in Korea may still circulate, one needs to note that development of sustainable infrastructure is an integral part of UNESCAP's suggested five track approach to green growth, together with eco-tax reform, demand-side management, greening the market, and developing eco-efficiency indicators. In Korea's case, sustainable infrastructure such as the Four Rivers Project will also be instrumental in flood management and mitigation, in particular to prepare for scenarios involving an aggravation of seasonal flooding due to climate change. Further, while green infrastructure development is a significant part of Korea's New Green Deal, the project also involves significant investment in renewable energy.

One of the accomplishments of the December 2010 UN climate conference in Cancun was a pledge of $100 billion a year from developed nations to the developing world to help pay for requisite emissions cuts and climate adaptation by the year 2020. Although the pledge is rather general and no practical steps towards implementation have yet been established, developed countries, including Korea, have good reason to like the deal, as it mentions the "mobilization" of the funds, thus suggesting a possible role for private sector money.[7] The agreement reached in Cancun includes a deal on technology transfer, which may give Korea an opportunity to establish itself as a "green middle power," by playing a prominent role in transferring low carbon technologies from developed to developing countries.

[7] Climate-change diplomacy: back from the Brink (18–31 December 2010). The Economist, p 121.

U.S.-Korea 'Green' Cooperation and Korea's Commitment to LGGC

U.S.-Korea memoranda of understanding and collaborative endeavors intended to initiate and sustain green U.S.-Korea cooperation constitute significant evidence of Korea's commitment to LGGC and the establishment of relevant international partnerships. The United States and Korea have engaged in both government-to-government contacts and bilateral private sector business arrangements to identify synergies and spur cooperation between the two countries in areas pertaining to Green Growth. At the inter-governmental level, U.S. Energy Secretary Steven CHU and Korean Minister of Knowledge Economy LEE Yoon-ho signed a Statement of Intent on June 16, 2009, to begin cooperation on research and development in the area of "smart power grid" technology. Furthermore, when the U.S.-led Major Economies Forum on Energy and Climate (MEF) convened in L'Aquila, Italy, on the sidelines of the G-8 Summit on July 9, 2009, Korea and Italy were the countries assigned to lead the acceleration of the deployment of smart grids.[8]

The Korean Ministry of Education, Science and Technology is engaged in several research projects related to nuclear energy with various national laboratories of the U.S. Department of Energy under the DOE's International Nuclear Energy Research Initiative (INERI). The U.S. DOE's National Energy Technology Laboratory (NETL) and the Korea Institute of Energy Research (KIER) have signed a memorandum of understanding to work together on Clean Coal and Carbon Capture and Storage (CCS) Technology. The DOE has also begun sharing information on gas hydrate research and development with the Ministry of Knowledge Economy (MKE), and state-run Korea National Oil Corporation will take part in the DOE Alaska North Slope Project on gas hydrate extracting and processing. Korea is a particularly active member of the Asia Pacific Partnership on Clean Development and Climate as APP Chair of the Renewable Energy Task Force[9] and of the Buildings and Appliances Task Force.[10] The Korea Energy Management Corporation (KEMCO) serves on the Secretariat for both task forces.

[8]"Smart power grid" technology is one of eight technologies identified by MEF countries as essential "transformational low-carbon technologies."

[9]The Renewable Energy Task Force deals with four main areas: (a) development of economic indicators for renewable energy distributed generation; (b) development of "smart energy solutions" based on renewable resources; (c) elaboration of studies on the supply of bio-diesel fuel for use in transportation; and (d) development of "dye-sensitive" solar cells.

[10]Korea is an active participant in 10 out of 13 projects under the Buildings and Appliances Task Force. Korea also takes part in a project under the Cement Task Force that is looking into the feasibility of concrete structures for sequestering CO_2 and is in a project under the Cleaner Fossil Energy Task Force that assesses technologies to recover methane captured in coal beds and coal mines.

An 18-year-old joint high tech collaborative project between the Korea Center for Atmospheric Environmental Research (KCAER) and the U.S. National Oceanic and Atmospheric Administration (NOAA) has provided satellite imagery and greenhouse gas monitoring data that have been useful to scientists throughout the world, engaged in studying the extent and seriousness of global climate change. Korea's universities are also full participants in U.S.-Korean collaborative efforts on green projects. Seoul National University participates in a program to improve livestock waste management for methane recovery, as part of the Methane to Markets program spearheaded by the United States. LG International is also involved in programs researching the removal of methane from coalmines and landfills.

Moreover, the United States and Korea work together in a wide range of multi-lateral fora including: the International Energy Agency (IEA), the Asia Pacific Economic Cooperation (APEC) process, the Asia-Pacific Partnership on Clean Development and Climate (APP), the International Atomic Energy Agency (IAEA), and the Global Nuclear Energy Partnership (GNEP), among others.

In addition to government-driven efforts to initiate collaboration in green areas, the private sectors in the United States and Korea have also entered relevant agreements. For example, on July 8, 2009, the U.S. smart grid organization Grid Wise Alliance, chaired by IBM, and the Korea Smart Grid Association entered an agreement to share information on smart grid technologies and modernize electrical systems after the two umbrella organizations signed a Statement of Intent. General Motors Corporation chose LG Chemical as its sole supplier for lithium-ion batteries to be used in its plug-in hybrid electric car, the Chevrolet Volt, beginning in November 2010. LG Chemical is in the process of building a plant in Michigan to provide batteries to GM. On January 5, 2010, LG Chemical confirmed that its Troy, Michigan-based U.S. unit, Compact Power Inc. (CPI), would begin supplying lithium-ion batteries for commercial hybrid electric vehicles (HEVs) to Eaton Corp., a U.S. commercial vehicle components maker, beginning in November 2010. In July 2010, it was confirmed that CPI would supply lithium-ion battery packs for the electric 2011 Ford Focus the U.S. market.

On the research front, the Korea Institute of Energy Research and the University of South Carolina are carrying out joint fuel technology research through an exchange program. Hyundai and Kia Motors have joined the California Fuel Cell Partnership, and also participate in a partnership with Chevron Texaco and UTC Fuel Cell in a U.S. Department of Energy technology validation project for fuel cell electric vehicles. U.S. company UTC Fuel Cell is providing GS Power and Samsung Ever land 12 fuel cell systems to provide power to Anyang city in one of the largest fuel cell installations in the world, and the U.S. company Fuel Cell Energy is transferring technology and exporting equipment to POSCO Power Corporation to assemble Molten Carbonate Fuel Cells at POSCO's plant in Korea. Korea Parts and Fasteners has formed a joint venture with U.S.-based Plextronics to produce solar panels in Korea.

Technological Developments

Korea's technological competence augments its reputation as a potential "green middle power," bridging the gap between developed and developing nations in areas pertaining to clean energy. Technologically, Korea's shift toward eco-friendly solutions began a few years before the announcement of the New Green Deal. As MIT's Dr. SUH Nampyo assumed leadership of the Korea Advanced Institute of Science and Technology (KAIST), scholars who had worked on raising Korea's shipbuilding, automotive and electronic industries to superpower status, decided to transform Korea's engineering innovation powerhouse along the Energy, Environment, Water and Sustainability (EEWS) paradigm. President Lee's confidence in the New Green Deal may have been boosted by his first visit to KAIST in 2009 when he witnessed the debut of a new generation of electric transport. Operating on a lightweight battery supplemented by an underground electric cable, the hybrid electric car-train, called OLEV (on-line electric vehicle), is cheaper to operate than other battery-operated cars and does not require recharging stations. It was developed by KAIST in collaboration with a consortium, including Hyundai Heavy Industry, Daewoo Bus and SK Energy (Kwon 2009a). Hyundai Automobiles, LG Chemical, Samsung, SDI and LS Electric Cable have also stated their interest in this line of business (Lee 2009b).

OLEV meets U.S. and EU safety standards. This electric car has a lightweight lithium battery and a receiver on the bottom of the vehicle for a noncontact charge via airwaves from a magnetic strip embedded in the road. Scientists believe that at least 30% of roads will have to be equipped with the magnetic strips, placed mostly at intersections. The magnetic strips will be connected to charging stations powered by nuclear energy.

The government has invested KRW100 billion (USD83 million) in the OLEV cars as of late 2010, in addition to the KRW25 billion (USD21 million) spent on its initial development. OLEV infrastructure is cheaper than subway infrastructure, and the OLEV debuted publically in Daejon through a pilot bus program. On August 12, 2009, the Seoul Municipal Government decided to introduce and expand the use of OLEV public buses (Kwon 2009b). The vehicle may also be the next popular Korean product on the international market, as the U.S., Fiji, Australia, and Belgium have expressed interest in investing in Korean OLEV and hybrid cars. The OLEV company CT&T has been researching sites in Pittsburgh to build two plants creating 400 U.S. jobs, and some units have already been shipped to Japan (although the OLEV is not yet cleared to circulate in Korea). On the downside, OLEVs may cause a spike in the price of lithium, resulting in a sharp rise in their own price.

Korea has been making good progress on green infrastructure. On July 27, 2009, the MKE, the Korea Auto Industry Union, and 13 representatives of auto-part companies signed an MOU, agreeing to cooperate on green car R&D and the exchange of relevant technological and market information (Bae 2009). Some companies have already initiated preparations for the Electric Vehicle (EV) era. GS Park24 is currently promoting the installation of electric rechargers at public parking lots near large commercial areas and restaurants, while Samsung Engineering & Construction has applied for a patent on the Plug-in High Electric Vehicle Recharging Station System, aiming to install such a system in every apartment complex in Korea (Lee 2009b).

By late 2010, the first Korean tidal-power station was scheduled to become operational at Lake Shihwa in western Gyeonggi Province, generating a capacity of 254,000 kWh. In another green collaborative project, Hyundai Heavy Industries will cooperate with the North Jeolla provincial government to build a massive wind energy farm. The Green New Deal also aims to spur the recycling and renewable industry, in particular, through investment in facilities converting trash into fuels and bio-gas. With 75% of its renewable energy coming from waste, Korea already has a few years experience in this field, in particular through efforts by the Daesung Group, which has been recycling methane gas drawn from dump sites. Daewoo Engineering has already built Korea's first bio-gas power plant, which became operational in 2008.

Korea's automakers are also bound to play a key role in the Green New Deal. Hyundai Motor technicians have already test driven a hydrogen fuel cell-powered Tucson hybrid across the United States and Hyundai intends to begin commercializing a hydrogen fuel-powered car by 2012. The government has announced plans to raise fuel-economy standards in new cars, beginning in 2012. The plan mandates cars and vans with less than ten passenger seats must run an average of 17 km per liter (40 mpg) and produce less than 140 g (5 oz) of carbon emissions per km by 2015. This policy is aimed to lower carbon emissions and dependence on foreign energy. The government is working on additional incentives for the purchase of hybrid vehicles, giving out KRW2 million to people who trade their old vehicles for new hybrids.

Anticipating concerns regarding the infrastructure demands that hydrogen power will require, GS Caltex Corporation, one of Korea's top refiners, began testing a hydrogen fuel station in Seoul, in 2007. One distinctive feature of Korea's green auto industry is interest in LPG-electric, rather than gas-electric hybrids, due primarily to the fact that Korea has more LPG-fueled cars than any other country, having over a decade of experience producing and driving them. In addition to green automotive technology, the New Deal also targets energy-efficient LED illumination, production of solar cells for photovoltaic power generation and hydrogen fuel cells for heating and lighting. Even Korea Telecom (KT) has joined in this race towards the Green New Deal, unveiling its "Green KT, Green Korea" project that uses geothermal heating and air-conditioning systems for KT buildings and solar-powered base stations for KT wireless technology.

Further Opportunities to Engage in International Technological Cooperation

Korea's opportunities to engage in international environmental cooperation are ample and go beyond the regional scope of UNESCAP. The fact that the United States and Korea are on the verge of implementing a "Green Deal" should not be seen as Korea following the direction chosen by the current U.S. administration; both countries are opting to engage in cooperation in a whole array of energy and environment-related areas because of the clear potential. Building on General Motors and Ford's collaborative work with LG Chemical, Korea and the United States have an opportunity to explore ways to reduce carbon dioxide emissions through low-carbon technologies, standardize specifications and design safer plugs for the recharging of next-generation vehicle batteries.

Korean shipbuilders are expanding new niches in green industries as well. Experience producing propellers for massive cargo vessels can be leveraged in the production of turbines for wind and tidal energy generation. In March 2010, Korea's Daewoo Shipbuilding and Marine Engineering Ltd. (DSME) reached an agreement with the Canadian province of Nova Scotia, to establish a wind turbine tower and blade manufacturing facility in Pictou County. With a contribution of CD20.4 million, DSME will retain 51% of the common shares and will create hundreds of jobs in the coming year, in addition to the 120 jobs already created in 2010 (Tournemille 2010). This experience can prove useful in private sector couplings south of the border as well, in particular in Maine, the U.S. State nearest to Nova Scotia.

Through its Green New Deal commitment, Korea appears poised to strengthen its regional leadership in the promotion of environmentally sustainable growth. As green economic stimulus packages are being offered by governments worldwide, the UN Environmental Program (UNEP) estimates that South Korea is the only country that is currently spending enough of its stimulus on green investment in renewable energy, energy efficiency, and low-carbon technologies to reduce the costs of climate change later.

Critical Materials and Supply Chain Challenges

As it makes the transition towards a greener economy, Korea will have to be prepared to hedge against supply chain and geopolitical risks affecting those minerals essential to green industries. In Korea and other advancing economies, green IT as well as other technology-based industries' competitiveness and survival will depend on access to a stable supply of rare earth minerals. For example, rare-earth elements – such as magnesium, lithium, cobalt, and manganese – are essential raw materials in the production of secondary cells for green cars, including electric vehicles. Indium, gallium, molybdenum, selenium and platinum group elements are used in the production of solar and fuel cells. Manufacturers of system semiconductors need a constant supply of indium, germanium, niobium, strontium, and rare-earth elements. Over the past few years, as the middle class in emerging countries, including China and India demanded more LCD TVs, mobile telephones and automobiles, demand for rare minerals has been on the increase. Clean energy technologies account for about one-fifth of global consumption of such critical materials. As green technologies are further developed, their share is likely to grow. Permanent magnets, batteries, photovoltaic (PV) thin films and phosphors, all essential to green technologies, depend on materials at risk of supply disruptions over the next 5 years, although such risks are estimated to decrease over the medium and long term (5–15 years), as more rare earth mineral and other critical material deposits are identified and exploited (U.S. Department of Energy 2010).

Anticipating growing future demand in Korean industries, the MKE recently designated many of these rare metals as critical to Korea's green, IT, and high-tech industries, including second generation batteries for hybrid and electric vehicles, automobiles, mobile phones, and LCDs (e.g., lithium, chrome/chromite, manganese, molybdenum, cobalt, tungsten, indium, rare-earth elements, magnesium and titanium/titanium dioxide). The challenge for Korea is that these elements, just like all other needed natural resources, will be increasingly difficult to secure for its economy (Kim 2009).

One significant hurdle is that most of these so called "industrial vitamins" that will fuel the green and IT industries of the future are located in China, Africa, and Latin America, in regions where political instability, corruption, low levels of government transparency, lack of economic freedoms, and government interference are likely to distort markets and disrupt supply. A recent study by the Samsung Economic Research Institute (SERI) indicates that significant uncertainty could create problems for Korea's supply of eight rare metals of great importance to its manufacturing, four of them not included in the MKE list: niobium, bismuth, vanadium and germanium.

Furthermore, two of Korea's direct competitors, Japan and China, have engaged in an active attempt to secure a steady supply of rare metals heightening the contest for these resources. The Japanese government has pushed to secure global resources through their "four major strategies (pillars) to secure rare metals" movement. In the case of China, the world's largest holder of rare mineral reserves, the country has resorted to restricting the export of rare-earth minerals to 35,000 t per year for a 6-year period, beginning in 2009. Additionally, China also imposed an export duty of 25% on rare-earth minerals (Kim 2009). Currently, over 95% of the production of rare earth minerals is in China (U.S. Department of Energy 2010). A petition filed by United Steelworkers in the United States, claiming that China's green technology practices violate WTO rules, alleged that "China uses export quotas, taxes and licensing procedures to restrict exports of these [rare earth] minerals to users in the U.S. and other countries."[11]

To address this growing challenge, one strategy to secure access to resources may be to harmonize South Korea's international development efforts with its quest for natural resources. In other words, the government would provide development assistance to countries that hold key resources for tomorrow's green future, e.-g., Bolivia, the Democratic Republic of Congo, Gabon, Malawi, Mongolia, Nigeria, Serbia, South Africa, Tanzania and Zambia.

Korea's international efforts in these countries would build on its own developmental experience to empower local communities, to enhance central and local government transparency and accountability to reduce corruption-induced market distortion, and to improve the knowledge base of mining engineers and other relevant technical experts. In particular, long-term training and higher education degree programs offered by Korean universities, Korean foundations and private entities, could be a very efficient way of creating a cadre of technical experts and government officials familiar with Korea.

Assistance with geological surveys and exploration for rare minerals could be included in these comprehensive ODA packages, especially for countries that may possess untapped rare minerals, including Mongolia, the Democratic Republic of Congo and Zambia (the Copper belt). For more industrialized economies, Korea can offer technology transfers and enhanced green-technology collaborations.

Although Korea continues to be heavily dependent on outside natural resources to fuel its economy, its newly acquired status as an economically successful and politically responsible middle power has provided the country an opportunity to employ its resource diplomacy and experience to become a game maker in the exploration, extraction and distribution of the rare minerals that hold the key to the future.

Korea's Four Green Fronts

The greening of an economy has to develop in four essential areas: energy generation, industrial output, transportation, and household use. In South Korea's case, the limited land space restricts the degree to which the country can rely on alternative sources as the mainstay of energy production such as wind, tidal and solar power generation. While efforts are under way to develop clean alternative energy, it appears that, at least over the medium run, the most effective answer to the challenge of greening Korea's power generation and decreasing dependence on energy imports is more nuclear power generation.

[11] United Steelworkers' Section 301 Petition Demonstrates China's Green Technology Practices Violate WTO Rules. Petition Summary. 1. http://assets.usw.org/releases/misc/section-301.pdf. Accessed 6 Dec 2010.

The greening of South Korea's industry is perhaps the most difficult challenge ahead. Although South Korea and Japan have the highest levels of manufacturing industry energy efficiency among OECD regions – followed by Europe and North America – large energy intensive industries, such as chemical, petrochemicals, and steel, still remain at the core of Korea's economic growth. Nevertheless, the next driver of Korea's economic dynamism may be the development of new energy efficient technologies.

In transportation, although Korean car makers are striving to catch up with foreign competitors who have made greater progress in next-generation automobiles, Korean consumer attitudes towards hybrid vehicles appear to be out-of-sync with the growing popularity of the green movement among the government, business and the population at large. Sales of hybrid vehicles in Korea, including the hybrids Hyundai Avante, Kia Forte and Toyota Prius, amount for only about 0.5% of Korea's auto market, despite tax incentives amounting up to KRW3 million (USD 2,650) and additional discounts. More than a lack of commitment to a greener Korea, the main reasons for the hybrids' sluggish sales are their rather steep prices: a Hyundai Avante costs KRW17.5 million after the tax rebate and additional discounts, significantly more than the regular version of the same model (KRW11 million). Also the use of personal vehicles is lower and the distances driven shorter in Korea than in the markets where hybrids are more popular, such as the United States.

South Korea's Green Growth drive is likely to score its earliest successes inside Korean homes, despite the results yielded by a "green life capability index" developed by Samsung Economic Research Institute (SERI). According to that index, of the 29 OECD member nations surveyed, Korea ranked 24th in terms of "green life capability," and local community capability in Korea came in at 28th.

While the government's green capability includes relevant regulations and technologies, local communities have to be in charge of establishing rules of behavior that encourage a green lifestyle and restrict energy overconsumption. One significant hurdle that Korea's green communities will have to overcome is the confrontational culture fostered by the myriad of competing environmental activist groups – the same reason a robust grassroots South Korean environmental movement has been delayed.

Koreans were found to emit 11.1 t of CO_2 per person annually, ranking 17th among OECD members in terms of individual CO_2 emissions.[12] The Framework Act on Low Carbon, Green Growth calls for the need to reduce per capita CO_2 emissions through behavior change PR campaigns that call for the "boosting of home-based working, virtual conference[s], remote education, and remote medical treatment[s]."[13]

[12]The equivalent of 55 Seoul-Pusan trips in a 2,000 cm^3 Hyundai Sonata.

[13]Framework act on low carbon. Green Growth. Act No. 9931, 13 January 2010. Article 27(2). The Presidential Committee on Green Growth. http://eng.me.go.kr/board.do?method=list&bbsCode=law_law_law&categoryCode=00. Accessed 13 Oct 2010.

Green living campaigns in Korea are calling for switching off lights in buildings and turning the country into a "bicycle-friendly nation." In 2009, the Ministry of Public Administration and Security announced a plan to complete a 3,114 km bike-path network by 2018 that will cost KRW1.25 trillion. The national government will cover 70% of the cost and local governments will pick up the rest. Close to 8,300 jobs, including both temporary construction and permanent service jobs, are expected to be generated by the project.

On the local level, incentives to reduce CO2 emissions have also been created. The Seoul Metropolitan Government has established an incentive system for local residents to voluntarily lower their carbon emissions. Any citizen, business, or school is eligible to receive KRW50, 000 through the new eco-mileage system if they successfully reduce CO2 emissions by at least 10% for an average of 6 consecutive months.

Due to the 2010 summer heat wave, electricity usage reached dramatically high levels: on July 22, daily electricity usage reached 67.34 million KW, the second highest in South Korea's history. In July, regulations were enacted to cap the maximum allowable temperature inside government offices, department stores, banks and other large public spaces at 26°C (78.8 F). To secure sustainable energy reductions, such regulations will have to be augmented by increasing the energy efficiency of buildings and by public awareness campaigns aimed at the greening of lifestyles.

Green Challenges and Opportunities

Anyone who knows what Koreans can accomplish when they set their minds to something should not be surprised if the country that sacrificed its environment for rapid industrialization and accelerated economic growth undergoes another huge transition from one of the world's leading carbon emitters into a role model for countries attempting to implement eco-friendly policies.

Infrastructure programs are instrumental for creating new green jobs and for elevating Korea's profile as a "green country," thus building internal good will toward LCGG and boosting Korea's global image as an environmentally friendly country. Public awareness campaigns can play an important role, particularly in reducing individual carbon footprints in Korea.

However, the ultimate litmus test of the success or failure of Korea's LCGG policy is green technology innovation and development. In the grand scheme of things, Korea has become a "Newly Advanced Economy" (NAE), a concept proposed by Prof. Werner Pascha of the University Duisburg-Essen, Germany. As an NAE, Korea is very close to having exhausted the options for extensive economic growth through employing more capital and labor. Massive investment in infrastructure works will play a role in averting recession and creating jobs over the short and medium term, but further strong growth is only possible through intensive growth, that is through raising productivity. Korea's catching-up style growth model has to be modified to allow for innovation and technological change at the frontier of global knowledge. Green growth provides precisely such an opportunity by placing Korea in a leadership role in technological innovation pertaining to clean energy.

Nevertheless, as the country takes the requisite steps along a green path leading away from catch-up style growth, Korea also has to be aware of some of the obstacles and perils it is likely to encounter. As the global economy is slow to recover, if demand-side incentives for renewable energy such as feed-in tariffs (FiTs) are introduced – resulting in increased energy costs – Korea and other governments could feel increasingly compelled to link such demand-side mandates to preferential supply-side policies that promise to create green jobs and boost economic competitiveness over the long run. However, subsidies aimed at job creation and increased competitiveness could be seen as breaching trade laws and cause tension among trading partners (Ma and Pearson 2010a).

On October 15, 2010, the U.S. Trade Representative (USTR) decided to conduct a 90-day investigation into claims included in a United Steelworkers (USW) petition against China's alleged protectionist policies in the clean-energy sector. Following that investigation, on December 22, 2010, the United States requested formal consultations with the Government of China, with respect to China's Special Fund for Wind Power Manufacturing. USTR estimates that grants offered under this program since 2008 are likely to amount to several 100 million dollars, and alleges that the program amounts to WTO-prohibited import substitution, as the grants extended under the program are contingent on the use of domestic parts and components over foreign-made parts of components by Chinese wind power equipment manufacturers.[14]

If, after the mandatory consultation period of 60 days during which the complaining party and the responding party are required to attempt to reach a satisfactory resolution on the matter, the USTR decides to submit a subsequent case to the World Trade Organization (WTO), this may signal the beginning of an age of tension for green industries worldwide.[15] The USW petition claimed that some of China's green subsidies, contingent on export performance or domestic control, constitute a violation of Article three of the WTO Agreement on Subsidies and Countervailing Measures. In addition to the grants that form the subject of consultation between the USTR and the Government of China, Chinese subsidies include: the "Ride the Wind" program, which grants access to subsidized loans; priority connection to the power grid for wind power projects that use domestically produced equipment; and the extension of export credits to Chinese green

[14]United States Requests WTO Dispute Settlement Consultations on China's Subsidies for Wind Power Equipment Manufacturers. Press Release. Office of the United States Trade Representative. http://www.ustr.gov/about-us/press-office/press-releases/2010/december/united-states-requests-wto-dispute-settlement-con. Accessed 16 Mar 2011.

[15]Ibid, p. 2.

technology contractors through China's ExIm Bank.[16] The petition also claims that China's discrimination against imported goods and foreign firms – in the form of the local content requirements for wind and solar power plants, bidding preference extended to domestic wind companies, and exclusion of foreign firms from access to carbon credits – is a violation of Article III:4 of the General Agreement on Tariffs and Trade (GATT), requiring WTO member states to grant imported goods treatment at least equal to that extended to domestic goods.[17]

While Korea must avoid Chinese-style subsidies that could reinforce protectionist barriers in markets around the world and become breeding grounds for trade disputes, it is a fact that most countries that have initiated a green-energy drive have resorted to linking demand-side mandates to job creation: the U.S. biodiesel sector has been targeted in dumping claims made by Australia and the EU, while Japan has challenged Ontario's domestic content for renewable energy mandates (Ma and Pearson 2010b). Korea has the advantage of having put in place a coherent national energy policy, a commitment underscored by the enactment of effective green legislation, the Framework Act on Low Carbon, Green Growth, which clearly incentivizes renewable development.

In order to become instrumental in Korea's success as an NAE, its new Low Carbon, Green Growth development paradigm should incorporate clean-energy innovation as a means to graduate from the catch-up growth model; spur green growth and the creation of green jobs, while avoiding overly aggressive demand-side incentives or exaggerating job creation; hedge against potential disruptions affecting the supply chain of critical materials needed by green industries; and seek an increasingly active role as a "green middle power" spurring public and private sector international cooperation among developed nations, and between developed and developing nations.

References

Bae M-S (2009) MOS on greening vehicle. Energy J. http://www.ejnews.co.kr/news/articleView.html?idxno=887. Accessed 17 Jan 2010

Cho M-R (2009) Green growth and green society. Kuk-to 327:62–70

Choi Y-H (2009) MB style green growth is "Concrete" based growth. Newsis. http://news.naver.com/main/read.nhn?mode=LSD&mid=sec&sid1=001&oid=003&aid=0002818170. Accessed 15 Apr 2010

Kim H-N (2009) Rare metals and industrial competitiveness: efforts to secure supply and response measures. Korea Econ Trends. Weekly Insight. Samsung Research Institute. 13. Accessed 3 Mar 2010

[16]United Steelworkers' Section 301 Petition Demonstrates China's Green Technology Practices Violate WTO Rules. Petition Summary. p 2–3. http://assets.usw.org/releases/misc/section-301.pdf. Accessed 30 Nov 2010.

[17]Ibid, p. 3.

Kwon Y-E (2009a) EV runs in Seoul from 2011. Asia business. http://www.ajnews.cokkr/uhtml/read.jsp?idxno=200908121530116610087&section=S1N5&section2+S2N68. Accessed 3 May 2010

Kwon Y-E (2009b) EV runs in Seoul from 2011. Asia business. http://www.ajnews.cokkr/uhtml/read.jsp?idxno=200908121530116610087&section=S1N5&section2+S2N68. Accessed 3 Oct 2009

Lee S-H (2008) Characteristics and problems of low carbon, green growth policy. Environ Life (Hwangyong wa Saengmyung) 58:110

Lee G-H (2009a) Opposition berated green growth policy. Digital Times. http://www.dt.co.kr/contents.html?article_no=2009100702103516610003. Accessed 23 Feb 2010

Lee J-S (2009b) EV recharging business as a new blue ocean. Hankook Kyungje daily. http://www.hankyung.com/news/app/newsview.php?aid=2009100473621&sid=&nid=920&type=4. Accessed 27 Jan 2010

Ma D, Pearson W (2010a) Green collar politics: perils and opportunities for South Korea in the clean tech race. KEI Academic Paper Series 5(12):1–2. http://www.keia.org/Communications/Ma_Pearson_Paper.pdf. Accessed 20 Dec 2010

Ma D, Pearson W (2010b) Will green collar politics: perils and opportunities for South Korea in the clean tech race. KEI Academic Paper Series 5(12):11. http://www.keia.org/Communications/Ma_Pearson_Paper.pdf. Accessed 20 Dec 2010

Park Y-R (2009a) DP criticizes green growth. INEWS 24:1006. http://itnews.inews24.com/php/news_view.php?g_serial=447428&g_menu=020200. Accessed 7 March 2010

Park Y-R (2009b) DP criticizes green growth. INEWS 24:1006. http://itnews.inews24.com/php/news_view.php?g_serial=447428&g_menu=020200. Accessed 23 Feb 2010

Tournemille H (2010) Nova scotia and Daewoo announce wind turbine plant. Energy Boom. http://www.energyboom.com/wind/nova-scotia-and-daewoo-announce-wind-turbine-plant. Accessed 7 May 2010

U.S. Department of Energy (2010) Critical materials strategy summary. www.energy.gov. Accessed 7 Dec 2010

Chapter 16
Long-Run Protection: Determining Key Features of Growth and Sustainability in Northeast Asia

Matthew A. Shapiro

Introduction

This chapter attempts to advance understanding of the crucial links among growth, technological change, human capital accumulation, and openness and their impact on greenhouse gas (GHG)-related research and development (R&D) output in East Asia, specifically China, Japan, Korea, and Taiwan. Looking at the 1952–2004 period, I test not only for the impacts of a science and technology (S&T) and innovation set of GHG-output related inputs, but also test for the presence of the Environmental Kuznets Curve (EKC) in terms of the GDP-CO2 emissions relationship. This research builds upon the works of various important growth theorists, including Solow (1956, 1957), Nelson and Phelps (1966), Mankiw et al. (1992) and especially Benhabib and Spiegel (1994, 2002). The latter demonstrated the role of human capital on technological development (or more specifically technology catch-up) between 72 and 86 nations over extended periods of time, 1965–1985 and 1960–1995, respectively. In all of these studies, however, there is a limited role for openness to trade, capital, or technology in explaining variations in growth rates across countries. Nevertheless, and although certainly still controversial, many of the studies that have been done on East Asia and its so-called "economic miracle" argue that it is greater openness to trade, capital flows, and technology that have distinguished the East Asian experience from that of Latin America, Africa or other regions of developing countries.

What is remarkable about Japan, Korea, and Taiwan, and what is a grand qualifier for their "advanced economy" status, is that they have taken a proactive role to establish themselves as high-technology suppliers. In terms of GHG-related R&D, there is a vested interest for these countries to distribute knowledge and products to their neighbors, as the technology-based mitigation of GHGs and airborne pollutants has positive externalities for surrounding environs, particularly

J. Mahlich and W. Pascha (eds.), *Korean Science and Technology in an International Perspective*, DOI 10.1007/978-3-7908-2753-8_16,

if it positively affects economic growth.[1] Technology-oriented partnerships are advanced through opportunities to gain knowledge from more technologically-advanced countries, but technology transfer is typically limited in nations like China which has lax intellectual property laws, weak governance, and a preponderance of corruption. China is a particularly important case because of its current and prospective levels of GHG emissions. Given that climate change is a direct result of GHG levels in the atmosphere, the entire world but especially the countries of the Northeast Asian region can facilitate China in its application of the most advanced technology to mitigate GHG emissions.

There is a disincentive for China, and other developing countries for that matter, to seek out solutions for its environmental pollution and increasing GHG emissions. The EKC model, for example, describes how environmental clean-up efforts limit growth initially. But, the shape of the EKC is affected by a number of factors, including incoming flows of technology from abroad, which can temper the short-run costs of environmental clean-up, if not offset them altogether. Openness to technology from abroad can also limit growth if countries become excessively dependent on trade partners and if manufacturing is emphasized rather than innovation, but all four of the East Asian countries examined in this chapter have policy goals which incorporate innovation and provide the opportunity for sustainable, long-term growth.

To provide evidence of sustainable growth, this paper extends the Nelson-Phelps model of catch-up and shows how Japan, Korea, Taiwan, and potentially China are newly advanced economies. Our revised catch-up model accounts for and highlights the effects of innovations which increase a nation's potential for environmental sustainability and, thus, sustainable growth, rather than adapting general technologies from abroad. The impact of these environmentally-oriented S&T transfers is not different from the original model: a curvilinear increase in a nation's capabilities to address environmental sustainability, creating a catch-up effect in which a country rapidly acquires capabilities in the near-term and at a diminishing rate over the medium- to long-terms, as the gap with the leading countries is narrowed.

The most important contributor to curvilinear increases in capabilities is the simultaneous growth of a nation's innovative capacity in the area of environmental sustainability with the receipt of such transfers from abroad. In the absence of such a mechanism, there would simply be a level increase of the existing linear trajectory. Of course, a nation's innovative capacity is not predicated strictly on incoming technology transfers but involves domestic efforts to address existing deficiencies and establish targets which were not previously attainable.[2] These domestic efforts typically include a more rigorous focus on the national innovation system,

[1] I do not model the minutiae of the decision-making process for the providers or receivers of this technology.

[2] The Pascha and Mahlich (2007) volume discusses these domestic efforts in Korea and Northeast Asia.

including government funding of key education programs, R&D facilities (both public and private), and the establishment of international economic relations with more technologically advanced countries as well as countries which possess factors of production (labor and capital) to contribute to the generation of the targeted innovations (Nelson 1993). There is room for improvement in the Northeast Asian countries examined here, but strong evidence is offered below showing that innovative capacity combined with inflows of foreign R&D investment are sufficiently mitigating GHGs.

To address these issues of growth, environmental sustainability, and the Northeast Asian region, we turn next to the EKC model and the extent to which endogenous measures, the international transfer of technologies, and regional cooperation contribute to its shape. The relevant data, empirical methods, and empirical results are then presented, and a concluding section offers policy prescriptions and a call for even more regional integration. As this is a relatively new area of research, a review of the relevant literature is incorporated throughout the chapter.

The Environmental Kuznets Curve

The original Kuznets Curve (Kuznets 1955) reveals the relationship between economic growth and inequality, hypothesizing that inequality first increases with economic growth, levels off, and then decreases. The EKC applies a similar logic to explain environmental quality in an inverted U-shaped relationship with growth. That is, a country's initial development is coupled with decreases in air and water quality and environmental degradation. Over time and with further development, changes are made to limit pollution and environmental degradation, for the sake of the country's citizenry or in response to regional/international pressure. Support for the efficacy and applicability of the EKC is by no means universal.[3]

In the existing empirical literature, there is limited evidence of the EKC. Dinda's (2004) overview and survey of the EKC does not support its widespread presence, although local air pollutants seem to point to an inverted-U relationship between emissions and economic growth. Huang et al. (2008a) address some of the cross-country differences at the macro-level by differentiating between income level. Focusing explicitly on energy consumption, which is a legitimate proxy for environmental degradation and/or GHG emissions, the overall picture indicates a positive correlation between energy consumption and income for the middle income group and a negative correlation for the high income group. Huang et al. (2008a) attribute this difference to greater energy efficiency for countries in the high income group, but it must also be acknowledged that energy efficiency is

[3]See Dasgupta et al. (2002) and Millimet et al. (2003), which offer support, while Stern (2004), Copeland and Taylor (2002), and Arrow et al. (1995) argue against use of the EKC.

the result of S&T advances which are much more likely to occur in the high income group.

In their model of the effects of a carbon tax policy in a rich, open economy, (Bruvoll and Foehn 2006) note that the EKC can be explained in terms of the redistribution of more polluting industries from rich to poor, neighboring countries. In the case of Northeast Asia, such a relocation effort has been occurring, namely from Japan, Korea, and Taiwan to mainland China. A call for transfers of technology to China from these countries can be justified on normative grounds, thus. International agreements attempt to address this problem, but they must account for the role of region-based technology transfer.

The EKC dovetails with the neoclassical economic approach to assess the determinants of growth, assuming a constant returns-to-scale production function,

$$Y_i = K^{\alpha}(AH)^{1-\alpha}. \tag{16.1}$$

This attempt to incorporate the EKC with the Solow growth model is most consistent with Brock and Scott's (2004) "green Solow model," in which emissions intensities and abatement costs are incorporated into the analysis. They do not acknowledge, however, the role of openness. We pursue two theoretical avenues: (1) a simultaneous examination of the determinants of economic growth and the determinants of environmental sustainability; (2) a claim that the EKC is present in Northeast Asia. The two forms of development are represented by i, with economic growth measured by GDP levels and sustainable development measured by CO2 emissions.

Equation 16.1 presents constant, A, capital stock, K, and educational attainment, H, where the capital share is set at one-third and the labor share (AH) is set at two-thirds.[4] As it is commonly expressed in the literature, A represents exogenously growing labor-augmenting technology. Educational attainment, H, is generally expressed in one of two forms: as a measure of school enrollment or as a measure of average years of schooling attained. Barro and Lee (2000), based on their study of educational attainment, have the relevant data for both of these measures, but our basis of educational attainment is more centered on tertiary enrollment rates and the number of researchers, which is more appropriate when focusing on S&T output.

Capital stock, K, is shown in the standard capital accumulation equation,

$$\dot{K} = I_K Y_i - \delta K, \tag{16.2}$$

where the change in physical capital stock is the difference between I_K, the amount of investment in physical capital, and δK, the depreciation of capital. If we divide both sides of Eq. 16.1 by AL, defining y, k and h as Y/L, K/L and H/L, respectively, we have

[4]These estimates of α and $(1-\alpha)$ conform with Benhabib and Spiegel (2002).

$$\frac{y_i}{A} = \left(\frac{k}{A}\right)^{\alpha} h^{1-\alpha}. \tag{16.3}$$

The theoretical specification presented in Eqs. 16.1–16.3 treats technological change, A, as exogenously determined, implying nonexcludability and nonrivalry, as specified by Solow (1957). A, thus, receives no compensation and may be exploited without limits. Arrow (1962), on the other hand, claims that increases in capital goods, K, increase knowledge through "learning by doing."[5] As such, knowledge is treated similarly to that of the Solow model, but, as Romer (1990) indicates, fails to acknowledge the tendency for firms to intentionally invest in R&D. I acknowledge Romer's claims that technological change is endogenous in that it arises from intentional actions made in response to market conditions.

Jones (2002) develops a related theoretical model which replaces labor augmenting technology, A, with individual skill level, h. Defining "skill" as "the range of intermediate goods that an individual has learned to use,"[6] advanced capital goods are utilized as

$$\dot{h} = \mu e^{\psi s} A^{\gamma} h^{1-\gamma}, \tag{16.4}$$

where A^{γ} is the world frontier of technology. Building on the work of Eaton and Kortum (1994), Jones (2002) describes the parameter μ as the productivity of a country using education to learn to use new ideas. In the traditional sense, high values of μ may reflect a high quality education system. Here, however, I expand this parameter to include the productivity of a country based on its degree of openness. If we substitute Eq. 16.4 into our original production function, it will be shown that growth is now not just a function of the ability to use new ideas, but also the ability to productively use incoming capital and technology, especially that which mitigates GHG emissions.

Theorizing in the endogenous growth framework, we are faced with interpretation problems. Econometric estimates, such as Kim and Lau's (1994) analysis of the meta production function (MPF), identify the fact that production knowledge is imperfectly available and requires large amounts of tacit knowledge.[7] Second, developed countries' firms may fear that the communication of technology to counterparts in developing countries will lead to future competition in a Schumpeterian sense. Third, a considerable part of learning is local, meaning that

[5]"Learning by doing" is the education process which occurs during production. This education may occur in a training facility (college- or firm-based) separate from the production floor, or it can happen in the S&T case by reviewing publications and patents.

[6]Jones (2002), p. 126.

[7]"Tacit knowledge" is defined here, in accordance with queryLangloi and Foss (1997), p. 17), as knowledge which "can be acquired only through a time-consuming process of learning by doing." (For a definition of "learning by doing", see fn. 11)

knowledge transfer occurs across sectors.[8] Finally, openness measures, embodied by μ in Eq. 16.4, are meant to account for differentiating policies between countries, particularly the rules and regulations pertaining to intellectual property rights (IPRs) and FDI, although IPRs can have different effects, depending on a country's stage of development.

Mansfield (1995) claims that there is a direct relationship between the strength and weakness of a country's IPR regime and the kinds of technology transferred to that country, particularly with regard to high-technology industries. The strength of IPRs is also a focus of Yang and Maskus (2003), who claim that stronger IPRs may discourage innovation and reduce international technology transfer in the preliminary stages.[9] Kim (2003), however, claims that technology transfer increases as the returns to innovation resulting from such IPRs become apparent. These increases are also dependent on international licensing, local wages, and other aspects of absorptive capacity.

There is no consensus as to the effects of openness, at least in the traditional growth literature. Harrison (1996) looks at multiple measures of trade openness, concluding that – for most of the measures –openness is correlated with higher growth. Edwards' (1998) seminal analysis of 93 countries uses nine indices of trade policy to address the complexities of international trade and also lends support for the claim that openness leads to increased growth. Wacziarg (2001) develops this line of discussion further, targeting the sources of gains from trade in a dynamic framework, while Rodriguez and Rodrik (2000) argue that the relationship between trade openness and growth is not settled, primarily because of the aforementioned endogeneity problem, although some (Lee et al. 2004) do attempt to minimize the potential for endogeneity through two-stage regression analysis.

The endogeneity issue is prevalent for trade openness, but analyses of capital openness typically conflate it with other forms of openness, most notably openness to technology. Findlay (1978), for example, models technology transfer as a function of FDI, among other things, while Wang (1990) shows that both human capital and technology diffusion are positively related to FDI inflows, given institutional shifts within the recipient country. Similar findings are presented by Borensztein et al. (1998), who examine and show that FDI flows from developed to developing countries over time, but that the degree and effect of technology transfers is dependent upon a country's existing human capital stocks.

Given the mutual effects and endogeneity of forms of openness, it is no surprise that the majority of the macro-level studies cited above make calls for further analysis at the micro-level, the literature of which provides a number of relevant

[8]For example, the growing share of employment in the service sectors may represent an institutional infrastructure requiring more technology-supporting services, such as the marketing, finance, and transportation of advanced technologies.

[9]Yang and Maskus also point out that the increase in technology transfer via inward licensing may occur with higher costs per license and possibly higher prices. Ultimately, the economic effects in terms of welfare are uncertain.

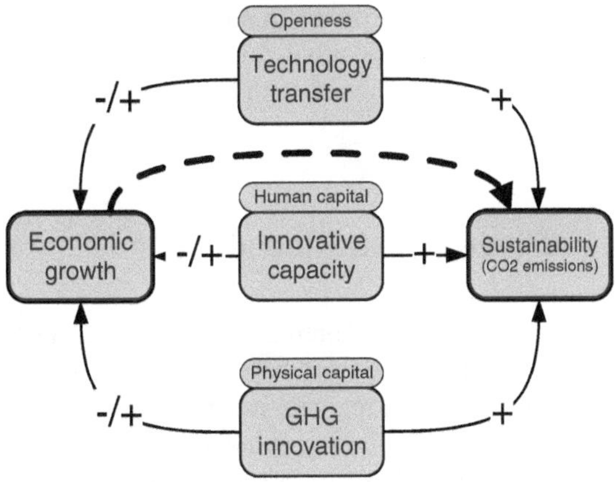

Fig. 16.1 Growth-EKC causal diagram and predicted effects

findings. Responding to and supporting Amsden's (1989) claims that growth models for late industrializers must incorporate imports of foreign technology rather than technological innovation, Zhang and Zou (1995) construct a model based on foreign capital imports. By dividing capital accumulation into home-produced capital and imported foreign technology, they show that economic growth rates and foreign technology imports are positively correlated and more significant than domestic accumulation.[10]

Given the aforementioned model and existing literature, the effects of openness, human capital, and physical capital upon economic growth and environmental sustainability are assumed to be positive, shown in Fig. 16.1 with the EKC framework (expressed with bold borders and arrows). Yet, as Fig. 16.1 shows, the traditional variables of the growth model are replaced with proxies relating specifically to sustainability and innovation. The proxies that are offered for openness, human capital, and physical capital are, respectively, technology transfer, human capital measured through innovative capacity, and physical capital measured by returns to investment in GHG-related innovation.

These proxies are not expected to impact economic growth or environmental sustainability uniformly, as technology transfer, innovative capacity, and GHG-based innovation restrain growth in the short-term and bolster growth in the long-run, as described above. This is likely the result of excessive costs from initial investments in the pursuit of these technologies or the orientation of labor away from certain profitable enterprises in less high-technology sectors to more uncertain but potentially rewarding S&T ventures. As well, technology transfer is expected to yield significant increases in growth, but a recipient country which lacks the requisite capacity may waste time and resources, so the impact can be negative in

[10]The distinction here, it should be noted, is in terms of capital goods imports, not FDI.

the short-run and positive in subsequent periods. These three variables may not have a uniformly positive effect on growth, but sustainability is expected to increase across the board, with the greatest yields occurring in the long-term.

Contributors to Sustainability and the EKC in Northeast Asia

We focus now on S&T output in Northeast Asia which has potential utility for other countries in the region and beyond. There is much promise for technology transfer through specialized regional integration, which is now in its relative infancy with regard to GHG emission reduction efforts. Gross differences in environmental governance, private sector responsiveness, and S&T output continues to present a significant hurdle for a smooth transition to East Asian cohesiveness, to which we now turn.

In Japan, the Ministry of the Environment (MOE) has supported a number of comprehensive measures to treat what it describes as "sustainable society," particularly the reduction of GHGs and the promulgation of a sound material cycle through reduced consumption and increased reuse and recycling of products (MOE 2007). Japan's S&T efforts currently reflect the third instalment of the Science and Technology Basic Law enacted in 1995.[11] One of the six goals of this law is sustainable development defined by a combination of economic growth and environmental protection. To this end, the largest share of the S&T budget allocations for 2009 are for low-carbon technologies, totalling 164 billion yen (Wada 2009).[12] The emphasis on low-carbon technologies is also indicative of Japan's long term approach to S&T, exemplified by the goals of the "Innovation 25" guidelines in 2025 and the "Cool Earth 50" proposal (from the 2007 G8 summit) of halving global GHGs by 2050.

Korea also conflates S&T-based efforts to reduce GHG emissions with economic growth. The three elements of "green growth" – minimize energy consumption while pursuing economic growth, minimize GHG emissions, and develop new growth engines – are grounded in commitments of 3.7 billion dollars of 23 billion dollars in government funding over the next 5 years to be invested in renewable technologies such as photovoltaics and wind turbines (Kim and SukJoon 2009). The chair of the Presidential Committee on Green Growth reported that the Lee Myung-bak administration has set a goal to reduce GHG emissions by 21–30% by 2020, in line with international goals to reduce GHG emissions 50% by 2050 (Na 2009).

[11]The First Basic Plan (1996–2000) targeted increases in government expenditures and a new R&D system; the Second Basic Plan (2001–2005) focused on increases in the knowledge base and increasing the competition for research funds.

[12]Innovative technologies received 52.3 billion yen, S&T diplomacy received 46.7 billion yen, regional (domestic) system promotion received 69.3 billion yen, and public-private R&D projects received 19.5 billion yen.

Taiwan's Agenda 21, the National Environmental Protection Plan, the Sustainable Development Action Plan (Republic of China Executive Yuan 2004), the Basic Environment Act of 2002 (Article 23), and the institution of the Ministry of Economic Affairs (Bureau of Energy) are all focusing on sustainability and efficiency as core goals of Taiwan's energy policy. The Taiwan Industrial Greenhouse Office (TIGO) under the Ministry of Economic Affairs was established specifically to reach a GHG-reduction goal of 10% (based on 2000 emissions) by 2015 (Chen 2008). This is for good reason, as Taiwan is very dependent on fuel imports (98% of sourced fuel) (Courtenay 2007). Despite the fact that Taiwan is not a signatory to the Kyoto Protocol, it adheres to the protocol's objectives and emphasizes renewable energy S&T to combat GHG emissions.

In China, there has been a strategy for sustainable development in place since 1996 (Rongping 2009), efforts to mitigate GHGs are secondary to air and water pollution corrections, as the Chinese Ministry of Environmental Protection (MEP) emphasizes reductions in chemical oxygen demand (COD) and SO2 emissions above all else. These two measures quantify water and air quality, respectively, but NO2 and especially air particulate (TSP) concentrations in China's urban areas are significantly higher than in neighboring countries, presented in Fig. 16.2. Major efforts are being made to address water and air quality, as eight cities have been banned from engaging in construction which results in increased COD or SO2 emissions (MEP 2009a). There is no indication, though, that CO2 emissions reductions targets are a primary concern for the MEP. As well, China maintains low energy efficiency and outdated technologies, and there are really no measures established to promote energy-saving S&T. Formal legislation, such as the Law on Science and Technology Progress (2007), does not sufficiently emphasize financial support and investment mechanisms (USAID and AECEN 2008).

However, the 17th National Congress of the CCP in October 2007 presented a plan to coordinate increases in GDP and sustainability (USAID and AECEN 2008). Subsequent commitments put forth in December 2007 at the 13th Conference of Parties of the UNFCCC amounted to a 20% energy savings target by 2012 and a viable GHG-related S&T policy. A proactive role is being played in Beijing to make this a reality, such as the Leading Coordination Committee on Global

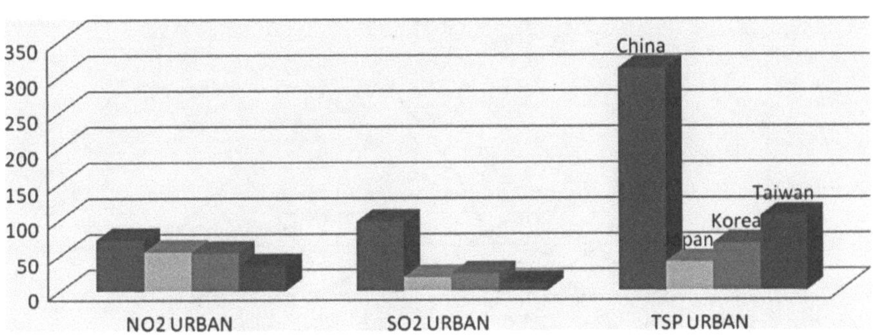

Fig. 16.2 GHG & pollutant emissions micrograms per cubic meter (Source: Esty et al. 2005)

Environmental S&T and the Law on Science and Technology Progress, the latter of which addresses IPRs (USAID and AECEN 2008). On the whole, this is a major effort to thread together the disparate agencies and ministries in China and unify the country's S&T goals.

Additional qualitative differences among these four Northeast Asian countries include, as of 2005, disparities in S&T infrastructures. Figure 16.3 exhibits the lower levels of innovation, tertiary enrollment rates, and number of researchers per 100 million people in China, relative to its neighbors. GHG-based R&D output for China is also a reflection of differences in S&T infrastructure, shown in Fig. 16.4. Data collection for patents was done through the online patent search function of the USPTO, with search parameters based on the presence of either "greenhouse effect" or "greenhouse gas" in the patent description or the article's topic. These keywords are by no means all-inclusive parameters to capture the degree of GHG-oriented innovation, but a cursory analysis of a number of keywords over the relevant time period confirms that these two terms are greatest in number and cover the widest area of industry classes. GHG-based patents in China have never really taken off and are noticeably lower than those of Japan, Korea, and Taiwan. Without sufficient technology inflows from these countries to China, the costs of effective CO2 reductions will remain high.

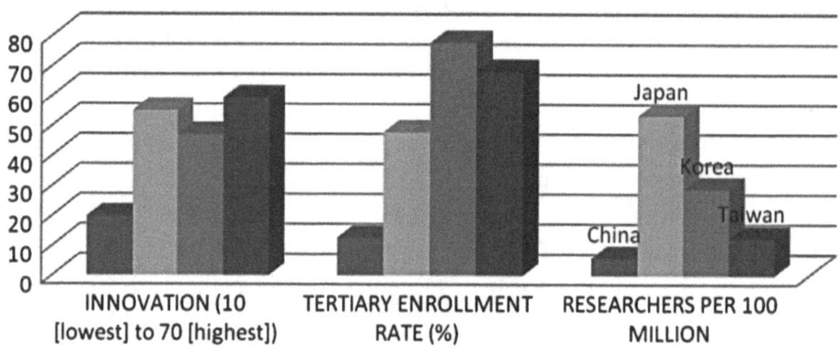

Fig. 16.3 Science & technology (Source: Esty et al. 2005)

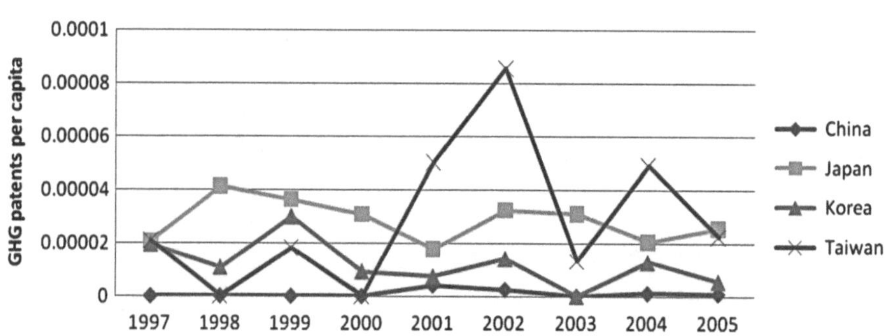

Fig. 16.4 Greenhouse gas R&D output (Source: Author's calculations)

Existing regional efforts to treat the absence or deficiency of national efforts to address GHG emissions include the Environmental Cooperation-Asia (ECO-Asia), under the United States Agency for International Development (USAID). This program encourages investment in clean energy technologies, among other sustainability-oriented goals, and it has been involved in China since 2007 to upgrade old coal-fired power plants and reduce CO2 emissions. A total of $109 million will help reduce at least 11 million tons of CO2 emissions over the lifetime of the affected power plants (USAID and ECO-Asia 2009). Similarly, the Asia Development Bank (ADB) recently approved an energy policy which provides reliable and affordable energy to all citizens of the region while simultaneously focusing on efficiency and renewable energy projects, such as wind power projects in China (ADB 2009). There are also efforts to tap the emerging carbon market and fund clean energy projects in Asia through the Asia Pacific Carbon Fund of the (ADB 2006). This can yield additional returns for those countries and areas which have highly inefficient or antiquated energy infrastructures, much like the rural areas of China.

Also at the regional level, efforts to deal with disparities in environmental governance are facilitated through the Asian Environmental Compliance and Enforcement Network (AECEN), which helps improve environmental compliance and enforcement. Since 2005, 13th Asian countries have worked in tandem with USAID, the ADB, the US EPA, the OECD, the World Bank, and other organizations to improve compliance through the exchange of policies and practices. The East Asian vision of a sound material cycle society is also embodied in attempts to synchronize efforts across the region, such as the Second Asia 3R Conference, held in Tokyo. This coordination effort extends beyond the reduction of material waste, stressing efficiency in resource use, S&T innovation, and international technology transfer (MOE and Institute for Global Environmental Strategies 2008). These are crucial advances but must continue at the same pace as sustainability-oriented inputs and technology transfers.

Data, Methods & Results

I first test for the EKC in China, Japan, Korea, and Taiwan by examining the relationship between GDP and CO2 emissions over the 1952–2004 period. GDP per capita data is drawn from the Penn World Table Version 6.2 (Heston et al. 2006), and CO2 data is measured in 1,000 metric tons (mtons) exclusively emitted from fossil-fuel consumption, available from CDIAC (2009). This specific account of CO2 acts as a control for any changes in its levels which might result from diminishing or expanding carbon sinks, agricultural shifts, and increases or decreases in livestock. Thus, the type of S&T infrastructure and output in each country affects our CO2 emissions data most directly.

While existing research may not necessarily validate the EKC, there are significant omissions in this literature. For example, the EKC does not fit the environmental efficiency patterns of 17 OECD countries, based on the 1980–2002 period (Halkos and Tzeremes 2009), but Korea, Taiwan, and China are excluded from the analysis. A similar study for the 1971–2004 period for 113 countries reaches a similar set of conclusions, although GHGs are not emphasized (Luzzati and Orsini 2009). When a modified functional form of the EKC is applied, specifically one which accounts for CO2 emissions among different GHGs,[13] the EKC is confirmed for OECD countries but not for non-OECD countries (Galeotti et al. 2006); yet, others have found that the EKC for CO2 is not well supported (Romero-Avila 2008) or that innovation and S&T to mitigate GHG emissions must accompany any analysis of the EKC, such as the Clean Development Mechanism (CDM) (Huang et al. 2008b). Failures to treat the EKC at the macro-level also indicate the need for case-specific analyses, and this chapter is one of several attempts to bridge this gap. Existing work on China examines city levels of SO2 pollution, supporting the EKC and concluding that China is able to "tunnel through" the EKC with application of S&T advancements (Brajer et al. 2008), which is consistent with Huang et al. (2008b).[14]

A first look at the EKC for these four countries in Fig. 16.5 largely reveals a standard rather than an inverted U-shaped curve. Japan is the exception, which I attribute to its position at the world's technology frontier and its lengthier exposure to CO2 emissions, as emissions in year-one were nearly 28,000 metric tons, while Korea and Taiwan only reached such levels in the early 1970s and mid-1980s, respectively. Despite clear confirmation of the EKC in Japan, I believe that China, Korea, and Taiwan are on the verge of plateauing. CO2 emissions in Korea in 2005 and 2006 (not presented in Fig. 16.5) show a 4% decrease and a negligible two-tenths of a percentage increase from 2004 to 2005, respectively. Emissions in China and Taiwan also decreased year-on-year 2004–2005 and 2005–2006.

Statistical analysis confirms that the inverted U-shaped EKC curve is present for these four countries. Table 16.1 presents regression results from a GLS fixed effects model, which is used to control for unexplained country-specific variation among these four countries. GLS modeling techniques control for heteroskedasticity (Greene 2002) which could be important in the present context. Models 2 and 3 represent quadratic and cubic functions, respectively, indicating that CO2 emissions increase with low levels of GDP and decrease with higher levels of GDP. The cubic function (Model 3) is particularly noteworthy given the second increase in CO2 emissions from even higher levels of GDP, although the coefficients are difficult to interpret given correlations among the explanatory variables.

---

[13]For example, SO2, NOx, and CO2.

[14]Others examine pollution in the form of sewage discharge in China, which is peripheral to our discussion of GHG emissions, but they conclude that technical progress tends to reduce the amount of industrial wastewater pollutants (Gu et al. 2009). Another China-based case study looks specifically at how the EKC in China also focuses on the country's first special economic zone (SEZ). In Shenzhen, for example, production-induced pollutants were found to support the EKC, although consumption-induced pollutants did not (Liu et al. 2007).

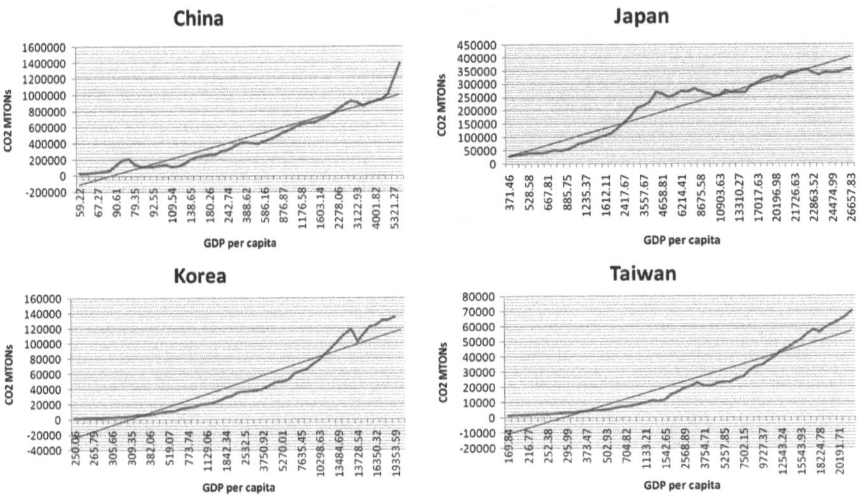

Fig. 16.5 The EKC in Northeast Asia: 1952–2004 (Source: Author's calculations based on Heston et al. (2006) and CDIAC (2009) data. GDP per capita measured in thousands of dollars)

Table 16.1 Regression (fixed effects) output: EKC

	(1) CO2	(2) CO2	(3) CO2
	8.702872***	39.21955***	141.9126***
GDP	(2.464924)	(7.812042)	(13.22694)
		−0.0011607***	−0.0104762***
GDP squared		(0.002842)	(0.0010971)
			2.36e–07***
GDP cubed			(2.73e–08)
	203,812.4***	76,176.94	−180,294.9***
Constant	(29,111.27)	(41,276.13)	(42,801.04)
N	100	100	100
r2	0.365	0.720	0.691
F-stat	12.47***	15.60***	43.55***

Standard errors in parentheses* $p < 0.05$; ** $p < 0.01$;
*** $p < 0.001$

Before turning to the endogenous growth/environmental sustainability models, an exploratory comparison of energy intensity data Energy Information Administration (EIA) (2009) and GDP per capita levels offers some important insight into the sustainability-related changes in these four countries. Energy intensity, measured in British thermal units (btu), quantifies the total primary energy consumption per dollar of GDP using purchasing power parities.[15] Energy intensity

[15]This measure is used as one of two proxies for GHG-based innovation in the subsequent regression analysis.

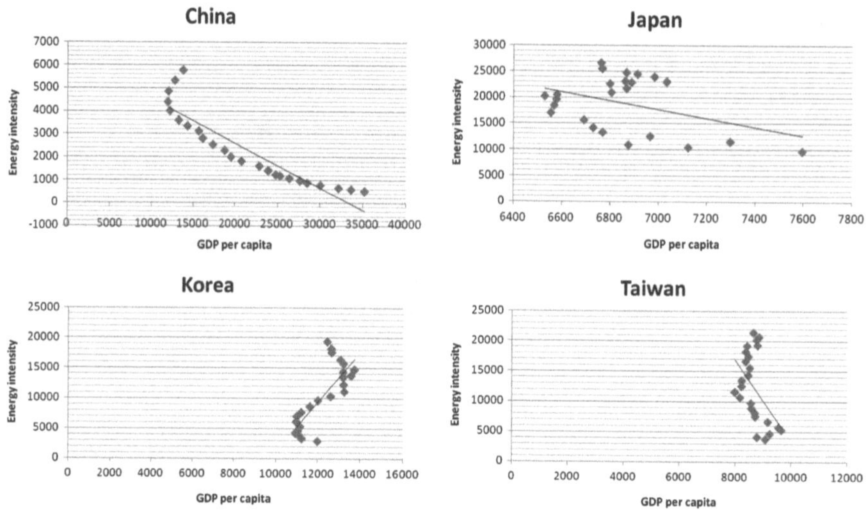

Fig. 16.6 Energy intensity and GDP: 1981–2004 (Source: Author's calculations based on Heston et al. (2006) and EIA (2009) data)

should be decreasing with increases in GDP per capita, given increases in S&T infrastructure and increasing economies of scale, which is the case in China, Japan, and Taiwan, shown in Fig. 16.6. In Korea, however, two periods of decreasing intensities border a period of increasing intensities from the late 1980s to the 1997–1998 financial crisis, creating a semblance of a backward S-shaped curve. The post-financial crisis period, represented by the uppermost scatter plot points, indicates a revitalized attempt to keep energy costs down. This backward S-shaped curve is partially identified in Taiwan, and, in China, there is a clear shift towards higher energy intensities in the most recent high-growth period, effectively establishing the bottom turn of the backward S-shaped curve.

When assessing the impacts of our proxies for openness, human capital, and physical capital, I again use the GLS model to control for heteroskedasticity and unexplained country-level variation. Our proxy for growth and sustainability is GDP per capita and carbon dioxide emissions, respectively. The proxy for openness is the share of investment in business R&D from foreign sources and the proxy for human capital is the number of full time researchers, both of which are taken from the OECD MSTI (2009) database. GHG-based innovation is measured in two ways, GHG-oriented patents and energy intensity, both of which have been described above. To test for the combined effects of incoming technology transfer and domestic capabilities to contribute to economic growth and environmental sustainability, a foreign R&D-FTE researchers interaction term is included. In total, eight fixed effects regressions are run to determine these variables' effects

Table 16.2 Summary statistics

Variable	Obs.	Mean	Std. Dev.	Min	Max
GDP	100	10,755.43	7,596.582	452.28	26,657.83
GDP squared	100	1.73e + 08	1.88e + 08	204,557.2	7.11e + 08
GDP cubed	100	3.24e + 12	4.46e + 12	9.25e + 07	1.89e + 13
CO2 emissions	108	315,310.1	349,345.3	20,504	1,664,589
GHG patents	36	0.0000176	0.0000188	0	0.0000859
Energy intensity	104	12,110.06	6,500.395	6,492.5	35,241
Foreign R&D funding	43	0.4727209	0.7630702	0.002	3.985
FTE researchers	56	435,783.1	311,279.1	45,778	1,223,756
For. R&D*FTE interaction term	43	327,311.8	657,410	119.3129	2,769,822

on economic growth and environmental sustainability, the results of which are presented in Table 16.3. Summary statistics for these variables (and the EKC model variables) are presented in Table 16.2.

Regression output for the endogenous growth model (Table 16.3, models 1–4) is largely consistent with our predicted effects described in Fig. 16.1. GHG-related innovation, measured first as GHG-related patents and then as energy intensity, has a positive effect on GDP per capita, although not at a statistically significant level. Innovation capacity, or the number of FTE researchers in each country, has a positive and statistically significant effect on growth, as does openness to technology from abroad, based on the amount of foreign funding for local firms' R&D. There is no significant effect, though, of the interaction term, which counters much of the existing literature. To recapitulate, both indigenous and foreign efforts play an important role for growth in Northeast Asia, but GHG-related innovation does not.

These results are partially reflected in those of the environmental sustainability model (Table 16.3, models 5–8), as an increase in the number of GHG-related patents decreases CO2 emissions from the consumption of fossil fuels. The coefficient, however, is not statistically significant and, more importantly, our second proxy for GHG-based innovation – energy intensity – actually increases CO2 emissions. This is likely due to the existence of a gap between energy consumption through fossil fuel generation and energy intensity improvements, a phenomenon which is present in the reverse S-shape relationship between growth and energy intensity. Regarding the other determinants of CO2 reductions, foreign R&D funding again contributes to sustainability by reducing CO2 emissions by 93,259 and 58,018 mtons for each percentage share increase to local firm R&D efforts. Expansion of human capital stock for R&D – the number of FTE researchers – also contributes to an expanding supply of CO2. This offers evidence in support of Esty et al. (2005) conclusion that S&T efforts and the ecological footprint are positively correlated. Finally, the combined effects of openness and domestic R&D capabilities significantly lower CO2 emissions in the model which accounts for GHG patents (Table 16.3, model 6).

Table 16.3 Regression (fixed-effects) output: Determinants of growth and sustainability

	(1) GDP	(2) GDP	(3) GDP	(4) GDP	(5) CO2	(6) CO2	(7) CO2	(8) CO2
	11,823,604.426	12,181,648.321			−1.572e + 08	−6.376e + 07		
GHG patents	(12,153,673.20)	(12,618,384.15)			(1.96e + 08)	(1.38e + 08)		
			0.744	0.669			29.358*	24.765
Energy intensity			(0.75)	(0.86)			(11.82)	(13.09)
Foreign R&D	2,982.417***	3,274.165	4,141.629***	3,762.336	−93,259.080***	−13,650.848	−58,018.957***	−78,118.480**
funding	(751.12)	(1,729.99)	(751.72)	(2,154.14)	(10,161.22)	(17,251.81)	(12,762.66)	(27,381.83)
	0.041***	0.041***	0.046***	0.047***	0.781***	0.558***	0.961***	0.937***
FTE researchers	(0.01)	(0.01)	(0.01)	(0.01)	(0.08)	(0.07)	(0.06)	(0.06)
For. R&D*FTE interaction		−0.001		0.001		−0.195***		0.030
term		(0.00)		(0.00)		(0.04)		(0.04)
	1,764.627	2,199.790	−8,600.328	−8,023.525	97,483.862*	195,184.694***	−321,730.771*	−268,018.291
Constant	(2,979.93)	(3,592.97)	(7,571.49)	(8,291.61)	(34,205.76)	(31,920.95)	(122,102.13)	(138,634.08)
N	26	26	35	35	30	30	43	43
r2	0.621	0.622	0.721	0.722	0.970	0.986	0.947	0.948
F-stat	10.40***	7.41***	24.17***	17.51***	246.33***	388.61***	214.96***	160.00***

Standard errors in parentheses

* $p < 0.05$; ** $p < 0.01$; *** $p < 0.001$

Conclusion

This discussion has confirmed that, for environmental sustainability in Northeast Asia, openness is more significant than local efforts to innovate and increase absorptive capacity. The number of researchers in a country does, however, play a significant role, while investments in GHG-related innovation do not. This has been shown through an expanded neoclassical growth model which acknowledges that economic growth or environmental sustainability are functions of the ability to use new ideas as well as the ability to productively use incoming capital and technology. There are indications, though, that local efforts and capabilities combined with incoming technology are important in reducing CO2 emissions, which is consistent with the claims of Borensztein et al. (1998). Also evident for Northeast Asia is a consistent and inverted U-shaped EKC. Absent from the above analysis, though, is a two-stage empirical procedure in which the growth function itself (including its determinants) is a function of CO2 emissions. This was not possible, however, given the correlation between CO2 emissions and the determinants of growth, as explained in the theoretical section above.[16] All the same, innovation and S&T in pursuit of solutions for GHG emissions is crucial, and there are a number of curvilinear effects involved with efforts to reduce GHG emissions. Our exploratory analysis of energy intensity, for example, revealed its non-linear relationship to economic growth. The reverse S-shaped curve indicates that China is following in the footsteps of Korea and, to a lesser degree, Taiwan.

China dominates the discussion of country-specific analyses of the EKC, as it is the veritable hinge upon which climate change policy swings. This is true in both the immediate and near-term sense, as India and other developing countries will ultimately follow international standards which address China's present and future GHG emissions. China is also a common case study for the EKC hypothesis in light of its rapid industrialization, large geographic size, problematic governance structures (also correlated with growth), and lax legislation to address GHG emissions. Given that technological openness is a positive predictor of both growth and environmental sustainability, China must establish closer ties within the region, particularly as Korea plans to devote nearly its entire basic R&D budget to green S&T pursuits.

A major challenge to China's adoption and application of technologies from within the region is the availability of such technology elsewhere. Recently, Vice Premier Li Keqiang met with U.S. Energy Secretary Steven Chu and Commerce Secretary Gary Locke to discuss bilateral cooperation on clean energy and related technologies (MEP 2009b), but it is likely that the relationship will be affected by China's adherence to the principle of "common but differentiated responsibilities."

[16]Future work on endogenous growth in an EKC framework should attempt to examine the determinants of growth which are uncorrelated with CO2 emissions. Such efforts may still be plagued with endogeneity problems, given policy efforts to tackle climate change.

This refers to the distinction between developed and developing countries in reducing GHG emissions: developed countries are largely responsible with the existing stock of GHGs and thus should shoulder the responsibility of GHG reductions more than countries of the developing world, where per capita emissions are still relatively low. On these grounds, China opposes unilateral adoption of GHG emissions reduction targets and, in response, the U.S. is unwilling to share key GHG-related S&T output with China without compensation. Whether or not future empirical research is able to verify the regional-versus-international benefits of openness to GHG-related technology transfer, laborious and lengthy negotiations at the international level can be avoided. These four countries are on track to coordinate, given revived emphasis on climate change-mitigating policies and innovation efforts.

References

Amsden A (1989) Asia's next giant: South Korea and late industrialization. Oxord University Press, New York

Arrow KJ (1962) Economic welfare and the allocation of resources for invention. In: Nelson RR (ed) The rate and direction of inventive activity. Princeton University Press, Princeton

Arrow KJ, Bert B, Robert C, Partha D, Carl F, Holling CS, Bengt-Owe J, Simon L, Karl-Goran M, Perrings C, David P (1995) Economic growth, carrying capacity, and the environment. Ecol Appl 15:91–95

Asia Development Bank (ADB) (2006) Carbon market initiative: the Asia Pacific carbon fund.

Asia Development Bank (ADB) (2009) New ADB policy targets secure, Clean energy for Asia.

Barro RJ, Lee J-W (2000) International data on educational attainment: updates and implications. CID Working Paper, No. 42

Benhabib J, Spiegel MM (2002) Human capital and technology diffusion. FRBSF Working Paper 2003-02

Benhabib J, Spiegel MM (1994) The role of human capital in economic development: Evidence from aggregate cross-country data. J Monetary Econ 34:143–173

Borensztein E, De Gregorio J, Lee JW (1998) How does foreign direct investment affect economic growth? J Int Econ 45:115–135

Brajer V, Mead RW, Xiao F (2008) Health benefits of tunneling through the Chinese Environmental Kuznets Curve (EKC). Ecol Econ 66:674–686

Brock WA, Scott TM (2004) The green Solow model. Social Science Research Institute Working Paper No. 2004–16. Available at SSRN: http://ssrn.com/ezproxy.gl.iit.edu/abstract=555262

Bruvoll A, Foehn T (2006) Transboundary effects of environmental policy: Markets and emission leakages. Ecol Econ 59:499–510

Carbon Dioxide Information Analysis Center (2009) Terrestrial carbon management data sets and analyses. http://cdiac.ornl.gov/carbonmanagement/

Chen S-H (2008) Elevated vision, enhanced illumination allowing energy policy to shine forth again in obscurity. http://www.tier.org.tw/eng2008/eng_analysis/EA200806_4.asp

Copeland BR, Taylor MS (2002) Trade, growth and the environment. J Econ Lit 16:147–168

Courtenay P (2007) Taiwan sets sights on clean energy. *Taiwan Today*

Dasgupta S, Laplante B, Wang H, Wheeler D (2002) Confronting the environmental Kuznets curve. J Econ Perspect 16:147–158

Dinda S (2004) Environmental Kuznets curve hypothesis: A survey. Ecol Econ 49:431–455

Eaton, Jonathan and Samuel Kortum (1994) International patenting and technology diffusion, NBER Working Paper No. 4931
Edwards S (1998) Openness, productivity and growth: What do we really know? Econ J 108:383–398
Energy Information Administration (EIA) (2009), International energy statistics. http://tonto.eia.doe.gov/cfapps/ipdbproject/iedindex3.cfm?tid=6&pid=29&aid=12&cid=&syid=1980&eyid=2008&unit=BKWH
Esty DC, Levy M, Srebotnjak T, de Sherbinin A (2005) 2005 Environmental sustainability index: Benchmarking national environmental stewardship. Yale Center for Environmental Law & Policy, New Haven
Findlay R (1978) Relative backwardness, direct foreign investment, and the transfer of technology: A simple dynamic model. Q J Econ 92:1–16
Galeotti M, Lanza A, Pauli F (2006) Reassessing the environmental Kuznets curve for CO2 emissions: A robustness exercise. Ecol Econ 57:152–163
Greene WH (2002) Econometric analysis. Prentice Hall, Upper Saddle River
Gu K, Liu J, Wang Y (2009) Relationship between economic growth and water environmental quality of Anshan City in Northeast China. Chin Geogr Sci 19:17–24
Halkos GE, Tzeremes NG (2009) Exploring the existence of Kuznets curve in countries' environmental efficiency using DEA window analysis. Ecol Econ 68:2168–2176
Harrison A (1996) Openness and growth: A time-series, cross-country analysis for developing countries. J Dev Econ 48:419–447
Heston A, Summers R, Aten B (2006) Penn world table Version 6.2. Center for International Comparisons of Production, Income and Prices at the University of Pennsylvania
Huang B-N, Hwang MJ, Yang CW (2008a) Causal relationship between energy consumption and GDP growth revisited: A dynamic panel data approach. Ecol Econ 67:41–54
Huang WM, Lee GWM, Wu CC (2008b) G.H.G. Emissions, G.D.P. Growth and the Kyoto Protocol: A revisit of environmental Kuznets curve hypothesis. Energy Policy 36:239–247
Jones CI (2002) Introduction to economic growth. W.W. Norton & Company, New York
Kim L (2003) Technology transfer & intellectual property rights: The Korean experience. In: UNCTAD-ICTSD (ed) Intellectual property rights and sustainable development. UNCTAD, Geneva
Kim J-I, Lau L (1994) The sources of growth of the east Asian newly industrialized countries. J Jap Int Econ 8:235–271
Kim, SukJoon (2009) Strategy for low-carbon green growth in Korea, AAAS annual meeting, Chicago, IL
Kuznets S (1955) Economic growth and income inequality. Am Econ Rev 45:1–28
Langloi RN, Foss NJ (1997) Capabilities and governance: The rebirth of production in the theory of economic organization. DRUID Working Paper No. 97–2
Lee HY, Ricci LA, Rigobon R (2004) Once again, is openness good for growth? J Dev Econ 75:451–472
Liu X, Helilig GK, Chen J, Heino M (2007) Interactions between economic growth and environmental quality in Shenzhen, China's first special economic zone. Ecol Econ 62:559–570
Luzzati T, Orsini M (2009) Investigating the energy-environmental Kuznets curve. Energy 34:291–300
Mankiw NG, Romer D, Weil DN (1992) A contribution to the empirics of economic growth. Q J Econ 107:407–437
Mansfield E (1995) Intellectual property protection, direct investment, and technology transfer: Germany, Japan, and the United States. International Finance Corporate Discussion Paper, No. 27
Millimet DL, List JA, Stegnos T (2003) The environmental Kuznets curve: Real progress of misspecified models? Rev Econ Stat 85:1038–1047
Ministry of the Environment (MOE) of Japan, Institute for Global Environmental Strategies (2008) Issues paper, The Second Asia 3R Conference, Tokyo, Japan

Ministry of Environmental Protection (MEP) PRC (2009) China pursues leadership in new energy solutions
Ministry of Environmental Protection (MEP) PRC (2009) MEP releases emission reduction results of all provinces and five major power groups
Ministry of the Environment (MOE) of Japan (2007) Becoming a leading environmental nation in the 21st Century: Japan's strategy for a sustainable society
Na J-J (2009) Korea to cut greenhouse gas emission by 30%. Korea Times, 4 August 2009
Nelson RR (1993) National innovation systems: a comparative analysis. Oxford University Press, Oxford
Nelson RR, Phelps ES (1966) Investment in humans, technological diffusion, and economic growth. Am Econ Rev 56:69–75
OECD (2009) Main science and technology indicators (MSTI). www.sourceoecd.org2009
Pascha W, Mahlich J (eds) (2007) Innovation and technology in Korea: Challenges of a newly advanced economy. Springer, Berlin
Rodriguez F, Rodrik D (2000) Trade policy and economic growth: A Skeptic's guide to the cross-national evidence. In: Bernanke B (ed) NBER macro annual 2000. National Bureau of Economic Research, Cambridge
Romer P (1990) Endogenous technological change. J Political Econ 98:S71–S102
Romero-Avila D (2008) Questioning the empirical basis of the environmental Kuznets curve for CO2: New evidence from a panle stationarity test robust to multiple breaks and cross-dependence. Ecol Econ 64:559–574
Rongping M (2009) Innovative policy for sustainable development in China. Paper presented at the AAAS Annual Meeting, Chicago
Solow RM (1956) A contribution to the theory of economic growth. Q J Econ 70:65–94
Solow RM (1957) Technical change and the aggregate production function. Rev Econ Stat 39:312–320
Stern DI (2004) The rise and fall of the environmental Kuznets curve. World Dev 32:1419–1439
USAID, and Asian Environmental Compliance and Enforcement Network (AECEN) (2008) Strengthening legal and policy frameworks for addressing climate change in Asia. AECEN Secretariat, Bangkok
USAID, and Environmental Cooperation-Asia (ECO-Asia) (2009) Financing secured through USAID's support will help Slash 17 million tons of CO2 emissions from Chinese and Indian power plants. USAID/ECO-Asia e-Newletter 4: 1
Wacziarg R (2001) Measuring the dyanmic gains from trade. World Bank Econ Rev 15:393–429
Wada T (2009) Recent developments in S&T policies after the lost decade. Paper presented at the AAAS Annual Meeting, Chicago, IL
Wang J-Y (1990) Growth, technology transfer, and the long-run theory of international capital movements. J Int Econ 29:255–271
Yang G, Maskus KE (2003) Intellectual property rights, licensing, and innovation. World Bank Policy Research Working Paper, No. 2973
Yuan, Republic of China (Taiwan) Executive (2004) Taiwan Agenda 21: Vision and strategies for national sustainable development. Approved at the 18th Council Meeting of the Council for National Sustainable Development, Executive Yuan
Zhang X, Zou H-F (1995) Foreign technology imports and economic growth in developing countries. World Bank Policy Research Working Paper, No. 1412

Chapter 17
Comment on M.A. Shapiro: "Long-Run Protection: Determining Key Features of Growth and Sustainability in Northeast Asia"

Reinhold Hofer

M.A. Shapiro's paper deals with core topics of economic thinking closely tied to the actual political debates on a "low carbon growth path". Therefore the paper "... attempts to advance the understanding of the crucial links among growth, technological change, human capital accumulation, and openness and their impacts on greenhouse gas (GHG)-related research and development (R&D) output in East Asia ..." (p. 1). The analysis of growth processes is an already longstanding economic issue. While studying conditions for sustainability is a somewhat more recent topic, it is nevertheless intensively discussed worldwide and high on the agenda of politics (e.g. UNIDO's (1993) Agenda 21; EU Commission's (2006) Green Paper).

Hence, distinctive features of the paper include interesting considerations tackling dimensions of growth together with the concept of sustainability. In this respect concentrating on indicators resulting from an innovation perspective is quite appropriate, and dealing with empirical results from East Asia provides an interesting contribution and a real value added. Nevertheless, what remains to be discussed is the understanding of the relations between openness, growth and sustainability as much as (human) capital stocks and their influence on growth and sustainability. Therefore, the following comments are addressing a few concerns regarding (1) the model structure and (2) the applied empirical indicators.

Endorsing generally the argument of innovative behaviour and the mobilisation of related capital (whether human or material) as key determinants for growth – and if steered towards "green investments" also for sustainability – its implementation in an economic model puts some questions in place.

The author draws his analysis from growth models enriched by elements of sustainability. Hence, he starts with the production function approach and discusses the implementation of openness. Lastly he concludes hypothesising positive effects from openness, human capital and physical capital upon economic growth and environmental sustainability. Nevertheless – similar to almost all growth models – "environmental capital" (i.e. non-renewable natural resources) is excluded by assuming equal stocks over time or negligible relevance of structural changes. But when dealing with sustainability issues this dimension may be of importance

J. Mahlich and W. Pascha (eds.), *Korean Science and Technology in an International Perspective*, DOI 10.1007/978-3-7908-2753-8_17,

according to growth strategies being based on different factors (e.g. Japan or Korea being less equipped with natural resources and have had to build their growth and development strategies on other factors, while in other countries like China endowment with natural resources, e.g. coal, is much larger and consequently growth and development strategies can at least partly follow other paths OECD 2009b). This certainly important constraint of the model is reflected by its assumption that other factors (beside openness, human capital and physical capital) had no significant influence on growth (GDP per capita) and sustainability (represented by GHG reductions and energy consumption). This implies also that institutional characteristics in the countries and historical developments have not created relevant differences. Again one may just think of Japan and Korea, which have been heavily depending on energy import and hence had huge incentives to reduce energy-intensity. Consequently related institutional settings, including supportive governmental initiatives appeared.

Furthermore, one would also have to have a closer look at the history of these countries. Several periods in Korea's catching-up process, for example, appear with their different treatment of (1) elements in the national innovation system (OECD 2005), (2) implemented institutional setting and (3) lastly, notice given to sustainability. Hence, the different periods organised along different development regimes (i.e. imitation in the 1960s and 1970s, transformation in the 1980s and innovation in the 1990s till now – with extremely large shares of business R&D compared to other countries and being based on chaebol industry structures) took sustainability differently. As the study is based on data from 1952 to 2004, these changes in accompanying R&D regimes may derive with breaks and adaptations causing influences on growth and sustainability, which are not considered in the analysis.

In addition, sustainability can be interpreted as a much broader concept and is – at least in the European context (see the European Sustainable Development Scoreboard).[1] It can include several dimensions (OECD 2008). Ignoring effects from involved elements, like social cohesion and its relation to growth, may put a too large emphasis on just technological change and ignore the important efforts in building up institutional frameworks supportive for a sustainable growth (i.e. implementing feasible regulations and educational structures, amongst others).[2]

An additional remark is to be made on the integrated characteristics of openness in the model, as the major results of the paper shows the importance of this dimension for growth and sustainability. But in case of the East Asian "growth

[1] Cf. EU Commission 2009 and Eurostat Statistical Book: "Sustainable development in the European Union". 2009 monitoring report of the "EU sustainable development strategy".

[2] Analyses of the success factors of Asian Tigers support the view that specific institutional arrangements in line with a somewhat even income distribution pattern may have been very supportive.

miracle"[3] openness in general is a much discussed element. Even when it initiates discussions towards a more interregional, cross-country perspective – hence, looking for the internalisation of external effects – it is not always fully clear, why imported technological change should work in this way. The arguments follow the assumption that technology can be (easily) transferred from abroad and that this technological change contains innovations producing capital with less green house gas emissions (or reduce energy intensity). But, technology transfer needs a set of conditions forming absorptive capacity and may in most cases not be serving the goal of reducing green house gas emissions – instead being determined by the overall production and consumption structure.

Other sources for discussions are the applied indicators of the variables. To start with, the representation of sustainability via CO2 emissions or energy intensities is not a comprehensive approach but delivers more focused results. In the paper, technological change is measured against its contributions to reduce CO2 and energy consumption related to GDP produced. Therefore, it is assumed these technological changes have no influence on other elements of sustainability (see above). On the other hand, less energy consumption (total primary energy consumption per dollar of GDP) does not always mean increasing sustainability (e.g. when thinking of developments by different industrial structures or shares of service sector, mobility patterns and labour division).

Taking this more general argument aside, one has also to recognise that CO2 as green house gas is not only reduced by less emissions via technological change but also by changes in land-use (i.e. intensified reforestation builds increasing carbon sink and vice versa). Ignoring structural effects like this may consequently cause biases in the analysis and bear the threat of overvaluing the contribution of technological change on GHG reductions. The approach taken, basing the calculations on emissions from fossil-fuel consumption does not consider e.g. household energy use and its CO2 emissions. Consequently, the presented relationship between GDP and CO2 emissions includes the assumption that all other CO2 emission (and their changes) has no correlation with GDP. Science and innovation in this respect are treated as if only to produce and implement substitutive new technologies in industries and activities that produce CO2 emissions as a consequence of "fossil-fuel consumption". In addition, efforts to adopt these new technologies may – at least partly – be encountered by the given subsidies on oil (e.g. in case of Korea).

Another issue to address related to the applied empirical indicators is a possible correlation between FTE researchers and foreign R&D funding – the former being taken as a representative variable for human capital, the latter for openness. The analysed East Asian countries are characterised by high shares of business R&D expenditures – with a significant share being financed from abroad. Subsequently, the highest shares of researchers are employed in foreign funded business R&D

[3] A "miracle" which has been intensely discussed (e.g. Krugman 1994) and seen mainly as a mobilisation effect of resources together with saving behaviours.

activities. This correlation creates a problem of causal relations, which may result in less clear interpretive power and hence may be responsible for the overvaluing of "openness". Therefore, indicators for human capital seemingly depending on national efforts would be helpful here.

Openness and technology transfer represented by foreign R&D funding may also have its shortcomings. First, the incentives behind these foreign R&D funds can widely differ, including a kind of capacity extension instead of transferring knowledge. Secondly, foreign funding of business R&D represents an engagement in R&D activities in the fundee's country but may not indicate a "green" technology transfer as such. An assumption has to be implemented here, stating that this kind of R&D activity has a close relation to R&D activities abroad that generate energy reducing or CO2 reducing technologies. Thirdly, the role of R&D in technology transfer is not exclusive. There exist other options like capital investments and embodied technology, which may act as a channel for technology transfer – all the more, if imported capital goods for manufacturing are seen as the main source for reducing energy use and CO2 emissions. Overall, foreign funded business R&D represents a rather specific indication of openness and is partly misleading when looking for its influence on sustainability.

Furthermore, technological change with an influence on climate change (reducing GHG) as the results of R&D, triggered by GHG patents is representing only a quite limited part of the story. First, there is the fact that patent indicators imply a lot of shortcomings (OECD 2009a), as they are typically an intermediate result and may not be followed by products and processes implemented in markets. Secondly, the transforming process from R&D activities via GHG patents towards related CO2 reductions is not fully clear. What type of R&D results in GHG patents and how does this R&D activity relate to CO2 reduction? In consequence, the influence of physical capital represented by GHG patents may be largely underestimated (just compared to the saving behaviour in those countries).

On a more technical level an estimation issue appears. While the arguments in the paper mainly stress growth, sustainability and its determinants, the performed estimations seem only to contain level figures but no growth figures (as is typical for the concept of EKC). Consequently, the estimations provide much more evidence of income levels and their determinants than of growth processes.

Beside all these caveats, the paper provides important insights for the East Asian context. It points to the need for an international perspective – one more serious about cross-border relations concerning sustainable growth paths with less GHG emissions and energy use – that follows a so-called "low-carbon growth path" (e.g. EU COM 2006). In general, such a sustainable growth path is perceived within reach by restraining energy demand growth and driving production towards low carbon sources[4] – and these arguments can be seen behind the given structure of the

[4] To take an example from Europe: amongst the most experienced countries of "low-carbon growth" is Denmark (total energy consumption has not increased since 1980, despite a 50% increase in GDP).

presented model, augmented by the important role of cross-border exchange and technology transfer (i.e. openness). Nevertheless, one would have to ask how robust these results are in other contexts and what to learn from these countries' experiences. While a role for cross-border cooperation in the case of CO2 seems meaningful in general, it is not this clear, if the openness (i.e. technology transfer) of the country is indeed the major explanatory factor for reductions. Just think on the pro-active role of governments, which in the Korean case appears again quite prominent, when a new "five-year" plan steers the efforts to transform the economy towards a green growth economy (e.g. East Asia Climate Forum)[5].

Compared to the results presented for "newly advanced economies" the European countries – as examples of "old advanced economies" – show a slightly different appearance. GHG was more or less stable in the last decades – with some fluctuations according to business cycles. This descriptive figure points to some different opportunities for "low-carbon path growth". Being on the technological frontier, already performing high productivity and energy-efficient production processes, other elements have to be stressed too. The social dimension becomes a lot more important – including need for new institutional adaptations. With a highly qualified human capital, individualisation and democratisation receive much more attention regarding their effects on a "low carbon growth path". Furthermore, these results call for stabilising society and dealing with social security and structural effects, like an ageing society. Overall, this would bear indications for differentiating between "old advanced economies" and catching-up "new advanced economies", when aiming for sustainability solutions.

Summing up, the analysis provided by M.A. Shapiro's paper attempted to advance understanding of links between technological change, human capital accumulation, openness, growth and impacts on GHG related R&D. The conducted empirical research indicates openness as the most significant determinant for growth and sustainability in case of Northeast Asia, meaning that imported technological change outweighs local efforts – a result which has to be treated with caution as it implies assumptions concerning foreign technology as perfectly matching the countries' absorptive capacities and needs. Major policy conclusions drawn on the empirical results include that countries with GHG problems should open their economy and have an incentive for close relationship with foreign technology suppliers to increase GDP per head and reduce CO2 at the same time. While there is indeed an argument for cross-border efforts to reduce CO2 (i.e. internalise external effects), it is not so clear why this so-called "low carbon growth path" should mainly be based on technology transfer from abroad, leaving aside all national structural changes in production and consuming behaviour and the countries' specific institutional contexts.

[5] http://www.iisd.ca/ymb/climate/eacf2/

References

EU Commission (2006) A European strategy for sustainable, competitive and secure energy. Green Paper. COM (2006)105 final, Brussels
EU Commission (2009) GDP and beyond: measuring progress in a changing world. COM (2009) 433 final, Brussels
Krugman P (1994) The myth of Asia's miracle. Foreign Affairs Vol. 73, Issue 6:62–78
OECD (2005) Economics Survey of Korea. Paris
OECD (2008) Sustainable development: linking economy, society and environment. Paris
OECD (2009a) Patent statistics manual. Paris
OECD (2009b) OECD reviews of innovation policy: China Paris
UNIDO (1993) Agenda 21: earth summit – the United Nations programme of action from Rio

Index

J. Mahlich and W. Pascha (eds.), *Korean Science and Technology in an International Perspective*, DOI 10.1007/978-3-7908-2753-8,

FSC
www.fsc.org
MIX
Papier | Fördert gute Waldnutzung
FSC® C083411

Kolibri 360 GmbH
Ferdinand-Jühlke-Straße 7
99095 Erfurt, Deutschland
produktsicherheit@kolibri360.de